A BASIC INTRODUCTION TO POLLUTANT FATE AND TRANSPORT

A BASIC INTRODUCTION TO POLLUTANT FATE AND TRANSPORT

An Integrated Approach With Chemistry, Modeling, Risk Assessment, and Environmental Legislation

FRANK M. DUNNIVANT
Whitman College
Walla Walla, Washington

ELLIOT ANDERS
Educational Solutions, LLC
Mohawk Trail Regional School District,
Shelburne Falls, Massachusetts

WILEY-INTERSCIENCE

A JOHN WILEY & SONS, INC., PUBLICATION

Published by John Wiley & Sons, Inc., Hoboken, New Jersey
Published simultaneously in Canada

For general information on our other products and services or for technical support, please contact our
Customer Care Department within the United States at (800) 762-2974, outside the United States at (317)
572-3993 or fax (317) 572-4002.

Wiley also publishes its books in a variety of electronic formats. Some content that appears in print may
not be available in electronic formats. For more information about Wiley products, visit our web site at
www.wiley.com.

Library of Congress Cataloging-in-Publication Data:

Dunnivant, Frank M.
      A basic introduction to pollutant fate and transport : an integrated approach with chemistry, model-
ing, risk assessment, and environmental legislation / Frank M. Dunnivant, Elliot Anders.
          p.   cm.
      Includes bibliographical references and index.
      ISBN-13 978-0-471-65128-4
      ISBN-10 0-471-65128-1 (cloth)
      1. Pollution–Mathematical   models. 2. Pollutants–Environmental   aspects. 3. Environmental
chemistry. 4. Environmental risk assessment. 5. Environmental policy. I. Anders, Elliot. II.
Title.
    TD174.D86 2005
    628.5'2—dc22

                                                                                      2005012262

Printed in the United States of America
10 9 8 7 6 5 4 3 2 1

To the chemists who have had a positive influence on my career:
Loretta McLean, John Coates, Alan Elzerman, Philip Jardine, and
Rene Schwarzenbach. Thank you.
Frank M. Dunnivant

To my grandmother, who would understand very little of the
technical details of this book, but who
appreciates hard work and family more than anyone I know.
Elliot Anders

CONTENTS

PART VI *POLLUTANT CASE STUDIES*

CHAPTER 12 *CASE STUDIES OF SELECTED POLLUTANTS* 399

PREFACE

While there are hundreds of books on fate and transport phenomena, this is the first truly introductory textbook on the subject. All of the books that I have reviewed require a working knowledge of linear algebra or differential equations, which effectively eliminates a majority of the people working in fate and transport. This book presents and integrates all of the aspects of fate and transport: chemistry, modeling, risk assessment, and the environmental legal framework. We approach each of these topics initially from a conceptual perspective, and then we explain the concepts in terms of the math necessary to model the problem. The only prerequisites for understanding the concepts covered in this book are a basic knowledge of algebra and first-year college chemistry. In addition, for the fate and transport modeling chapters (Chapters 4–9), we have included a simple, user-friendly simulator, Fate®, which uses basic models to predict the fate and transport of pollutants in lake, river, groundwater, and atmospheric systems. Fate® can be an effective teaching and learning tool, as discussed in the "How to Use Fate®" section of the introductory materials.

This book is the result of a challenge I made to one of my senior chemistry students. I challenged Elliot Anders, the co-author of this book, to create a new version of EnviroLand, the precursor to Fate®. If he did so, I told him, I would write a book to accompany it. To my surprise, Elliot finished the software in a few months and to meet my end of the bet, I had to write a textbook around the software. We feel that this textbook provides a very unique instructional tool for students and environmental professions who lack the rigorous mathematical backgrounds to be able to derive the governing fate and transport equations, but nonetheless require an understanding of the subject. This book can be used to teach a variety of classes, from a new type of hydrology or environmental chemistry course to new fate and transport courses for support personnel who want to work in the environmental arena. I use the book to teach environmental chemistry to undergraduate students majoring in chemistry, geology, and biology. These students usually have sufficient background to work in environmental remediation, but lack the basic engineering knowledge to be truly effective in this area. I have had great success in expanding the academic horizons of science students to areas of modeling, risk assessment, and environmental legislation. In addition, there is no reason that this book cannot be used in a graduate course in fate and transport, since it provides an especially extensive and complete conceptual development of fate and transport modeling. In this case, the professor can use the book as a conceptual guide while teaching the derivation portion of the course in the classroom. We hope you enjoy our approach to environmental fate and transport.

TO THE INSTRUCTOR

The material in this textbook has been used for five years to teach an undergraduate course in environmental chemistry to chemistry, biology, and geology students who have not completed linear algebra or differential equations. Students from these disciplines tend to have excellent skills and knowledge in environmental studies but lack an understanding of how to apply this knowledge to environmental applications such as fate and transport modeling and risk assessment. Students feel a strong sense of empowerment as they come to understand how to approach an environmental pollution event and gain an appreciation of the many scientific and political factors controlling the remediation of contaminated sites. A software package is integrated into the textbook, and we have suggested important technical papers for class discussion for many of the chapters. In addition, we have designed laboratory exercises specific to the topics covered in the lecture material. This textbook would be ideal for an introductory course in environmental chemistry, pollution science, environmental science, environmental engineering, or pollutant fate and transport.

A set of laboratory exercises has been designed for this textbook. As with all laboratory experiments, it will be instrumental that the instructor test the procedures prior to using them. This will ensure the success of your laboratory class. A complete set of instructions for solution preparations, suggested level of difficulty, suggested lab time requirements, apparatus construction guidelines, and hints for success can be obtained from Wiley-Interscience after you have adopted the textbook for your class.

TO THE STUDENT

This is a new type of textbook, in the sense that it combines many disciplines to craft a practical discussion of environmental pollution. We first start with an overview of the chemistry of fate and transport processes. Next, we introduce you to how professionals model a pollution event in the real world and then present you with a conceptual understanding of the models that these professionals use. At the end of each fate and transport chapter, we incorporate our conceptual ideas into two simple and common models that can be used to predict the pollutant concentration in environmental systems. Next, we use the results from fate and transport modeling as a basis for our risk assessment calculations to determine whether a hazardous situation is present. The final two chapters include environmental legal aspects and case studies of common pollutants that have been spread around the world.

TO THE ENVIRONMENTAL PROFESSIONAL

While most environmental engineers with find this book to be a basic review, many of other environmental professionals who do not understand the derivation of complex environmental models will find this book to be a valuable asset for developing a conceptual understanding. Our goal is to bridge the knowledge gap between the engineers and mathematicians and the many important workers associated with a SuperFund site. The only prerequisites for understanding the subjects covered in this book are a working knowledge of algebra and college-level general chemistry. We provide the background and explain everything else. If you are an engineer or modeler working in the environmental arena, you will find this book to be a must for your colleagues.

HOW TO USE THE BOOK WITH FATE® AND ASSOCIATED SOFTWARE

This book comes with a CD-ROM containing Fate® and the pC–pH Simulator®. Fate® is certainly the most important tool, since it can be used in a variety of ways. First, Fate® can be used as an in-class tool to illustrate how each of the fate and transport models work. It enables the instructor to quickly and easily show how changing model input parameters affect the resulting pollutant concentration. Students and instructors will find Fate® to be an invaluable resource in working homework problems. All of these problems require lengthy, multistep calculations, and each step can be checked with Fate®. This decreases the need for students to rely on a tutor or their instructors, who in turn would have to manually examine through the students' work to find errors. We suggest the following approach. First, work the homework problem manually, consulting Fate® only when you do not understand which calculation step to complete next. Check your manually calculated answers at each step in the process against Fate® to see if you are correct. If you do not understand why you are doing a calculation, then certainly consult your instructor. The pC–pH Diagram Simulator® works in a similar manner to Fate®, but is used only in Chapter 3 to understand acid–base equilibrium and buffers. The final piece of software is The Water and Wastewater Tutorial, available from the Wiley ftp site at ftp://ftp.wiley.com/public/sci_tech_med/pollutant_fate/ which can be viewed on your own time or in class to illustrate how modern water and wastewater treatment facilities work. You will need to download the free Flash Player® in order to use the three software packages.

ACKNOWLEDGMENTS

First and foremost I would like to thank my editor, Samantha Saalfield, for her turning my vague, poorly worded concepts into a readable manuscript. Howard Drossman (Department of Chemistry, Colorado College) and Brennan Jordan (Department of Geology, Whitman College) served as reviewers of the manuscript and significantly added to the clarity, technical content, and integration of the sciences into the theme of pollutant transport. This textbook would have been impossible without the efforts and patience of many students who served as guinea pigs during the classroom testing of the material. The students include members of my

2004 Environmental Chemistry class: Mary P. Ashby, Bjorn E. Bjorkman, Isaac R. Emery, Marni E. Hamack, Brian T. Logan, Megan M. Murray, Robert M. Olsen, Viet V. Phan, Mark-Cody J. Reynolds, Rishi Shrivastava, Kristen J. Swanzy, Kathryn L. Taylor, and Emily J. Welborn. Robert Olsen also designed the hydrology experiments for this book. I thank all of you for a job well done.

FRANK M. DUNNIVANT

Walla Walla, Washington
September 2005

SYMBOLS

A	activity
ABS	the absorption factor (unitless) that accounts for desorption of the pollutant from the soil matrix and absorption of the pollutant across the skin
AF	soil-to-skin adherence factor
AT	average time period of exposure
BW	body weight
c	speed of light
C	concentration of pollutant C
CA	pollutant concentration in the air
CDI	cumulative daily index, the mass of pollutant taken up per unit body weight and unit time
CF	conversion factor
CF	pollutant concentration factor in the fish or shellfish
CR	contact rate
CS	pollutant concentration in the soil
$C(t)$	concentration as a function of time traveled from a source
$C(x)$	concentration as a function of distance from the source
CW	pollutant concentration in the drinking water
d	depth of a system
e^-	concentration of electrons
E	longitudinal dispersion (eddy) coefficient for streams
E	energy
ED	the exposure duration
EF	exposure frequency
E_H	reduction potential
FI	fraction of soil ingested from the polluted site; the fraction of the daily fish intake from the polluted source
g	gravitational acceleration constant
h	Planck's constant
H_r	height of the release (length)
IR	ingestion rate of water
k	first-order rate constant
K_d	water–solid distribution coefficient
K_{DOM}	water–natural organic matter partition coefficient
K_H	Henry's law constant
K_{OC}	octanol–water partition coefficient
K_p	water–solid partition coefficient
K_w	ionization constant for water
M_0	total mass of pollutant in a syste
NOM	natural organic matter
PC	chemical-specific dermal permeability constant
$p\varepsilon$	negative log of the molar concentration of electrons

xix

pH	negative log of the molar hydrogen ion concentration
pOH	negative log of the molar hydroxide ion concentration
Q_e	volumetric flow rate out of a system
Q_i	volumetric flow rate into a system
Q_m	pollutant source (mass/time)
s	slope of a stream system
SA	surface area
t	time
v	velocity
u	wind velocity (length/time)
w	width of a system
W	rate of continuous discharge into a system
Z	charge of an ion
λ	wavelength of a photon of light
μ	ionic strength
σ_x	horizontal dispersion coefficient (length)
σ_z	vertical dispersion coefficient (length)
γ	activity coefficient

GLOSSARY

Adsorption: The process of accumulating an excess of a chemical on a surface. As used in this text, it refers to the concentrating of pollutant on a mineral or NOM-coated mineral surface.

Advection: The transport of pollution in the direction of flow.

Anaerobic: An aquatic or atmospheric system that does not contain oxygen. This term is the same as anoxic.

Anoxic: Devoid of oxygen.

Abiotic reaction: A chemical reaction that takes place without the aid or in the complete absence of microorganisms.

Basel Convention: An international treaty regulating the reporting, disposal, and transport of hazardous waste. The United States is currently not a member of this treaty.

Biota: Any living organism in an ecosystem.

Biotic reaction: A chemical reaction that occurs due to a microbial process (enzymes in the microbes cell).

BOD: Biochemical oxygen demand, the amount of dissolved oxygen required by microorganisms to oxidize organic matter present in the water.

Cation exchange capacity (CEC): The concentration of sorbed cations that can be readily exchanged for other cations.

Confined aquifer: An underground body of flowing water that is located below an layer of strata that is impermeable to water.

Conservative tracer: A chemical tracer that does not degrade and is not sorbed by anything in the water. The chemical moves freely with the water.

Contaminant: A chemical that is out of its proper place. In this text we will use the term contaminant and pollutant interchangeably.

Dispersion: A mixing process resulting from advection which always dilutes the concentration of pollutant.

DNAPL: Dense nonaqueous phase liquid. An example is carbon tetrachloride.

DO: Dissolved oxygen; the concentration of dissolved oxygen in water, usually from 0 to 14 mg/L.

DOM: Dissolved organic matter. These are components of NOM that are soluble is water.

E: The longitudinal dispersion (eddy) coefficient for streams.

Environmental Impact Statement: A study required by the National Environmental Policy Act to attempt to determine if any adverse affects will occur from a governmental action such as the building of a building or plant or even a remediation effort.

Empirical: A relationship based on experiment data.

Epilimnion: The upper region of a stratified lake.

Eutrophication: An overproductive aquatic system. Excess alga growth occurs during the day due to the presence of excess nutrients, but during nighttime hours the oxygen is depleted below levels that can support aerobic life.

Explanative modeling: A model that attempts to explain how something happened. For example, modelers can use mathematical relationships to explain how a pollutant moved to where it is, where it originally came from, or how much pollutant was originally released.

Head: The height of a water column.

Hypolimnion: The lower or bottom region of a stratified lake.

LNAPL: Light nonaqueous phase liquid. An example is gasoline.

Longitudinal dispersion: The mixing of pollution in water or air in the direction of flow.

Modeling: An attempt to explain a process in a simpler form. Models can take on several forms including physical models, which are usually small-scale versions of the real thing. We will limit our discussions in the text to mathematical models that are very simple mathematical relations of more complicated processes. The more terms we include in our models, the more processes we account for in our model and theoretically the more accurate our model will predict the real system.

NAPL: Nonaqueous phase liquid. An example would be oil or gasoline.

Nitrobenzene: A benzene molecule with an attached nitro (NO_2) group. Nitrobenzenes can also have other functional groups such as methyl (CH_3), chloro (Cl), and many other chemical groups attached to the benzene ring.

NOM: Natural organic matter. NOM results from the accumulation of degradation products from plants and animals in soil and water.

Non-point pollutant source: A source of pollutant that cannot be identified as a specific location. Examples would be runoff from a large area of land such as a parking lot or agricultural setting and atmospheric inputs.

Partitioning: Very similar to adsorption but partitioning does not involve a site specific reaction. It is more of a solvation or dissolving of a pollutant into NOM.

pC: A way of representing concentration units on the log scale. p stands for the negative log of anything. C stands for the concentration (in any units) of any chemical species.

pH: The negative log of the hydrogen ion concentration.

Point pollutant source: A source of pollution that can be pinpointed to a specific location. For example, the output pipe from an industrial process or sewage treatment plant would be a point source.

Predictive modeling: A model that attempts to predict what will happen at a future time. For example, in this text we are concerned with predicting the concentration of pollutant at some point (location or time) in the future.

Pulse or instantaneous release of pollutant: A release of pollutant that occurs over a very short time scale and contains a finite volume or mass of pollutant. This type of release is in contrast to a step or continuous pollutant release.

Refractory pollutant: A pollutant that does not readily degrade (degrades slowly or not at all).

Remediate: To clean up a waste site to acceptable pollutant concentrations.

Residence time: The average time a chemical spends in an environmental compartment. This can be obtained by dividing the volume of the compartment or mass of a chemical in a compartment by the outflow from the system (volume/volume per time = average residence time).

Sensitivity analysis: An iterative process of testing a model where one parameter at a time (volume, rate constant, etc,) is systematically increased or decreased while the result (pollutant concentration) is recorded.

Sorption: A generic term referring to adsorption and partitioning.

Stratification: A process that results in two distinct layers of water in a lake system. Stratification results due to heating of the surface water and cooling of lower waters by the Earth that sets up a density difference in the two bodies of water. The cool water settles to the bottom of the lake, while the warmer water is present at the surface.

Surface aquifer: The groundwater closest to the land surface.

Step or continuous pollutant release: A release of a pollutant that occurs over a long time scale. Examples include the constant release of sulfide from a pulp mill and the release of nitrate from a sewage treatment plant.

Vadose zone: The portion of the ground that is unsaturated with respect to water.

Variable: A symbol representing a mathematical term or the term of interest in a mathematical expression. For example, velocity is an important term in fate and transport, and it is represented by the variable, v.

Watershed: The drainage area of land surrounding and feeding water into a lake or river basin.

PART *I*

INTRODUCTION

"Through the history of literature, the guy who poisons the well has been the worst of all villains."

—Author unknown

SOURCES AND TYPES OF POLLUTANT, WHY WE NEED MODELING, AND HISTORICAL CONTAMINATION EVENTS

1.1 INTRODUCTION

A good starting point for our discussions and this book is to ask, "Is there a common sequence of events leading to the identification, characterization, and remediation of a hazardous waste site?" There are a variety of answers to this question, but a general order of events often occurs as follows.

- First, a pollutant is observed to be present or the potential of a pollutant release from a proposed industrial site is identified. This can result from routine monitoring of a pollutant's concentration at the site, through the known manufacturing of the pollutant at the site, through research identifying the cause of an illness or cancer cluster in the community of a site, or during an environmental planning assessment (also called an environmental impact assessment, EIS).

- Second, the source of the pollutant is identified at the hazardous waste site or a theoretical release can be simulated.

- Third, pollutant fate and transport modeling is completed to determine what pollutant concentrations will result at specific points at the site over time (referred to as receptor sites where humans are the receptors of the pollutants).

- Fourth, the results of the pollutant fate and transport modeling are used in risk assessment calculations to estimate health risks.

- Fifth, a decision or plan of remediation is negotiated between the local citizens, local and federal governments, and the party responsible for the hazardous waste site.

These steps also outline our approach in this book. In this chapter, we will look at several historical hazardous waste release events and describe types and sources of potential pollutant releases. In Chapter 2 and 3, we will look in-depth at the chem-

A Basic Introduction to Pollutant Fate and Transport, By Dunnivant and Anders
Copyright © 2006 by John Wiley & Sons, Inc.

istry associated with fate and transport phenomena in environmental media. In the next set of chapters (4–10), we will develop and learn to use pulse and step fate and transport models for rivers, lakes, groundwater, and the atmosphere. In Chapter 11, we will use pollutant concentrations from the fate and transport modeling to perform basic risk assessment calculations to estimate risk to human health. Environmental legislation for the United States and Europe will be covered in Chapter 12. Finally, in Chapter 13 we will look at the history of several pollutants that are present at undesirable concentrations across the globe.

1.2 THE NEED FOR MODELING OF POLLUTANTS IN ENVIRONMENTAL MEDIA

Many pollutants have been found to be ubiquitous in nature; that is, every environmental compartment that has been tested has shown some level of contamination. Two of the chemicals that fall into this category are PCBs (polychlorinated biphenyls) and DDT (1,1'-(2,2,2-trichloroethylidene)bis[4-chloro-benzene]). These compounds accumulate especially well in the environment due to their refractory behavior (they are not easily chemically or biologically degraded). Modelers like to divide nature into more easily mathematically described boxes, such as the atmosphere, rivers, lakes, groundwater, and biota (living organisms). In this book we simplify this further by only considering one section of a river, a small portion of a groundwater aquifer, or a small portion of the atmosphere.

There are two basic goals in pollutant modeling: to explain how a pollutant got where it is (a form of thermodynamic equilibrium) and to predict how fast a pollutant will move through an environmental compartment in the future (a form of kinetics). Later in this chapter we will look at several examples of major pollutant release events; the approach taken to these reflects the first goal of modeling (explanation of how a pollutant got where it is). Although this approach is very important to understanding how pollutants move in the environment, a more common use of modeling today is to predict if and/or how fast a pollutant will move once it is accidentally released. In the United States this is included in an Environmental Impact Statement. Most developed countries require a variety of industries to perform some type of environmental impact modeling in the unlikely and unfortunate event of an accidental release. This book will concentrate on predictive modeling, but by understanding the processes involved in the predictive modeling approach, it is relatively easy to appreciate how the explanation type of modeling is completed.

In order to predict where a pollutant will go, we use a set of fate and transport equations that describe the chemical and physical processes occurring in the environmental compartment under evaluation, such as mixing, outflow through the effluent of the system, evaporation, volatilization, and chemical or biological degradation. The same modeling equations and concepts are used for the explanation type of modeling (where pollution already exists), but a reverse process is used. In explanation modeling, the modeling equation is fit to pollutant concentration data from field measurements in order to obtain estimates of mixing and degradation rates in the specific system. Both modeling approaches will become clearer when we reach the modeling chapters in this text, especially Chapter 4.

1.3 POLLUTION VERSUS CONTAMINATION; POLLUTANT VERSUS CONTAMINANT

Is there really a difference between the terms pollutant and contaminant? Some would argue no, while others will argue vehemently yes. In the United States, environmentalists and those working for Environmental Protection Agency (EPA) prefer the terms pollution and pollutant. But what makes a chemical a pollutant? *Basically this distinction is determined by where the chemical is located and how much of the chemical is present.* For example, a bottle of mercury chloride is typically not considered a pollutant when it is sitting on the shelf in a chemical storeroom. But pour that same chemical down the drain, and it immediately becomes a pollutant, even at extremely low concentrations. This is the logic behind usage preferred by other branches of the United States government, such as the Department of Energy (DOE). The DOE prefers to use the terms contamination and contaminant. If a chemical is in its proper place it is not considered a problem and is thus not a contaminant. Likewise, if a chemical is present in an environmental media below the concentration deemed to be a problem (which is subject to a variety of views and laws), then the chemical is not a problem and contamination is not present. Thus, the legal definition usually specifies two factors, where the pollutant is and its concentration. For example, the first author of this book worked on a project for the Idaho Engineering and Environmental Laboratory (INEEL), one of many U.S. DOE sites, in which we intentionally placed a rapidly degrading radioactive substance in a water-filled lagoon in order to characterize the movement of water and radionuclides in the subsurface media. As long as the chemicals stayed within the controlled boundaries of the lagoon (the approved experimental area) they were not considered a contaminant. But one day a violent storm blew a small amount of the water and sediment a few feet out of the lagoon. The DOE safety officials went crazy. Although the vast majority (I would speculate over 99.99%) of the chemical was still present in the lagoon, the extremely small amount of chemical "released" (amounting to a few specks of dirt) was officially classified as a contamination event. So, when you use these terms be aware of who you talking to. In this book we will use the more common terms pollutant and pollution.

1.4 POLLUTION CLASSIFICATIONS

There are perhaps as many ways to categorize types of pollution as there are pollutants. One broad way is to categorize by physical phase: solid, liquid, or gaseous. This categorization is useful from a standpoint of treatment and disposal. Many treatment technologies are based on the physical phase of the pollutant, such as bag filtration houses for atmospheric particulates or filtration of particles in liquids. Another general but more chemical means of categorization is the inorganic, organic, and radioactive nature of the waste. Mixed waste, a very difficult waste to treat, usually refers to organic and radioactive waste being present in the same waste product. Inorganic waste can be further broken down into nontoxic and toxic metals, metalloids, and nonmetals. Metal waste can be subdivided into heavy metals and transition metals. Toxic wastes can be grouped as carcinogens, terratogens, and mutagens (discussed in Chapter 10). Compounds that are subjects of heightened public awareness, such as PCBs, are divided out even further. Other wastes are listed as hazardous based solely on their origin from a specific industrial process (for example, metal plating wastes) or when a specific chemical is present in the waste stream (for example, the presence of PCBs). As you see, there are many ways to categorize or list a waste, and each country will have their own system for classifying pollutants.

Another interesting way to categorize waste is by the risk it posses. For example, say you have a hazardous waste site that needs to be remediated and there are 20 pollutants of interest. But, from a risk standpoint, 5 of the 20 pose a significantly higher hazard based on risk assessment calculations. In some countries it is customary to focus the remediation effort on these 5 chemicals and disregard the other 15 "less hazardous pollutants." Such an approach is used by the U.S. Environmental Protection Agency to calculate health risk based on source, and an overview of the approach is shown in Figure 1.1.

1.5 SOURCES OF POLLUTION

As with types of waste, there are many ways of categorizing sources of waste (or potential pollutants). We will only mention a few of the more useful approaches and give examples of generated volumes from major industrialized countries. First, we will discuss point and nonpoint sources.

Sources of pollutants are commonly divided into one of two types, point or nonpoint sources. Point sources are well-defined sources such as the end of a pipe, smoke stack, or drain. Nonpoint sources are less well defined, but contain all sources where you cannot directly pinpoint the emission. The distinction becomes a bit more vague, depending on physical scale. For example, if you are studying a large watershed (the land surface that drains into a stream or lake), a cattle holding lot could be considered a point source. Obviously the farm is the source of the cattle waste. But on a smaller scale, what is the point source? The answer is possibly each cow, or the entire holding lot, since the pollution will be spread over its entirety. But, in general the terms point and nonpoint are used to describe sources.

Table 1.1 lists sources of waste by 10 categories: agricultural, chemical industry, mining industry, energy industry, municipal and hazardous landfills, medical industry, food processing, domestic waste, municipal government, and federal government. General wastes for each source are also listed in the right-hand column. This list is certainly not complete, but contains the major sources of waste generated and provides a starting point for discussion of waste sources.

Since this book deals mostly with the transport of toxins, we will be especially concerned with hazardous waste. A summary of the amounts of hazardous waste

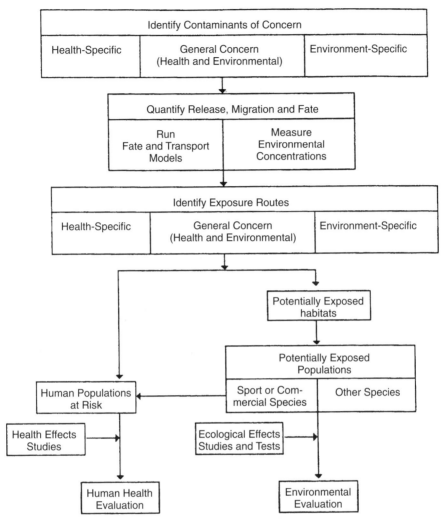

Figure 1.1.   An alternative method of assessing risk (U.S. Environmental Protection Agency, 1989).

generated by country members of the Basel Convention (an international treaty regulating the reporting, disposal, and transport of hazardous waste) is given in Table 1.2 for the year 2000. The magnitudes of these values are stunning, and, in general, the more developed the country, the more hazardous waste it generates. This correlation of waste to economical level is a homework exercise that you should conduct, and, by the way, it would make an excellent exam question. You will undoubtedly note the lack of data on two highly industrialized countries: Japan and Germany. Even without these two countries, the total amount of hazardous waste annually generated is an obscene 93,992,999 metric tons. Another country you will note absence

TABLE 1.1. Potential Sources of Chemical Emissions (Point and Nonpoint Sources)

Source	General Waste (Representative but not Exhaustive)
Agricultural	Field and chemical wastes
	Nutrients (fertilizers)
	Pesticides/herbicides
	Petroleum fuels
	Feedlot waste
	Dairy waste
Chemical industry	Metal products
	Metal sludges
	Nonmetal waste
	Electrical equipment waste
	Detergents/soaps/cleaners
	Petroleum-related waste
	Metal plating
	Cooling-tower additives
	Film-processing waste
	Solvents
	Wastewaters and off-specification wastes
	Pesticides
	Insecticides
	Chlorinated hydrocarbons
	Organophosphates
	Carbamates
	Microbial insecticides
	Pyrethroids
	Rodenticides
	Herbicides
	Petroleum oils
	Carbamates
	Triazines
	Phenoxy herbicides
	Amide herbicides
	Fungicides
	Precursors to smog (NO_x, hydrocarbons)
Mining industry	Mine tailings
	Mineral leaching (cyanide)
	Acid mine drainage
	Coal
	Smelting waste
	Atmospheric particulates
Energy industry	Petroleum-based waste
	Solvents
	Gas and vapor emissions
	Coal tars
	Boiler waste
	Nuclear waste
	Petroleum stored in underground storage tanks (gasoline stations)
	Precursors to acid rain and smog

(*Continued*)

TABLE 1.1. Potential Sources of Chemical Emissions (Point and Nonpoint Sources) (continued)

Source	General Waste (Representative but not Exhaustive)
Municipal and hazardous landfills	All chemicals disposed in the landfill
Incinerators	Incomplete combustion of feedstocks
	Combustion by-products
	Metals
	Particulates
Medical industry	Biodegradable waste (biohazards)
	Pharmaceutical waste
	Solvents
Food processing	Waste food products
	Rinsing waste
	Slaughterhouse waste
Domestic waste	Detergents/cleaners
	Pesticides, etc.
	Fertilizers
	Compose materials
	Paints/solvents
	Gasoline
Municipal governments	Chemicals associated with water and wastewater treatment
	Sewage (biochemical oxygen demand)
Federal governments	Weapons-related waste (conventional and nuclear)
	Nuclear waste
	Petroleum-based waste

from the list is the United States, since it has not signed the Basel Convention. For comparison purposes, the United States generated 40,821,481 tons in the year 2001 (EPA, 2001), which is 43.4% of the world total shown in Table 1.2. A summary of this waste for the United States is given in Table 1.3 by state and territory, and in Table 1.4 by generator (U.S. Environmental Protection Agency, 2001). Review Table 1.4 and see if you recognize any of the company names. An additional homework exercise would be to conduct an Internet search of a company and determine what are their waste chemicals.

From the standpoint of modeling, the subject of this text, we would like to be able to express our waste emission or "sources" in mathematical terms. Two mathematical source terms we will use throughout this text will be the "pulse" and the "step" inputs of pollutant. The pulse and step inputs represent both extremes of the duration of pollutant inputs to a system. You should familiarize yourself with these terms. These two terms can be slightly confusing, mostly because of the use of multiple names for them. A pulse release, also commonly referred to as an instantaneous release, is one that occurs over a small time scale (an instant in time, hence instantaneous), but may contain large or small amounts of pollutant. Examples of pulse releases would be the immediate release of one gallon of acetone or antifreeze into

TABLE 1.2. Total Hazardous Waste Generated in Year 2000 by the Parties to the Basel Convention

Country	Total Hazardous Waste Generated (Metric Tons)
Austria	980,558
Bulgaria	755,677
Canada	No data
China	8,300,000
Croatia	25,999
Cuba	1,023,638
Czech Republic	2,603,337
Denmark	287,491
Ecuador	85,859
Egypt	170,000
Estonia	5,965,750
Finland	1,203,000
France	9,150,000
Georgia	93,000
Germany	No data
Hungary	3,392,628
Iceland	12,550
Israel	279,987
Italy	No data
Japan	No data
Jordan	17,390
Kyrgyzstan	6,087,869
Luxembourg	96,526
Malaysia	344,550
Morocco	987,000
Netherlands	2,722,828
New Zealand	No data
Norway	650,000
Oman	242,098
Poland	1,627,143
Portugal	194,724
Qatar	280
Republic of Korea	2,756,984
Republic of Moldova	7,122
Romania	860,892
Russian Federation	12,800,000
Singapore	121,500
Slovakia	1,600,000
Slovenia	128,395
Spain	3,293,705
Sri Lanka	40,617
Sweden	1,100,000
Ukraine	2,613,400
United Kingdom	6,296,043
Uzbekistan	15,074,459
Total	93,992,999

Source: www.basel.int/natreporting/index.html, accessed December 8, 2003.

TABLE 1.3. Rank Ordering of States Based on Quantity of RCRA Hazardous Waste Generated and Number of Hazardous Waste Generators, 2001

State	Hazardous Waste Quantity			Number of Generators			Reported Status	
	Rank	Tons Generated	Percentage	Rank	Number	Percentage	LQG	Non-LQG
Texas	1	7,555,402	18.5	6	879	4.6	874	5
Louisiana	2	3,883,563	9.5	13	462	2.4	420	42
New York	3	3,534,261	8.7	2	1,992	10.5	1,990	2
Kentucky	4	2,686,583	6.6	20	316	1.7	316	0
Mississippi	5	2,165,734	5.3	32	162	0.9	157	5
Ohio	6	1,889,067	4.6	3	1,071	5.6	955	116
Minnesota	7	1,662,632	4.1	24	259	1.4	256	3
Kansas	8	1,571,587	3.8	26	223	1.2	208	15
Alabama	9	1,569,675	3.8	22	270	1.4	266	4
Illinois	10	1,412,100	3.5	4	955	5.0	954	1
Indiana	11	1,127,542	2.8	9	625	3.3	491	134
Massachusetts	12	1,121,752	2.7	14	435	2.3	430	5
New Mexico	13	962,808	2.4	46	41	0.2	38	3
Oklahoma	14	887,643	2.2	30	169	0.9	135	34
Arkansas	15	857,910	2.1	28	201	1.1	185	16
California	16	807,297	2.0	1	2,544	13.4	2,520	24
Georgia	17	760,043	1.9	17	363	1.9	362	1
Michigan	18	649,207	1.6	8	786	4.1	571	215

(Continued)

11

TABLE 1.3. Rank Ordering of States Based on Quantity of RCRA Hazardous Waste Generated and Number of Hazardous Waste Generators, 2001 (continued)

State	Hazardous Waste Quantity			Number of Generators			Reported Status	
	Rank	Tons Generated	Percentage	Rank	Number	Percentage	LQG	Non-LQG
Tennessee	19	629,834	1.5	15	396	2.1	393	3
New Jersey	20	586,210	1.4	5	892	4.7	891	1
North Dakota	21	574,614	1.4	52	15	0.1	13	2
Hawaii	22	464,857	1.1	47	36	0.2	31	5
Florida	23	400,107	1.0	16	376	2.0	355	21
Pennsylvania	24	398,403	1.0	7	868	4.6	868	0
North Carolina	25	329,721	0.8	12	473	2.5	443	30
Wisconsin	26	294,754	0.7	11	489	2.6	489	0
Nevada	27	277,258	0.7	39	78	0.4	78	0
Washington	28	240,795	0.6	10	506	2.7	506	0
Idaho	29	214,409	0.5	43	48	0.3	31	17
Virginia	30	209,447	0.5	23	265	1.4	264	1
Puerto Rico	31	176,555	0.4	36	84	0.4	84	0
South Carolina	32	142,510	0.3	19	319	1.7	290	29
Missouri	33	101,782	0.2	21	298	1.6	285	13
West Virginia	34	101,195	0.2	35	131	0.7	131	0
Arizona	35	96,544	0.2	29	193	1.0	190	3
Utah	36	88,664	0.2	36	84	0.4	83	1
Colorado	37	66,791	0.2	33	144	0.8	135	9

38	Connecticut	62,524	0.2	18	360	1.9	357	3
39	Oregon	49,945	0.1	27	206	1.1	206	0
40	Iowa	47,147	0.1	30	169	0.9	153	16
41	Wyoming	37,566	0.1	48	27	0.1	22	5
42	Nebraska	31,382	0.1	38	83	0.4	79	4
43	Maryland	17,577	0.0	53	14	0.1	14	0
44	Delaware	17,512	0.0	41	66	0.3	65	1
45	New Hampshire	12,269	0.0	25	231	1.2	166	65
46	Rhode Island	9,435	0.0	34	132	0.7	128	4
47	Trust Territories	8,999	0.0	55	3	0.0	2	1
48	Montana	6,877	0.0	45	44	0.2	38	6
49	Maine	6,168	0.0	40	77	0.4	70	7
50	Alaska	5,094	0.0	42	50	0.3	48	2
51	Vermont	4,099	0.0	44	47	0.2	46	1
52	District of Columbia	2,113	0.0	50	18	0.1	18	0
53	Virgin Islands	1,971	0.0	56	1	0.0	1	0
54	South Dakota	950	0.0	51	16	0.1	16	0
55	Guam	412	0.0	49	26	0.1	12	14
56	Navajo Nation	188	0.0	54	6	0.0	6	0
	Total	**40,821,481**	**100.0**		**19,024**	**100.0**	**18,135**	**889**

Note: Columns may not sum due to rounding.

Reporting requirement changes for the 2001 National Biennial will make cursory comparisons of the 2001 National Biennial Report to National Biennial Reports prior to 2001 misleading. Refer to the introduction for a complete explanation.

Source: National Biennial RCRA Hazardous Waste Report: Based on 2001 Data.

TABLE 1.4. Fifty Largest RCRA Hazardous Waste Generators in the United States, 2001

Rank	EPA ID	Name	City	Tons Generated
1	TXD008080533	BP Products North America Inc.	Texas City, TX	2,039,862
2	LAD008213191	Rubicon Inc.	Geismar, LA	1,856,429
3	KSD007482029	Vulcan Materials Co.	Wichita, KS	1,530,780
4	LAD008175390	Cytec Industries Inc.	Waggaman, LA	1,444,351
5	NYD002080034	GE Silicones, LLC	Waterford, NY	1,420,189
6	MSD096046792	E. I. Du Pont DE Nemours and Co.	Pass Christian, MS	1,287,978
7	TXD059685339	Diamond Shamrock Refining and Marketing	Sunray, TX	1,257,357
8	NMD048918817	Navajo Refining Company	Artesia, NM	956,611
9	OHD042157644	BP Chemicals Inc.	Lima, OH	913,555
10	TXD083472266	Lyondell Chemical Company	Channelview, TX	880,382
11	TXD008123317	E. I. Du Pont De Nemours and Company	Victoria, TX	832,632
12	MND006172969	3M Company	Cottage Grove, MN	718,536
13	TXD008079642	E. I. Du Pont De Nemours and Company	Orange, TX	610,350
14	NDD006175467	Tesoro-Mandan Refinery	Mandan, ND	573,556
15	OKD000829440	Zinc Corporation of America	Bartlesville, OK	525,938
16	KYD006373922	Atofina Chemicals, INC.	Carrollton, KY	507,659
17	ARD043195429	Great Lakes Chemical Corp Central	El Dorado, AR	486,514
18	NYD000707901	IBM Corporation-E Fishkill Facility	Hopewell Junction, NY	467,961
19	HID056786395	Tesoro Hawaii Corporation Refinery	Kapolei, HI	464,076
20	ILD042075333	Cabot Corp.	Tuscola, IL	442,406
21	ALD046481032	Sanders Lead Company, Inc.	Troy, AL	427,000
22	MSD008186587	Morton International A Rohm & Haas Co.	Moss Point, MS	405,135
23	IND003913423	Bethlehem Steel Corp. Burns Harbor	Burns Harbor, IN	393,494
24	MSD033417031	First Chemical Corporation	Pascagoula, MS	357,825
25	TXD008081697	BASF Corporation	Freeport, TX	333,344
26	NYD000824482	Occidental Chemical Corporation	Niagara Falls, NY	319,693

27	ALD004009320	Hunt Refining Company	Tuscaloosa, AL	316,146
28	MID006013643	Pfizer Inc. Parke-Davis & Co.	Holland, MI	309,388
29	OKD987072006	Norit Americas Inc., Pryor Facility	Pryor, OK	306,348
30	FLR000068007	K. C. Industries, LLC, Mulberry, Florida	Mulberry, FL	287,413
31	GAD003275252	International Paper Co.	Savannah, GA	279,828
32	NVT330010000	U.S. Ecology, Inc.	Beatty, NV	270,654
33	CAR000081422	Brite Plating Co. Inc	Los Angeles, CA	265,205
34	MND006253801	Superior Plating Inc.	Minneapolis, MN	259,434
35	NYD002245967	Reynolds Metals Company	Massena, NY	253,969
36	KYD006371314	Noveon Inc.	Louisville, KY	238,881
37	KYD006384531	Harshaw Chemical Co.	Louisville, KY	211,777
38	IDD070929518	FMC Idaho LLC	Pocatello, ID	209,532
39	TXD087491973	Asarco INC.	Amarillo, TX	199,005
40	MAD007325814	Texas Instruments INC.	Attleboro, MA	190,643
41	NJD002385730	Dupont Chambers Works	Deepwater, NJ	189,594
42	MAD000189068	M/A-Com INC., A Division of TYCO Interna	Burlington, MA	186,953
43	NYD002232304	ALCOA Inc.	Massena, NY	181,477
44	NCD047368642	E. I. Du Pont & Co.—Fayetteville Works	Fayetteville, NC	181,379
45	KYR000011718	Guardian Automotive Morehead Plant	Morehead, KY	175,726
46	GAD063152573	SAFT America, INC	Valdosta, GA	174,139
47	KYD048878805	EPT Drives & Components Div. Operations	Maysville, KY	160,321
48	TND003376928	Eastman Chemical Company, Tennessee OPE	Kingsport, TN	154,410
49	ARD006354161	Reynolds Metals Company	Arkadelphia, AR	154,134
50	ALD000608216	Hager Companies—Montgomery	Montgomery, AL	151,263
Total				**26,761,230**

Note: column may not sum due to rounding.
Reporting requirement changes for the 2001 National Biennial will make cursory comparisons of the 2001 National Biennial Report to National Biennial Reports prior to 2001 misleading. Refer to the introduction for a complete explanation.

Source: National Biennial RCRA Hazardous Waste Report: Based on 2001 Data.

a river, or the same into a groundwater well, or the flushing of one toilet of human waste into a stream. We use the terms pulse and instantaneous to stress that the release occurs immediately, travels through the system as a pulse of pollution (a narrow band), and is not a prolonged release.

In contrast, a step or continuous release occurs over longer periods of time. For modeling purposes we like to assume that the release occurs indefinitely. Examples of step releases would be the constant release of a pollutant from an industry at "acceptable" or "nonacceptable" levels, the continuous release of leachate from a landfill into the groundwater system, and the 24-hour outflow from a sewage treatment plant. Pollutant release concentrations can be high or low, but as long as they are constant, we can model them using relatively simple transport equations.

Many pollutant sources fit into these two simple categories, but you might think of them as end-members because there are also other pollutant sources that do not fit neatly into these categories. For example, suppose an industry legally emits a pollutant for 5 years, but installs better waste treatment technology or shuts down the process completely. How could we treat this? One approximation would be to use a step model for the first five years and then switch to a pulse model. To be more accurate, we would derive a specific model to fit the exact emission scenario. As noted earlier, the pulse and step inputs are very common, and since this is an introductory text on fate and transport, we will only use models based on these two types of inputs. As long as there is an equation describing the pollution input, mathematicians can modify the final fate and transport equation to model the pollution as it moves through an environmental compartment.

1.6 HISTORIC EXAMPLES OF WHERE FATE AND TRANSPORT MODELING ARE USEFUL

There is no lack of unfortunate and disastrous pollution events that fate and transport models have been applied to in an effort to better understand and be able to predict pollutant movement in the environment. These are divided into surface water, groundwater, and atmospheric events. You may recall many of the events, discussed below, from news reports or case studies that you have covered in other classes.

1.6.1 Surface Water

The Release of Biochemical Oxygen Demand (BOD) Waste to Streams.
One of society's greatest accomplishments has been the treatment and disposal of our sewage waste. However, you may be shocked to learn that as late as the 1980s, several major coastal cites in the United States only partially treated their sewage before dumping it into estuaries, bays, and oceans. Many major cities across the world are still modernizing their treatment facilities, and huge strides have been made in this effort. Since the installation of sewage treatment plants, developed countries have eradicated many diseases associated with human waste. But from the standpoint of nature, human pathogens are of little concern and the main problem is the large amount of biodegradable organic waste, known as biochemical oxygen

demand (BOD), that is released into natural systems. Water contains relatively small amounts of dissolved oxygen (DO) and when untreated sewage is released into a stream, the oxygen needed to oxidize these organic wastes far exceeds the oxygen present in the stream water. This causes the stream to become anaerobic (devoid of dissolved oxygen), and the biological organisms dependent on this DO are killed. In addition, the stream smells like an open sewer, which is essentially what it is.

The modeling of the consumption of DO by organisms oxidizing BOD is relatively simple. The governing equation, known as the Streeter–Phelps equation, will be discussed in Chapter 6. For now, we will only concern ourselves with a plot of DO downstream from a sewage treatment plant. Figure 1.2 shows the DO profile of a stream starting at the input of untreated sewage and downstream from the inlet point. Note how the DO immediately plummets as waste enters the streams and microbes consume the organic matter. In fact, the stream modeled here becomes anaerobic from 180 to 430 km downstream from the waste input (an incredibly long distance) and only slowly recovers as the stream undergoes natural re-aeration and as the BOD is oxidized by microorganisms.

Now note the effect of adding a sewage treatment plant to the system where sewage waste is treated prior to release to the stream (Figure 1.3). In this modeling effort, we assume that 95% of the BOD is removed, which is easily achievable in modern treatment facilities. The DO of the stream and its associated wildlife are not noticeably affected by the oxidized organic matter (treated waste). Thus we cannot only predict the effects of adding sewage to a stream, but we can also evaluate the effects on the stream by adding treatment systems.

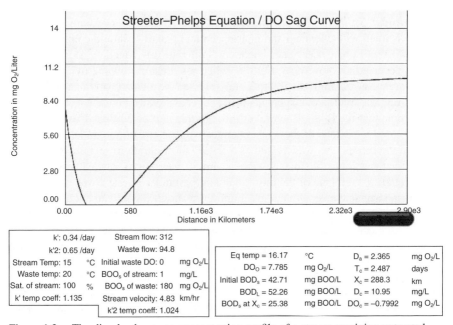

Figure 1.2.   The dissolved oxygen concentration profile of a stream receiving untreated sewage waste. (Results from Fate®.)

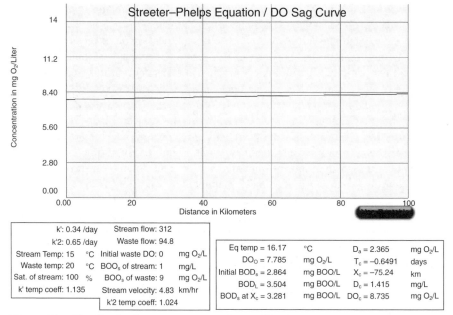

Figure 1.3. The dissolved oxygen concentration profile of a stream receiving 95% treated sewage waste. (Results from Fate®.)

A Chemical Spill on the Rhine River.

The Rhine River system in Western Europe is lined with chemical and pharmaceutical plants. It should come as no surprise that there is an occasional chemical spill into this aquatic system. A major release occurred on November 1, 1986, when a storehouse owned by Sandoz Ltd. near Basel, Switzerland caught fire and released pesticides, solvents, dyes, and various raw and intermediate chemicals (Capel et al., 1988). Figure 1.4 shows a map of the route that the Rhine River takes through this part of Europe. This release has been referred to as "one of the worst chemical spills ever" (Anonymous, 1987). For modeling purposes, this was treated as a pulse (short duration) release. Monitoring points for the various chemicals were set up at the stations labeled in Figure 1.4 (Capel et al., 1988). One of the most abundant pesticides released was disulfoton, a thiophosphoric acid ester insecticide. Figure 1.5 shows the movement and flushing of disulfoton through the Rhine (Capel et al., 1988). Note the bell or Gaussian shape of the concentration profile, which is characteristic of a pulse release. As the pulse of disulfoton moved downstream, it was diluted and degraded as indicated by the broader peaks and lower concentrations shown in Figure 1.5. This is a case where the model can be fit to the data to better understand how pollutants move through the system and thus can be used to predict downstream concentrations of later releases of pollutants. Of course this does not help the wildlife affected by a disaster, but it can help to minimize the impact on drinking water systems that can be shut down as the peak concentrations of pollution pass.

Map of the Rhine River with its monitoring stations[a]

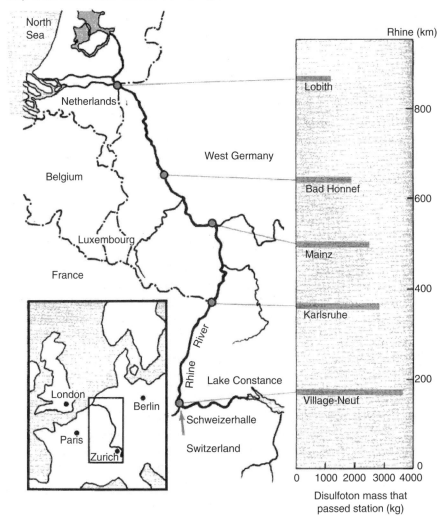

[a]The dashed lines are international borders. The graph on the right side shows the
mass of disulfoton that passed the five monitoring stations.

Figure 1.4.   The route of the Rhine River in Western Europe. [Reprinted with permission
from Capel et al. (1988). Copyright 1988 American Chemical Society.]

A Chemical Spill on the Tisza River Bordering Hungary and Romania.   On
January 30, 2000 a mine tailing dam, owned by the Aurul SA Baia Mare Company,
broke and released a cyanide-rich waste into the Somes River which subsequently
drained into the Tisza and Danube before entering the Black Sea (UN, 2000). It has
been estimated that nearly 100,000 cubic meters of aqueous and suspended cyanide
and heavy metals were released that resulted in the death of 500 tons of fish. This

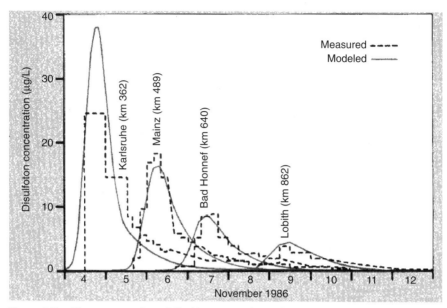

Figure 1.5. The concentration of disulfoton at each monitoring station on the Rhine river. [Reprinted with permission from Capel et al. (1988). Copyright 1988 American Chemical Society.]

pollution event, like the one of the Rhine discussed above, illustrates the international effort needed in controlling release events and the politics of such a release. While the company in Switzerland bore the full responsibility (liability) for the Rhine incident, Romania took the stand that they had no international treaties with Hungary and that they were therefore not responsible for damages to the fishing industry (Schaefer, 2000). Liable or not, if an environmental impact study (utilizing fate and transport modeling) had been conducted showing the possible extent of damage to the affected rivers, more stringent safety measures would certainly have been in place.

1.6.2 Groundwater

Polluted groundwater systems are undoubtedly the most difficult and costly to remediate (clean up). Currently, many rural and urban citizens around the world depend on groundwater for their source of drinking, industrial, and agricultural water. This is becoming more and more common as the global population increases and many surface water resources become unfit for use as drinking water. Yet, in the past, placement of liquid waste in a "deep" well was considered a perfectly acceptable and legal means of disposal. "Deep" is a relative term, but it was understood that the layer of groundwater (aquifer) that received the waste was not a surface aquifer (not confined by an impermeable layer of geologic media). There are a number of deep-well injection sites across the globe, and while many developed nations have made this practice illegal, it may still be practiced in some parts of the world.

INEEL Test Area North Deep Well Injection.   The deep-well injection project that the authors are most familiar with was located on the Idaho National Engineering and Environmental Laboratory (INEEL) at the Test Area North (TAN) site. This U.S. government owned and operated hazardous waste site was placed on the U.S. EPA's National Priorities List (NPL) on November 21, 1989. Injection well TSN-05 was drilled in 1953 to a depth of 93 m (305 ft) with a diameter of 30.5 cm (12 in.). The groundwater surface at the site is approximately 63 m (206 ft) below the land surface. During its years of operation (not given in the report), the well received approximately 133,000 L (35,000 gallons, 193,000 kg) of liquid and dissolved trichloroethylene (TCE), organic sludges, treated sanitary sewage, metal filing process waters, and low-level radioactive waste streams. Basically the well was the means of disposal for any liquid or semi-liquid waste (domestic and hazardous) produced at the site. Although several pollutants of concern were disposed of and detected in the groundwater at TAN, we will only show the results for TCE.

Figure 1.6 shows iso-concentration circles (isopleths; lines of equal TCE concentration) for the TAN that were obtained by fitting a step-model to field measurements of TCE concentrations in the groundwater (explanatory modeling). Since TCE is suspected of being present in the aquifer as a dense nonaqueous phase liquid (DNAPL), it solubilizes slowly and the input of TCE can be considered a continuous source (step input over an extended period of time). You will note that near TSN-05 injection well the isopleth is for 1000 parts per billion (ppb) while the lowest concentration shown in this figure is 3 ppb (the maximum allowed drinking water concentration). If no remedial action was to be taken in 1994 (the proposed year of remediation if any was to be attempted) and the DNAPL continued to release TCE to the groundwater, another step model predicted that the TCE plume would expand to the one shown in Figure 1.7 for the year 2044.

So, how do we decide what is to be attempted in the way of remediation? Pollutant fate and transport modeling can be one of many tools to evaluate this question. For example, what if, through extensive remediation of the site, the source of TCE could be removed. How would this affect the TCE concentrations for year 2044 (a date selected by DOE and EPA)? These results are shown in Figure 1.8. As you can see, there is significant improvement in the groundwater quality. The source of the contamination has been removed, and the plume of contamination has migrated down gradient and has been significantly diluted. Still, pollutant fate and transport modeling cannot be the only tool in the decision process. Other factors, such as the cost of this magical cleanup and the associated health risk of drinking the contaminated groundwater and of remediation practices, must also be considered.

The Release of Nitrobenzene-Based Solvents at the Sondermulldeponie Landfill in Kolliken, Switzerland.   This site is a bit more complicated since it involves a lot of chemistry to understand the pollution scenario. Around 1978, Sondermull disposed of nitroaromatic liquid wastes in metal drums in a shallow pit (Colombi, 1986). The disposal pit was lined with sawdust in the event that the drums would someday leak. Due to flooding from changes in the groundwater table, the metal drums rusted and eventually leaked their contents. In 1985, groundwater wells located down-gradient ("downhill" in a groundwater system) from the disposal area

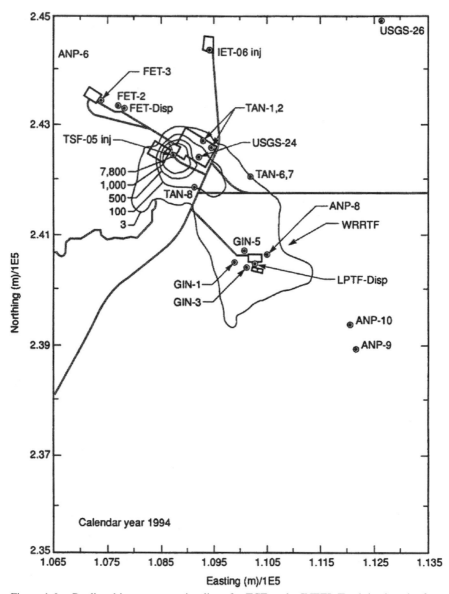

Figure 1.6.   Predicted iso-concentration lines for TCE at the INEEL Tan injection site for the year 1994 [INEEL, Department of Energy (Dunnivant, et al., 1994)].

were found to contain reduced aromatics, such as chloroanilines, but no nitrobenzene compounds. Anyone who knows the Swiss knows that they are meticulous record keepers and always have been. The company that disposed of the waste produced records showing that they had not disposed of any aniline compounds, and thus they asserted that they were not the ones responsible for the contamination. So, an investigation was conducted to attempt to explain this dilemma.

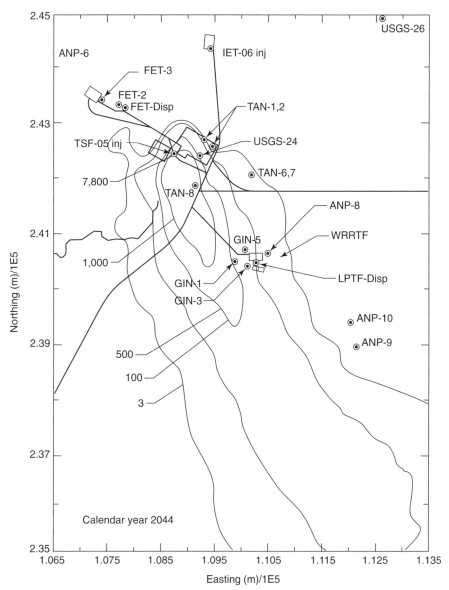

Figure 1.7.   Predicted iso-concentration lines for TCE at the INEEL Tan injection site for the year 2044 without remediation [INEEL, Department of Energy (Dunnivant et al., 1994)].

Groundwater modeling could have easily been conducted using a step model if you were only modeling the transport of nitrobenzene compounds. But this was not the issue since the monitoring well down-gradient from the point source (the landfill) only contained anilines. Thus, the importance of considering the chemistry occurring in the landfill and groundwater became apparent. Schwarzenbach et al.

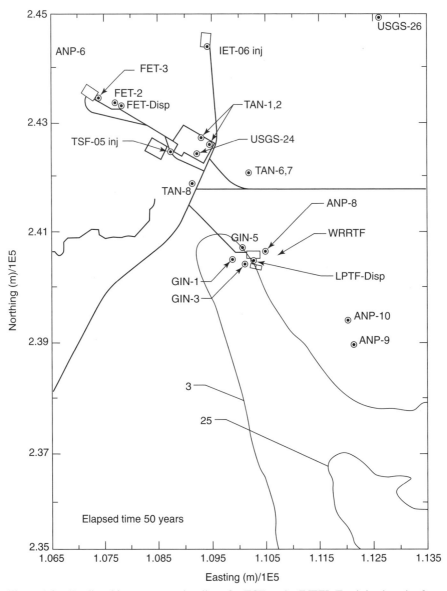

Figure 1.8. Predicted iso-concentration lines for TCE at the INEEL Tan injection site for the year 2044 after remediation of the immediate area around the injection well (INEEL, Department of Energy, Dunnivant et al., 1994).

(1990) and Dunnivant et al. (1991) proposed that as the water table changed, it could have allowed water to enter the landfill and allowed oxidation of the meal drums. After the drums ruptured, the leaking nitrobenzene compounds were subjected to reducing conditions from oxidation of the sawdust and, in the presence of reduced humic acid, the nitrobenzene compounds were reduced to their respective anilines

(Dunnivant et al., 1991). This case demonstrates the importance of record keeping, as well as the importance of the integration of chemistry and modeling in assessing the history of and in predicting the affects of a pollution event. This is why we will be discussing the role of chemistry as part of the modeling approaches in this book.

1.6.3 Atmosphere

Most of the models given in this textbook concern the fate and transport of pollutants in aquatic systems. Our medium of life is the atmosphere, however, and we are constantly exposed to pollutants through inhalation. Our exposure may be minimal in rural settings or more concentrated in urban areas, and there are many chronic health problems associated with this exposure. The most common fate and transport model used to study atmospheric emissions is concerned with a constant (step) release from a factory or an immediate (pulse) release from an industrial or transportation accident. We will look at two globally reported pulse emission accidents.

Methyl Isocyanate Release in Bhopal, India. In the early hours of December 23, 1984, a pesticide manufacturing plant in Bhopal, India accidentally released methyl isocyanate (MIC). The event was a pulse release (short-term) resulting in estimated concentrations that range from 13 ppm (Dave, 1985) to 100 ppm (Varmer, 1986). Death estimates range from 2000 (Dave, 1985) to 10,000 (Shrivastava, 1987, p. 65), while approximately 300,000 injuries were reported (Shrivastava, 1987). Although records indicate that concerns were raised before the accident, pollutant fate and transport modeling, as well as risk assessments, should have been conducted before the plant was even built. If the citizens of Bhopal had known of such modeling results, they may have been willing to surrender the small economic benefits from the factory in order to avoid the deaths and injuries caused by the accident. We will look at the Bhopal accident in more detail at the beginning of Chapter 10.

Accident at the Nuclear Power Plant at Chernobyl, Ukraine. A number of radioactive releases have occurred since our development of nuclear power plants. England experienced a major release at the Windscale site as early as 1957 (Bailey et al, 2002), and the United States' nuclear industry suffered the accident at Three Mile Island, Pennsylvania in March 1979. The world was shaken by the serious accident at the Chernobyl Unit 4 in the Soviet Union (present-day Ukraine) in 1986. Radionuclides were spread across both Eastern and Western Europe, thereby contaminating dairies in Austria, Hungary, Poland, and Sweden, not to mention the hazards posed by eating fresh fruit and fish from the immediate area. Released radionuclides were subsequently detected in the atmosphere over Canada, Japan, and the United States (Bailey et al., 2002). Thirty workers at the plant were directly killed from accident (20 from radiation exposure), another 209 were treated for acute radiation poisoning, and countless others were exposed to lower doses of radiation with unknown affects (http://www.uic.com.au/nip22.htm). In the end, 210,000 people had to be resettled (http://www.uic.com.au/nip22.htm). Modeling of such a large, widespread area, such as that affected by the Chernobyl accident, is also possible but less accurate, given the scale of the system and dynamic nature of weather systems. Still, you can rest assured that worst-case scenarios have now been modeled by governing agencies of nuclear power plants in developed countries.

1.7 ENVIRONMENTAL LAWS

Each country has developed a set of environmental laws governing the pre-manufacture testing, use, and appropriate disposal of chemicals. Some countries have extensive rules, while others are still in the process of developing these laws as their chemical industry develops. Environmental movements and the major environmental laws for the United States and Europe are discussed in Chapter 11. If the reader is from a different country, you may wish to research the laws for your country and compare them to those for the United States and Europe.

Concepts

1. What is the difference between explanatory and predictive modeling?
2. Explain the difference between a pollutant and pollution.
3. Define and give three examples of non-point source pollution from the Internet.
4. Define and give three examples of point source pollution from the Internet.
5. Define and give three examples of a step pollution event.
6. Define and give three examples of a pulse pollution event.

Exercises

1. Research a pollution event on the Internet. Write a one- to two-page summary including as many of the following as possible: the location, the company responsible for the site, the pollutants, the concentration of pollutants, the extent (surface area or volume) of pollution, the health effects of the pollutants, and the planned remediation.
2. Become socially active and research one of the chemical companies given in Table 1.4 using the Internet or by telephone or email. What chemicals do they use? What are their wastes? How do they dispose or destroy their wastes?

REFERENCES

Anonymous. *Environ. Sci. Technol.* **21**, 5 (1987).

Bailey, R. A., H. M. Clark, J. P. Ferris, S. Krause, and R. L. Strong. *Chemistry of the Environment*, 2nd edition, Academic Press, New York, Chapter 14, 2002.

Capel, P. D., W. Giger, R. Reichert, and O. Warner. Accidental input of pesticides into the Rhine River, *Environ. Sci. Technol.* **22**, 992–997 (1988).

Colombi, C. Sondermulldeponie Kolliken, wie weiter? *Phoenix Int.* **1**, 10–15 (1986).

Dave, J. M. The Bhopal methyl isocyanate (MIC) incident: An overview. In: *Proceedings of an International Symposium, Highly Toxic Chemicals: Detection and Protection Methods*. Schiefer, H. D., ed. Saskatoon, Saskatchewan, Canada, pp. 1–38, 1985.

Dunnivant, F. M., R. P. Schwarzenbach, and D. L. Macalady. Reduction of substituted nitrobenzenes in aqueous solutions containing natural organic matter. *Environ. Sci. Technol.* **26**, 2133–2141 (1991).

Dunnivant, F. M., G. J. Stromberg, A. H. Wiley, C. M. Hamel, and C. A. Leon. *Feasibility Study Report for Test Area North Groundwater Operable Unit 1-07B at the Idaho National Engineering Laboratory*, EFFF-ER-10802, January 1994, Idaho Falls, ID, 1994.

Internet 2: Chernobyl Accident. http://www.uic.com.au/nip22.htm. Accessed November 1, 2003.

Schaefer, A. Diasatrous cyanide spill could spawn liability reforms. *Environ. Sci. Technol.* **34**, A-203 (2000).

Schwarzenbach, R. P., R. Stierli, K. Lanz, and J. Zeyer. Quinone and iron porphyrin mediated reduction of nitroaromatic compounds in homogeneous aqueous solution, *Environ. Sci. Technol.* **24**, 1566–1574 (1990).

Shrivastava, P. *Bhopal: Anatomy of a Crisis*, Ballinger Publishing Company, Cambridge, MA, 1987.

United Nations Environment Programme, UNEP/Office for the co-ordination of Humanitarian Affairs, OCHA. *Spill of Liquid and Suspended Waste at the Aurul S. A. Retreatment Plant in Baia Mare*, 23 February–6 March 2000, Geneva, March 2002.

United Nations (UN), 2000, Secretariat of the Basel Convention, United Nations Environment Programme, Country Fact Sheets, available from http://www.basel.int/natreporting/index.html.

U.S. Environmental Protection Agency, 1989, Risk Assessment Guidance for Superfund Volume II, Environmental Evaluation Manual, Interim Final, EPA/540/1–89/001, Office of Emergency and Remedial Response, Washington, DC, p. 4.

U.S. Environmental Protection Agency, 2001. National Analysis, The National Biennial RCRA Hazardous Waste Report (based on 2001 data), available from http://www.epa.gov/epaoswer/hazwaste/data/brs01/index.htm.

Varmer, D. R. Anatomy of the methyl isocyanate leak on Bhopal. In: *Hazard Assessment of Chemicals*, Saxena J., ed, Hemisphere Publishing Corp., New York, pp. 233–299, 1986.

PART II

CHEMISTRY OF FATE AND TRANSPORT MODELING

"The noblest of the elements is water."

—Pindar, 476 B.C.

A Basic Introduction to Pollutant Fate and Transport, By Dunnivant and Anders
Copyright © 2006 by John Wiley & Sons, Inc.

CHAPTER *2*

BASIC CHEMICAL PROCESSES IN POLLUTANT FATE AND TRANSPORT MODELING

2.1 THE LIQUID MEDIUM: WATER AND THE WATER CYCLE

There are several factors responsible for life on Earth, including the Earth's ideal distance from the sun and a stable planetary rotation, which result in hospitable conditions. A view of the Earth from space almost tells the complete story as to why life as we know it thrives; the planet's surface consists of approximately 71% water. Water vapor rising from the oceans replenishes the landmasses with essential freshwater and, as a greenhouse gas, warms the Earth. As we will see in this textbook, water acts not only as the medium of life but also as a transport medium for chemicals that can threaten life. Whether we are discussing water in rivers, lakes, or groundwater, or as it leaves the oceans and passes through and cleanses the atmosphere, the unique properties of water will play an important role in understanding transport phenomena.

Water occurs on Earth in distinct settings that we generally refer to as compartments: atmosphere, land, groundwater, rivers, lakes, and oceans. The water cycle is the exchange between these compartments and is a highly dynamic system. Figure 2.1 is a simplified representation of the global water cycle, showing each major compartment with its respective mass of water, water's average residence time in each compartment, and the flux between compartments. Residence time refers to the average amount of time a molecule of water (in this case) spends in the compartment of interest. Of course the largest reservoir of water is the oceans, which are also associated with the longest residence time (40,000 years). Some molecules of water falling on the surface of the oceans are almost immediately volatilized (evaporation) and return to the atmosphere to fall across landmasses. Other molecules are taken deep into the ocean, not to see the light of day for tens of thousands of years. Water that is volatilized from the ocean surface usually only spends a short time in the atmosphere, about nine days on average. Most of the water returns to the oceans, but some also enters the terrestrial water cycle. As shown in Figure 2.1, water falling on the surface of land makes up glaciers, lakes, rivers, and land runoff. Surface water

A Basic Introduction to Pollutant Fate and Transport, By Dunnivant and Anders
Copyright © 2006 by John Wiley & Sons, Inc.

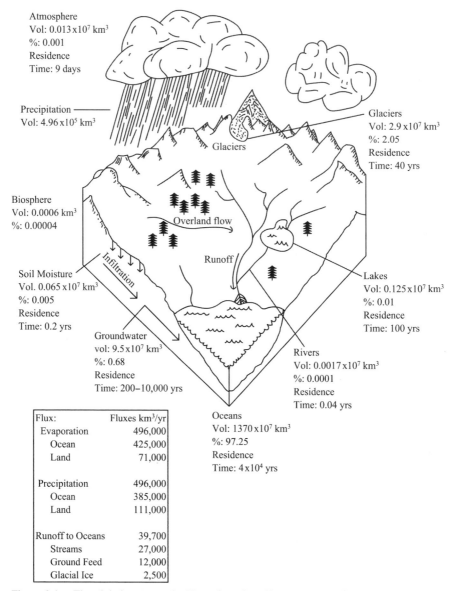

Figure 2.1. The global water cycle. [Data from http://www.geog.ouc.bc.ca; http://web.sfc.keio.ac.jp and Stumm and Morgan (1996).]

also percolates into the ground and forms two types of water: water in the unsaturated zone (vadose zone) and water that makes up aquifers, or groundwater. Only a small fraction of the freshwater is actually available to terrestrial life, since glaciers and deep groundwater are difficult to access and water in these compartments has relatively long residence times.

It is an understatement to say that water is not distributed evenly across the planet. Table 2.1 shows the total renewable water resources, water withdrawals, percent of renewable resources, and average percent used, by sector, for major geographic regions of the world. Perhaps the most important column is the percent of renewable resources used, which is an indicator of future crisis areas of the world. As this percentage approaches 100, the population in the area is using water faster than it can be supplied. The most important areas of concern are Pakistan, Turkmenistan, and Uzbekistan in Asia and most of the countries of the Middle East and Northern Africa. Several countries in the Middle East and Northern Africa are supplementing their freshwater with desalination plants fueled by their reserves of fossil fuels, but this cannot last forever. It is also interesting to note the use of freshwater by agriculture and industry, given in the last two columns in Table 2.1. Water from these sectors is almost always subject to contamination, and the fate and transport of these polluted waters is the subject of this textbook. Water scarcity will be an increasingly important issue in the future and has been listed as a major security issue by world leaders.

2.2 UNIQUE PROPERTIES OF WATER

Most people, even most college science students, do not realize how incredibly unique water is. In fact, as we will see later, if water behaved as a chemist would predict based solely on chemical structure, life as we know it would not exist on Earth. But first, let's discuss the general nature of water. Table 2.2 lists 38 unique properties of water that you might not predict if you just looked at water as a typical molecule (from http://www.sbu.ac.uk/water/anmlies.html). So, why is water so unique? All of the unique properties listed in Table 2.2 can be traced back to hydrogen bonding between individual water molecules. You should recall from first-year chemistry that hydrogen bonding is a special form of dipole–dipole interaction, reserved for intermolecular bonding between hydrogen and the nonbonding (lone pair) electrons of N, F, and O (the lone pair of electrons in the Lewis dot structure). Hydrogen bonding in water is especially important, since each water molecule has two bonded hydrogen atoms and two lone pairs of electrons. This allows for a complete alignment and bonding between each lone pair of electrons and hydrogen. In ice, virtually all of the nonbonding electrons on the oxygen atoms are bonded to adjacent hydrogens of another water molecule (see Figure 2.2). As ice melts to liquid water, approximately 15% of these hydrogen bonds between water molecules are broken; the net result is a shrinking of the volume of water (an increase in density). This accounts for the fact that water shrinks upon melting and that ice floats on liquid water, the first unique property of water listed in Table 2.2. All other inorganic and organic liquids have the reverse relationship; their solid phase is more dense and sinks in their liquid.

Two other unique properties that result from hydrogen bonding in water are its unusually high melting and boiling points. You will recall from general chemistry that elements in the Periodic Table are arranged based on periodicity, which is the term used for the relationships between atomic structure and behavior. If we take

TABLE 2.1. The Global Water Budget by Region

Region/Country	Total Renewable Water Resources (km³/year)	Total Water Withdrawals (m³/year)	Per capita (m³/person)	Average % of Renewable Resources	Average % Used by Agriculture	Average % Used by Industry
World	43,219	3,414,000	650	—	71	20
Asia (excluding Middle East)	11,321	1,516,247	1,028	29	79	10
Cambodia	121	520	60	0	94	1
Pakistan	52	155,600	1,382	100	97	2
Turkmenistan	1.4	23,779	5,801	116	98	1
Uzbekistan	16	58,051	2,598	132	94	2
Europe	6,590	367,449	503	9	25	48
Bulgaria	21	13,900	1,573	58	22	75
Middle East & North Africa	518	303,977	754	423	80	5
Egypt	1.8	66,000	1,055	127	82	25
Israel	0.8	1,620	287	108	54	7
Jordan	0.7	984	255	151	75	3
Kuwait	<1	538	306	3,097	60	2
Libyan Arab Jamahiriya	0.6	4,500	870	801	84	3
Oman	1	1,223	658	181	94	2
Saudi Arabia	2.4	17,018	1,056	955	90	1
United Arab Emirates	0.2	2,108	896	1,614	67	9
Yemen	4.1	2,932	253	123	92	1
Sub-Saharan Africa	3,901	72,556	147	5	66	7
North America	4,850	512,440	1,720	14	27	58
Central America & Caribbean	1,186	105,741	513	9	65	14
South America	12,246	156,948	614	1	76	7
Oceania	1,693	16,730	318	1	45	15

TABLE 2.2. Unique Properties of Water (http://www.sbu.ac.uk/water/anmlies.html)

Properties discussed in text
Water shrinks on melting (ice floats on water)
Unusually high melting point
Unusually high boiling point
Unusually high surface tension
Unusually high viscosity
Unusually high heat of vaporization
Unusually high specific heat capacity

Other properties dependent on hydrogen bonding
The solubilities of nonpolar gases in water decrease with temperature to a minimum and then rise
Specific heat capacity has a minumum at 36°C
A high density that increases on heating
The number of nearest neighbors increases on melting
The number of nearest neighbors increases with temperature
Pressure reduces water's melting point
Pressure reduces the temperature of maximum density
D_2O and T_2O differ from H_2O in their physical properties much more than might be expected
Unusually large viscosity increase as the temperature is lowered
Viscosity decreases with pressure
Unusually low compressibility
Compressibility drops as temperature increases down to a minimum
Low coefficient of expansion
Thermal expansivity decreases increasingly at low temperatures
The speed of sound in liquid water increases with temperature
Liquid water has over twice the specific heat of steam or ice
NMR spin-lattice relaxation are very small at low temperatures
Solute have varying effects on properties such as density and viscosity
No aqueous solutions even approach thermodynamic ideality
X-ray diffraction shows unusually detailed structure
Supercooled water has two phases and a second critical point
Liquid water may be supercooled, in tiny droplets, down to about −70°C
Solid water exists in a wider variety of stable crystal and amorphous structures than do other
    materials
Hot water may freeze faster than cold water
The refractive index of water has a maximum value at just below 0°C
At low temperatures, the self-diffusion of water increases as the density of pressure increases
The termal conductivity of water rises to a maximum at about 130°C and then falls
Proton and hydroxide ion mobilities are anomalously fast in an electric field
The heat of fusion of water with temperature exhibits a maximum at −17°C
The dielectric constant is high
Unusually high critical point

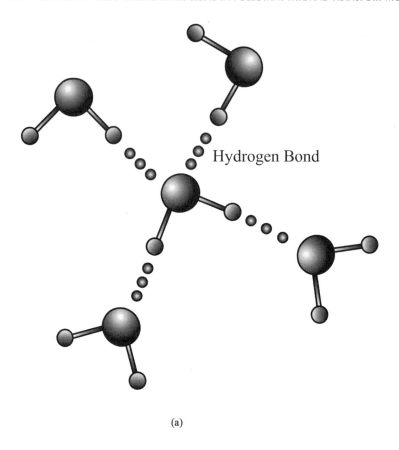

(a)

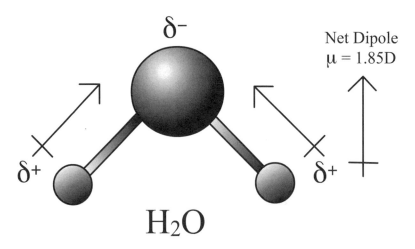

(b)

Figure 2.2.   Electronic structure of water. (a) Hydrogen bonding, (b) the dipole in a water molecule.

the hydrides formed by the elements in Group VI, the group containing oxygen and sulfur (hence H_2O, H_2S, etc.), and plot the melting and boiling points as a function of molecular mass of these hydrides, we observe a directly proportional relationship (refer to Figures 2.3 and 2.4). As the molecular weight of the hydride increases, so do the melting and boiling point. Thus, based on this trend and theoretical calcula-

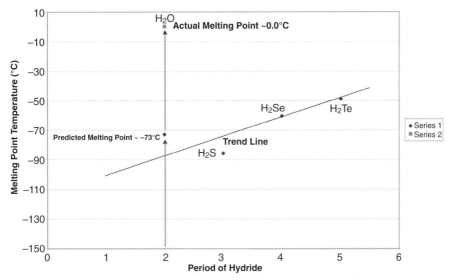

Figure 2.3. Melting points of Group VI hydrides. (Data from *The Handbook of Chemistry and Physics*, 1980.)

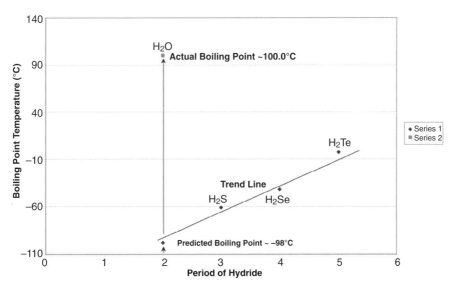

Figure 2.4. Boiling points of Group VI Hydrides. (Data from *The Handbook of Chemistry and Physics*, 1980.)

tions, water should boil at −73°C and freeze at −98°C, which would leave a very small thermal window for liquid water. As you know, fortunately this does not occur; water boils at 100°C and freezes at 0°C. The reason for this discrepancy is again due to hydrogen bonding. First, let's look at the observed melting point of water ice. Since water can undergo hydrogen bonding, there is a stronger-than-expected attraction between individual water molecules; in order to melt solid ice into gaseous water, you must break 15% of these stronger intermolecular bonds. To do this, we must put in more heat, manifested by an increase in temperature. Thus the melting point of water rises from the predicted −98°C to 0°C. In order to volatilize water, in turn, we must break the remaining intermolecular hydrogen bonds. Again, the strength of these bonds raises the melting point of water from the expected −73°C to the observed 100°C. Also note that not only have the absolute boiling and melting points of water increased, but the range between the two has increased from a difference of 25°C to a difference of 100°C. This also allows liquid water, and therefore aquatic life, to exist over a broader range of temperatures on Earth.

Hydrogen bonding between water molecules is also responsible for the unusually high viscosity and high surface tension of water. In order to understand surface tension, you must picture in your mind the three-dimensional nature of water. Each water molecule is completely surrounded by other water molecules, and each is hydrogen bonded to four other water molecules (minus the 15% of bonds not present in liquid water). In order for water to flow, and have relatively low viscosity, this three-dimensional bonding must be elastic and allow the relatively free movement of adjacent water molecules. Since the hydrogen bonding is stronger than expected, the water molecules are less fluid than expected, resulting in a high viscosity. Now take the three-dimensional structure of water and remove the upper layer so that you have an exposed isolated layer of water. Water does not like this, and the surface molecules cling very strongly to the water molecules below the surface. This gives rise to a very strong surface tension in water. Surface tension is related to the force it takes to break the surface of the water. An illustration of this high surface tension can be seen when you gently place a metal paper clip on the surface of water. It will float even though the metal is more dense than water. However, if you add a surface active agent, a surfactant or soap, the surfactant molecules will align themselves between the surface water molecules, break the surface tension (hydrogen bonds), and allow the paper clip to sink. Surface tension and viscosity will be important when we study water flow and pollutant transport in groundwater systems.

Two final unique properties of water are its unusually high heat of vaporization and high specific heat capacity. The high heat of vaporization is directly related to water's unusually high boiling point. In the absence of hydrogen bonding, water would more freely evaporate and go from a liquid to a gas. However, hydrogen bonding is a strong force holding water molecules together, and significant heat must be put into the system in order to separate liquid water molecules into gaseous molecules. Likewise, the strength of the bonds makes a lot of heat necessary to increase the average kinetic energy, or temperature, of water (its specific heat is 4.184 J/g-°C). This ability to absorb large amounts of heat without significantly changing temperature is important to life of Earth. Water, in the form of lakes, rivers, and oceans, serves to store heat on Earth's surface and greatly affects the weather. The climates of landmasses near large water bodies are significantly affected by the temperature of these water bodies. You have most likely heard of microclimates and the lake effect. In colder climates, cities located downwind from a large body of water receive large amounts of snow in the winter. Dry cold air moves across the relatively warm water body, thereby evaporating water; upon reaching the shore and elevated landmasses, the moist air mass rises with the topography and snow forms. However, when the source of the water vapor is closed off, after the water body cools and freezes later in winter, the snowfall greatly diminishes or completely ends. Water is a very important thermal regulator on a global scale, and its ability to only slightly change temperature with absorption of heat is a property directly related to hydrogen bonding.

2.3 CONCENTRATION UNITS

In general chemistry, a variety of concentration units are normally introduced, including grams/liter, moles/liter, mole fraction, molarity, and molality. Some general chemistry textbooks even briefly mention mg/L (parts per million). Chemists tend to use molarity (moles/L) more often, while engineers, hydrologists, and researchers dealing with pollutant fate and transport prefer mg/L (parts per million, ppm) or µg/L (parts per billion, ppb). We will generally use ppm and ppb in this textbook, since we are mostly dealing with pollutant modeling, except in Chapters 2 and 3, where we discuss chemistry. But where do the terms ppm and ppb come from? It is assumed that pure water contains 1.00 million parts of water per million parts of water, or 100%. We typically use a density of 1.00 g/mL for water and assume (incorrectly sometimes) that pollutants have a similar or identical density. So, for example, how many parts per million is a pollutant with a concentration of 1000 mg/L?

$$1000 \text{ mg pollutant}/\text{L of } H_2O \left(\frac{1.00 \text{ g pollutant}}{1000 \text{ mg}} \right)\left(\frac{1.00 \text{ L } H_2O}{1000 \text{ g } H_2O} \right)$$

$$= \frac{1000 \text{ g pollutant}}{10^6 \text{ g } H_2O} = 0.001 \text{ g}/\text{g} = 1000 \text{ parts per million}$$

Parts per billion (ppb) is defined as µg/L, and parts per trillion (ppt) is ng/L. These are the three basic units we use in pollutant monitoring. As shown above, in water we use a mass per mass calculation, while for gas we use volume per volume calculations. When we are dealing with radioactivity, we will use the milli, micro, and nano prefixes with the radioactivity unit of curies (Ci). A curie is the radioactivity of 1.00 g of pure ^{226}Ra, which is equal to 3.7×10^{10} disintegrations per second. Another unit commonly used unit (the SI equivalent) in the study of radioactivity is the becquerel (Bq). One becquerel is equal to 2.703×10^{-11} Ci.

You should recall from general chemistry that molarity (M) is another concentration unit, meaning moles of chemical per liter of solution (moles/L). This unit will be used when we are dealing with problems or situations that involve stoichiometry. If you are unfamiliar with this unit, you should review a general chemistry textbook. The molarity of a chemical species is often represented using brackets: [A] denotes the mol/L of species A.

$$\text{Molarity(M)} = \frac{\text{Mass (g) of substance}}{\dfrac{\text{Molar mass of substance (g/mol)}}{\text{Volume of solution (L)}}} = \frac{\text{Moles of solute}}{\text{L of solution}}$$

2.4   CHEMICAL ASPECTS OF ENVIRONMENTAL SYSTEMS

2.4.1   pH

You learned in general chemistry that pH is the negative log of the molar hydrogen ion concentration:

$$pH = -\log[H^+] \tag{2.1}$$

Ions in water are always surrounded by water molecules that partially cancel out the charge of the ion. The hydrogen ion, like all ions in solution, is surrounded by waters of hydration. For ease of writing, we normally represent the hydrogen ion as H_3O^+, suggesting one water of hydration, although in reality it can have 5 to 9. In fact, it is a cluster of water molecules with one extra proton.

The pH of a water solution is considered a master variable. By this, we mean that the pH of the solution can be the determining factor in a variety of parameters, especially chemical speciation, which we will spend considerable time discussing in Sections 2.5.1 and 3.2.2. Therefore, pH is one of the most common parameters measured for a water sample.

Although we commonly relate pH to hydrogen concentration [Eq. (2.1)], the pH of a water is almost always measured with an electrode, and electrodes measure activity instead of concentration. Activity is discussed in the next section.

Most natural waters have a pH between 5.5 and 9, but extreme pH values have been observed in natural settings such as geothermal water and eutrophic (organic-rich) systems. When hazardous waste enters natural environments, any pH value is possible, given the vast amount of acidic and caustic wastes that the chemical industry produces. Other topics related to pH that you should review from general chemistry are buffer solutions and the Henderson–Hasselbach equation, which we will look at closely in Section 2.5.1.

2.4.2   Activity

You will find, as you take more chemistry, that what you learned in general chemistry was a simplification of reality. This is true for units of concentration. For example, molar units are only appropriate in what are called "ideal solutions," in which the molar concentration of an ion or compound is equal to its activity. Activity is a measure of the effective concentration of an ion, accounting for its interactions with other ions that can mask it. Therefore, an ideal solution is one containing very little dissolved salt. Activity (A), the concentration that the solution "sees," is expressed by

$$A = \gamma[C] \tag{2.2}$$

where γ is the activity coefficient (almost always equal to or less than 1.00; for extreme conditions it can be greater than 1.00), and [C] is the molar concentration of the chemical or pollutant. The activity coefficient (γ) is a direct function of ionic strength, or the amount of other salts in the solution, which helps explain why we must be concerned with activity instead of concentration. If you picture an ion in solution, it is not present as a free cation or anion, but is surrounded by water molecules (waters of hydration) and by ions of opposite charge, which serve to balance out the charge of the ion of interest. A representation of such a configuration is shown in Figure 2.5 for ions of $CaSO_4$.

This balancing of charges tends to make the ion less active in solution than expected based on its concentration (by changing the ions mobility), and activity accounts for this canceling out of concentration. As noted above, the activity of a pollutant is equal to the activity coefficient times the molar concentration, and the activity coefficient is related to the salt content of the water. An activity coefficient of near 1.00 for dilute solutions yields an activity equal to the concentration. As the salt concentration increases, the activity coefficient decreases from 1.00, thus lowering the activity.

So, what is the practical reason for using activity rather than concentration? Remember that our end goal in pollutant fate and transport modeling is to be able to predict risk, and risk is based on toxicity. It has been almost universally found that water containing higher concentrations of nontoxic salts have lower toxicity for the same pollutant concentration. Toxicity is thus governed by activity instead of concentration.

So how can we quantify the effect of activity? As noted earlier, concentration is related to activity by the activity coefficient (γ), which is related to the total ionic strength (μ). We can calculate the ionic strength of any water, if we know the anion/cation composition of the water, by

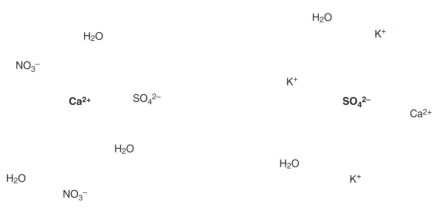

Figure 2.5.   Illustration of ions of Ca^{2+} and SO_4^{2-} in aqueous solution.

$$\mu = 0.500 \left(C_1 Z_1^2 + C_2 Z_2^2 + C_3 Z_3^2 + C_4 Z_4^2 + \cdots \right) = 0.500 \sum C_i Z_i^2$$

where C is molar concentration and Z is charge.

Here is an example calculation: What is the total ionic strength of (a) 0.100 M $NaNO_3$ and (b) 0.100 M Na_2SO_4.

(a) $\mu = 0.500 \left[(0.100 \times 1^2) + (0.100 * 1^2) \right] = 0.100\,M$

(b) $\mu = 0.500 \left[(0.100 \times 2 \times 1^2) + (0.100 \times 2^2) \right] = 0.300\,M$

From this trend we can develop the general rules shown in Table 2.3.

For mixtures of ionic salts, the empirically derived extended Debye–Hückel equation can be used to estimate γ:

$$\log \gamma = -\frac{0.512 Z^2 \sqrt{\mu}}{1 + \alpha \sqrt{\mu}/305} \qquad \text{at } 25°C \qquad (2.3)$$

where Z is the ionic charge you are calculating γ for, μ is the ionic strength, and α is the effective hydrated radius of the ion you are calculating γ (found in Table 2.4). Note the trend implicit in Eq. (2.3). Small, highly charged ions bind solvent molecules more tightly and have smaller hydrated radii than do larger or less highly charged ions. Also note some generalizations: (1) As ionic strength (μ) increases, the activity coefficient (γ) decreases, (2) as the ionic charge (Z) increases, the activity coefficient (γ) decreases, and (3) as the effective hydrated radius (α) decreases, the activity coefficient (γ) decreases.

Example. Calculate the activity coefficient and activity of Ca^{2+} in 0.0200 M $CaCl_2$.

$$\mu = 0.500 \left[(0.0200 \times 2^2) + (0.0200 \times 2 \times 1^2) \right] = 0.0600\ M$$

$$\log \gamma = -\frac{0.512(2^2)\sqrt{0.0600}}{1 + 600\sqrt{0.0600}/305} = -0.460$$

$$\gamma = 0.347$$

$$A = C\gamma = (0.0200)(0.347) = 0.00693.$$

Note the substantial difference between an activity of 0.00693 and a concentration of 0.0200 M. As noted, differences become more pronounced as the salt content increases. Waters where activity calculations are important include some groundwaters, inland salt lakes, estuaries, and oceans.

TABLE 2.3.   Estimation of Ion Strength Based on Salt Composition

Electrolyte	Molarity	μ
1:1	M	$1 * M$
2:1	M	$3 * M$
3:1	M	$6 * M$
2:2	M	$2 * M$

TABLE 2.4. Activity Coefficients and Hydrated Radii for Aqueous Solutions at 25°C (Data from Harris, 1999)

Ion	Hydrated Ion Radius (pm)	Ionic Strength of Solution				
		0.001	0.005	0.01	0.05	0.1
				Activity Coefficients		
Charge ± 1						
H^+	900	0.967	0.933	0.914	0.86	0.83
$(C_6H_5)_2CHCO_2^-$, $(C_3H_7)_4N^+$	800	0.966	0.931	0.912	0.85	0.82
$(O_2N)_3C_6H_2O^-$, $(C_3H_7)_3NH^+$, $CH_3OC_6H_4CO_2^-$	700	0.965	0.930	0.909	0.845	0.81
Li^+, $C_6H_5CO_2^-$, $HOC_6H_4CO_2^-$, $ClC_6H_4CO_2^-$, $C_6H_5CH_2CO_2^-$, $CH_2{=}CHCH_2CO_2^-$, $(CH_3)_2CHCH_2CO_2^-$, $(CH_3CH_2)_4N^+$, $(C_3H_7)_2NH_2^+$	600	0.965	0.929	0.907	0.835	0.80
$Cl_2CHCO_2^-$, $Cl_3CCO_2^-$, $(CH_3CH_2)_3NH^+$, $(C_3H_7)NH_3^+$	500	0.964	0.928	0.904	0.83	0.79
Na^+, $CdCl^+$, ClO_2^-, IO_3^-, HCO_3^-, $H_2PO_4^-$, HSO_3^-, $H_2AsO_4^-$, $Co(NH_3)_4(NO_2)_2^+$, $CH_3CO_2^-$, $ClCH_2CO_2^-$, $(CH_3)_4N^+$, $(CH_3CH_2)_2NH_2^+$, $H_2NCH_2CO_2^-$	450	0.964	0.928	0.902	0.82	0.775
$^+H_3NCH_2CO_2H$, $(CH_3)_3NH^+$, $CH_3CH_2NH_3^+$	400	0.964	0.927	0.901	0.815	0.77
OH^-, F^-, SCN^-, OCN^-, HS^-, ClO_3^-, ClO_4^-, BrO_3^-, IO_4^-, MnO_4^-, HCO_2^-, $H_2citrate^-$, $CH_3NH_3^+$, $(CH_3)_2NH_2^+$	350	0.964	0.926	0.900	0.81	0.76
K^+, Cl^-, Br^-, I^-, CN^-, NO_2^-, NO_3^-	300	0.964	0.925	0.899	0.805	0.755
Rb^+, Cs^+, NH_4^+, Tl^+, Ag^+	250	0.964	0.924	0.898	0.80	0.75

(Continued)

TABLE 2.4. Activity Coefficients and Hydrated Radii for Aqueous Solutions at 25°C (Data from Harris, 1999) (continued)

Ion	Hydrated Ion Radius (pm)	Ionic Strength of Solution				
		Activity Coefficients				
		0.001	0.005	0.01	0.05	0.1
Charge ± 2						
Mg^{2+}, Be^{2+}	800	0.872	0.755	0.69	0.52	0.45
$CH_2(CH_2CH_2CO_2^-)_2$, $(CH_2CH_2CH_2CO_2^-)_2$	700	0.872	0.755	0.685	0.50	0.425
Ca^{2+}, Cu^{2+}, Zn^{2+}, Sn^{2+}, Mn^{2+}, Fe^{2+}, Ni^{2+}, Co^{2+}, $C_6H_4(CO_2^-)_2$, $H_2C(CH_2CO_2^-)_2$, $(CH_2CH_2CO_2^-)_2$	600	0.870	0.749	0.675	0.485	0.405
Sr^{2+}, Ba^{2+}, Cd^{2+}, Hg^{2+}, S^{2-}, $S_2O_4^{2-}$, WO_4^{2-}, $H_2C(CO_2^-)_2$, $(CH_2CO_2^-)_2$, $(CHOHCO_2^-)_2$	500	0.868	0.744	0.67	0.465	0.38
Pb^{2+}, CO_3^{2-}, SO_3^{2-}, MoO_4^{2-}, $Co(NH_3)_5Cl^{2+}$, $Fe(CN)_5NO^{2-}$, $C_2O_4^{2-}$, Hcitrate^{2-}	450	0.867	0.742	0.665	0.455	0.37
Hg_2^{2+}, SO_4^{2-}, $S_2O_3^{2-}$, $S_2O_6^{2-}$, $S_2O_8^{2-}$, SeO_4^{2-}, CrO_4^{2-}, HPO_4^{2-}	400	0.867	0.740	0.660	0.445	0.355
Charge ± 3						
Al^{3+}, Fe^{3+}, Cr^{3+}, Sc^{3+}, Y^{3+}, In^{3+}, lanthanides[a]	900	0.738	0.54	0.445	0.245	0.18
citrate^{3-}	500	0.728	0.51	0.405	0.18	0.115
PO_4^{3-}, $Fe(CN)_6^{3-}$, $Cr(NH_3)_6^{3+}$, $Co(NH_3)_6^{3+}$, $Co(NH_3)_5H_2O^{3+}$	400	0.725	0.505	0.395	0.16	0.095
Charge ± 4						
Th^{4+}, Zr^{4+}, Ce^{4+}, Sn^{4+}	1100	0.588	0.35	0.255	0.10	0.065
$Fe(CN)_6^{4-}$	500	0.57	0.31	0.20	0.048	0.021

[a] Elements 57–71 in the periodic table.

Source: J. Kielland, *J. Am. Chem. Soc.* **59**, 1675 (1937).

As we continue in this textbook, we will almost always refer to concentration for simplicity, but remember that when you are dealing with a water with high salt content, it is better to work in activities.

2.4.3   Solubility

Solubility is defined as the maximum concentration of a chemical species that can be present in a solution at equilibrium. Since we will be dealing exclusively with water as our solvent, we are concerned with the aqueous solubility, given in moles/L or mg/L. We will divide our discussions into organic and inorganic pollutants. Although water is considered a universal solvent, it does not necessarily dissolve large quantities of all chemicals. Some organic pollutants, such as short-chained alcohols, acetone, acetonitrile, and a few other organic solvents, are miscible with water; that is, water and the solvent will completely mix in any proportion. Many other pollutants are incorrectly listed as insoluble in water, which really means that the solubility is very low. For example, the famous pollutants such as DDT and PCBs are listed as insoluble in many chemistry handbooks but are actually soluble in the ppb to ppm range. Their low solubilities contribute to these chemicals' toxicity to wildlife. In order to understand their toxicity, we must understand bioconcentration, or biomagnification.

Pollutants such as DDT and PCBs are hydrophobic; they do not like being dissolved in water. As noted above, their solubilities are in the ppb to ppm level in pure water. In the environment, they actively partition from (move out of) the water onto surfaces or into biological organisms, even at concentrations below their aqueous solubility. Microorganisms such as algae have a large surface area, and their hydrophobic cell surfaces will readily attract and sorb hydrophobic pollutants. Thus, these pollutants are concentrated into the algae or microbe cell mass. Organisms that feed on algae automatically concentrate more DDT and PCBs since they eat large quantities of algae over a longer lifespan. This bioconcentration of the pollutants continues up the food chain to the top predators, including birds of prey and humans. An example of bioconcentration in an ecosystem is shown in Table 2.5. As you see, the pollutant concentration in the water is only 0.00005 ppm, but it is bioconcentrated to ~25 ppm in biological species at the top of the food chain. This almost millionfold increase illustrates the dangers of these extremely low levels of pollutants in the environment.

The experimental determination of the exact solubility of a hydrophobic or any low solubility pollutant is very difficult and is wrought with analytical errors. The lower the solubility, the more error in the results and disagreement between laboratory determinations. Some solubility values for the same pollutant disagree by a factor of 10 to 100; however, these procedures do allow more precise estimates of the *relative* solubilites of hydrophobic pollutants. As a result of the disagreement in literature values, a modeling method to determine absolute and relative solubilities of compounds is becoming more accepted today; solubilities are estimated from highly sophisticated calculations using the chemical parameter program SPARC, developed by the U.S. EPA (http://ibmlc2.chem.uga.edu/sparc/). This program can

TABLE 2.5. Bioconcentration of DDT in Long Island Food Web (USA) (Woodwell et al., 1967)

Organism	DDT Residues (ppm)
Water	0.00005
Plankton	0.04
Silverside minnows	0.23
Sheephead minnows	0.94
Pickerel (predatory)	1.33
Needlefish (predatory)	2.07
Heron (feeds on small aquatic animals)	3.57
Herring gull (scavenger)	6.00
Osprey egg	13.8
Merganser (fish-eating duck)	22.8
Cormorant (feeds on large fish)	26.4

relatively accurately predict the aqueous solubility, vapor pressure, and Henry's law constant for any chemical with a known structure and melting point. Table 2.6 lists these parameters for several common pollutants that will be used in the modeling section of this textbook. This table and the SPARC model can be used in the modeling chapters to determine the source masses of pollutants in step and pulse models. The aqueous solubilities of different classes of pollutants are also compared in Figure 2.6. Note the wide range of pollutant solubilities.

While we will generally study organic pollutants in the aqueous phase, many sources of organic pollutants are pure liquids known as nonaqueous phase liquids (NAPLs). We discussed one example, from the Idaho National Engineering and Environmental Laboratory, in Chapter 1. The NAPL may be more dense than water and known as a dense nonaqueous phase liquid (DNAPL) or may be less dense than water and known as a light nonaqueous phase liquid (LNAPL). Neither of these liquids mix with water, but as water flows past the NAPL, compounds in the NAPLs slowly dissolve into the water phase. Although there are sophisticated models to predict the release rate of these "pools" of pollutants, one simple way to model them is to use the maximum aqueous solubility predicted from SPARC and use this as the input mass in a step transport model. We will discuss this in the groundwater modeling chapter.

The solubilities of inorganic pollutants (salts), at least as they are calculated in general chemistry, are much easier to determine than those of organics. You should remember the solubility product constant, or K_{sp}. Values of K_{sp} for several salts are given in Table 2.7. We will use an applied example from general chemistry to illustrate how solubilities are calculated from these constants. Note that in the following calculations we are concerned with the concentration of the dissolved metal ion in equilibrium with some solid phase.

Say that you have an industrial process that produces copper (I) bromide as a waste product. What will be the maximum Cu^+ solubility if this solid waste is contacted by rainwater? The K_{sp} for CuBr is 5.3×10^{-9}.

TABLE 2.6. Summary of Aqueous Solubility, Vapor Pressure, and Henry's Law Constants for Common Environmental Pollutants (Estimated by SPARC)

Classes of Pollutants	CAS #	Melting Point	MW	Solubility (mg/L)	VP (atm)	HLC (atm-L/mol)
C1–C2 Halocarbons						
Trichlorofluoromethane	75-69-4	−111°C	137.4	2,413	1.228	0.0569
Dichlorodifluoromethane	75-71-8	−158°C	120.9	523.3	5.957	0.231
Chlorodifluoromethane	75-45-6	−157.4°C	86.4	4,005	6.223	0.0216
1,2-Dichloromethane	75-71-8	−158°C	120.9	523.3	5.955	0.2311
Trichloroethene	79-01-6	−84.7°C	131.388	913.9	0.1773	0.02549
Tetrachloroethene	127-18-4	−22.3°C	165.83	127.3	0.0511	0.06657
1,1-Trichloroethane	71-55-6	−30°C	133.4	1,940	0.2032	0.01397
Methyl bromide	74-83-9	−93.7°C	94.94	23,970	1.831	0.003960
1,2-Dichloroethane	107-06-2	−35.5°C	98.96	7,903	0.1244	0.001558
1,1,2-Tetrachloroethane	79-72-1	−44°C	167.9	2,171	0.006077	0.004699
Hexachloroethane	67-72-1	187°C	236.74	12.95	9.931E-5	0.001816
Polychlorinated Biphenyls						
2-Chlorobiphenyl	2051-60-7	34°C	188.66	5.769	9.179E-6	0.0003001
4-Chlorobiphenyl	2051-62-9	78.8	188.66	1.915	2.245E-6	0.0002212
2,2'-DCB	13029-08-8	58°C	223.10	0.7528	1.304E-6	0.0003865
4,4'-DCB	2050-68-2	149.3°C	223.10	0.07793	5.732E-8	0.0001641z
2,2',3,3'-TCB	38444-93-8	123°C	291.99	0.008024	3.641E-9	0.0001312
2,2',5,5'-TCB	35693-99-3	84°C	291.99	0.03705	2.755E-8	0.0002171
2,3,4,5-TCB	33284-53-6	93°C	291.99	0.01478	5.235E-9	0.0001034
2,2',4,5,5'-PCB	37680-73-2	26°C	326.43	0.02945	1.266E-8	0.0001403
2,2',4,4',6,6'-HCB	33979-03-2	110°C	360.88	0.0009090	2.563E-10	0.0001018

(*Continued*)

TABLE 2.6. Summary of Aqueous Solubility, Vapor Pressure, and Henry's Law Constants for Common Environmental Pollutants (Estimated by SPARC) (continued)

Classes of Pollutants	CAS #	Melting Point	MW	Solubility (mg/L)	VP (atm)	HLC (atm-L/mol)
Carboxyate Esters (Pyrethrins)						
Pyrethrin I	121-21-1	Liquid at 25°C	328.45	3.303	8.517E-10	1.198E-7
Pyrethrins	8003-34-7	Liquid at 25°C	328.45	3.974	9.099E-10	1.141E-7
Pyrethrin II	121-29-9	Liquid at 25°C	372.46	12.37	4.053E-11	1.221E-9
Organophosphates (chemfinder.com)						
Parathion	56-38-2	6.1°C	291.26	3.064	3.498E-8	3.324E-6
Trimethyl phosphate	512-56-1	-46°C	140.08	7,750,000	0.02638	1.279E-7
Triphenylphosphate	115-86-6	50.5°C	326.29	38.74	2.666E-11	2.246E-10
Paraoxon	311-45-5	NA^a (assume liq.)	275.20	24,070	9.462E-8	1.082E-9
Methyl parathion	298-00-0	36°C	263.2105	58.51	3.033E-7	1.365E-6
Disulfoton	298-04-4	-25°C	274.41	0.4989	1.959E-7	0.0001078
Diazinon	333-41-5	~120°C	304.35	1.658	3.351E-8	6.152E-6
Dichlorvos	62-73-7	-60°C	220.98	761,700	8.389E-4	2.434E-7
Acephate	30560-19-1	88°C	183.17	5397,000	2.006E-7	6.134E-13
Carbamates						
Carbofuran	1563-66-2	151°C	221.26	44.88	2.203E-9	1.086E-8
Aldicarb	116-06-3	100°C	190.37	981.6	4.500E-7	8.723E-8
Carbaryl	63-25-2	145°C	201.22	17.59	9.532E-10	1.090E-8
Methomyl	16752-77-5	78°C	162.21	24,160	1.114E-6	7.479E9
Propoxur	114-26-1	87°C	209.24	262.5	6.175E-8	4.922E-8
Mexacarbate	315-18-4	85°C	222.29	218.4	8.370E-10	8.519E-10
Chlorinated Pesticides and Byproducts						
p,p—DDT	50-29-3	108.5°C	354.49	0.003758	2.185E-10	2.061E-5
p,p'-DDE	72-55-9	89°C	318.03	0.003216	5.141E-10	5.074E-5

o,p'-DDD	53-19-0	76°C	320.04	0.03473	1.010E-9	9.305E-6
p,p'-DDD	72-54-8	109.5°C	320.04	0.01293	3.133E-10	7.755E-6
Lindane (HCB)	58-89-9	113°C	290.83	10.14	4.059E-7	1.164E-5
PCP	77-10-1	46.5°C	243.39	114.3	4.934E-8	1.051E-7
Kepone	143-50-0	350°C (decomp.)	490.64	0.2108	1.290E-10	3.002E-7
Mirex	2385-85-5	485°C (decomp.)	545.54	2.516E-6	1.640E-10	0.03556
cis- or alpha chlordane	5103-71-9	106°C	409.78	0.06442	1.031E-8	6.556E-5
Heptachlor	76-44-8	95.5°C	373.32	0.1677	1.208E-7	0.0002688
Aldrin	309-00-2	104°C	364.91	0.01839	1.761E-7	0.003494
Dieldrin	60-57-1	175°C	380.91	0.1193	4.688E-9	1.498E-5
Endosulfan	115-29-7	106°C	406.93	1.211	1.947E-8	6.544E-6
Alachlor	15972-60-8	40°C	269.77	946.7	1.251E-8	3.566E-9
Aldicarb	116-06-3	100°C	190.37	981.6	4.499E-7	8.723E-8
Endrin	72-20-8	200°C (decomp.)	380.91	0.07949	3.125E-9	1.497E-5
2,3,7,8-Tetrachlorodibenzo-p-dioxin	1746-01-6	295°C	321.97	5.573E-5	2.426E-12	1.401E-5
Phenoxy Acid Herbicides						
2,4-DB	106-93-4	9.9°C	187.86	7,266	0.01260	0.0003258
Dalapon	75-99-0	166.5°C (decomp.)	142.97	69,300	3.125E-4	5.919E-7
Dicamba	1918-00-9	115°C	221.04	278.6	7.437E-7	5.901E-7
Dichlorprop	120-36-5	117.5°C	235.07	325.4	9.587E-8	6.925E-8
Dinoseb	88-85-7	40°C	240.22	2,120	1.305E-7	1.479E-8
MCPA	94-74-6	120°C	200.62	377.5	2.179E-8	1.158E-8
MCPP	93-65-2	93°C	214.65	494.8	1.559E-7	6.766E-8
2,4,5-T	93-76-5	153°C	255.48	63.47	1.714E-9	6.901E-9
2,4-D	94-75-7	140.5°C	221.04	257.7	1.315E-8	1.128E-8
Substituted Phenols						
4-Methylphenol	106-44-5	35.5°C	108.14	19,210	1.478E-4	8.317E-7
4-Ethylphenol	123-07-9	45.0°C	122.17	6,240	5.193E-5	1.017E-6
3-Methylphenol	108-39-4	11.8°C	108.14	25,090	1.969E-4	8.488E-7

(Continued)

49

Classes of Pollutants	CAS #	Melting Point	MW	Solubility (mg/L)	VP (atm)	HLC (atm-L/mol)
2-Chlorophenol	95-57-8	9.8°C	235.07	325.4	9.587E-8	6.926E-8
Phenol	108-95-2	40.9°C	94.11	58,840	4.148E-4	6.633E-7
4-Chlorophenol	106-48-9	42.7°C	128.56	27,890	9.998E-5	4.609E-7
4-Hydroxybenzoic acid	99-96-7	214.5°C	138.12	3,681	4.056E-10	1.257E-11
3-Chlorophenol	108-43-0	32.6°C	128.56	38,860	1.307E-4	4.322E-7
4-Hydroxyacetophenone	99-93-4	109.5°C	136.15	3,201	2.207E-8	9.378E-10
2-Hydroxybenzoic acid	69-72-7	158°C	138.12	2,190	6.139E-6	3.486E-7
4-Nitrophenol	100-02-7	113.8°C	139.11	7,514	1.028E-7	1.903E-9
Triazines						
Atrazine	1912-24-9	173°C	215.68	37.95	9.535E-9	5.419E-8
Eptam	759-94-4	Liquid at 25°C	189.32	152.4	3.357E-5	4.172E-5
Sutan	2008-41-5	Liquid at 25°C	217.37	18.35	1.502E-5	0.0001779
Vernam	1929-77-7	Liquid at 25°C	203.35	47.51	1.190E-5	5.096E-5
Tilam	1114-71-2	Liquid at 25°C	203.35	50.44	1.084E-5	4.372E-5
Propachlor	1918-16-7	77°C	211.69	648.7	1.547E-7	5.048E-8
Trifluralin	1582-09-8	49°C	335.28	2.121	2.858E-8	4.518E-6
Simazine	122-34-9	226°C	201.66	34.44	2.886E-9	1.690E-8
Propazine	139-40-2	213°C	229.71	12.24	8.861E-9	1.664E-7
Bromacil	314-40-9	158°C	261.12	913.7	2.437E-9	6.964E-10
Prowl	40487-42-1	56°C	281.31	1.044	7.028E-9	1.894E-6
Phthalate Esters						
Di- or bis(2-ethylhexyl) phthalate esters (DEHP)	117-81-7	−55°C	390.56	0.009507	1.046E-11	4.294E-6
Diethyl phthalate	84-66-2	−40.5°C	222.24	261.2	1.078E-6	9.169E-7
Dimethyl phthalate	131-11-3	5.5°C	194.19	2.383	5.391E-6	4.393E-7

Compound	CAS					
Di-n-butyl phthalate	84-74-2	−35°C	278.35	2.191	1.711E-8	1.631E-6
Di-n-octyl phthalate	117-84-0	25°C	390.56	0.003374	1.739E-12	3.096E-6

Polycyclic Aromatic Hydrocarbons

Compound	CAS					
Naphthalene	91-20-3	80.2°C	128.17	39.06	1.197E-4	0.0003926
Phenanthrene	85-01-8	99.2°C	178.23	0.9889	1.164E-7	2.098E-5
Anthracene	120-12-7	215.0°C	178.23	0.08508	8.126E-9	1.703E-5
Benzanthracene	56-55-3	84°C	228.29	0.06121	2.141E-10	7.987E-7
Pyrene	129-00-0	151.2°C	202.26	0.1055	6.148E-9	1.178E-5
Fluoranthene	206-44-0	107.8°C	202.26	0.2149	2.173E-8	1.280E-5
Fluorine	86-73-7	114.8°C	166.22	1.963	7.973E-7	6.750E-5
Chrysene	218-01-9	258.2°C	228.29	0.001461	6.278E-12	9.808E-7
Benzo(k)fluoranthene	207-08-9	217°C	252.31	0.003810	1.879E-12	7.258E-7
Benzo(b)fluoranthene	205-99-2	168°C	252.31	0.01237	6.849E-12	9.185E-7
Benzo(a)pyrene	50-32-8	176.5°C	252.31	0.003501	5.871E-12	9.356E-7
Benzo(e)pyrene	192-97-2	177.5°C	252.31	0.003605	6.871E-12	1.149E-6

Aliphatic Hydrocarbons

Compound	CAS					
C_5 (n-pentane)	109-66-0	−129.7°C	72.15	51.81	0.7393	1.029
C_6 (n-hexane)	110-54-3	−95.35°C	86.18	14.53	0.2245	1.331
C_7 (n-heptane)	142-82-5	−90.6°C	100.20	3.725	0.06942	1.867
C_8 (n-octane)	111-65-9	−56.82°C	114.23	0.9318	0.02128	2.608
C_9 (n-nonane)	111-84-2	−53.5°C	128.26	0.2349	0.006343	3.463
C_{10} (n-decane)	124-18-5	−29.7°C	142.28	0.05677	0.001925	4.824
C_{11} (undecane)	1120-21-4	−25.60°C	156.31	0.02252	5.839E-4	6.749
C_{12} (dodecane)	112-40-3	−9.66°C	170.34	0.01496	1.455E-4	9.555
C_{13} (tridecane)	629-50-5	−5.3°C	184.36	0.01030	4.376E-5	13.35
C_{14} (tetradecane)	629-59-4	5.82°C	198.39	0.007339	3.256E-5	18.76
C_{15} (pentadecane)	629-62-9	9.9°C	212.42	0.005402	4.345E-6	26.35
C_{16} (hexadecane)	544-76-3	18.12°C	226.12	0.004100	1.443E-6	37.12
C_{17} (heptadecane)	629-78-7	22°C	240.47	0.003184	4.814E-7	49.08

(Continued)

TABLE 2.6. Summary of Aqueous Solubility, Vapor Pressure, and Henry's Law Constants for Common Environmental Pollutants (Estimated by SPARC) (continued)

Classes of Pollutants	CAS #	Melting Point	MW	Solubility (mg/L)	VP (atm)	HLC (atm-L/mol)
C$_{18}$ (octadecane)	593-45-3	28.2°C	254.50	0.002490	1.603E-7	61.35
C$_{19}$ (nonadecane)	629-92-5	32.1°C	268.52	0.001998	5.748E-8	82.28
C$_{20}$ (icosane)	112-95-8	36.8°C	282.55	0.001631	2.139E-8	110.3
Substituted Nitrobenzenes						
2-Methylnitrobenzene	88-72-2	−10°C	137.14	764.6	2.299E-4	4.125E-5
3-Methylnitrobenzene	99-08-1	15.5°C	137.14	1,394	1.556E-4	1.529E-5
4-Methylnitrobenzene	99-99-0	51.6°C	137.14	869.1	8.833E-5	1.389E-5
2-Chloronitrobenzene	88-73-3	32.5°C	157.56	603.3	9.349E-5	2.442E-5
3-Chloronitrobenzene	121-73-3	44.4°C	157.56	900.3	1.139E-4	1.994E-5
4-Chloronitrobenzene	100-00-5	83.5°C	157.56	476.4	5.994E-5	1.982E-5
2-Acetylnitrobenzene	577-59-3	28.5°C	165.15	4,715	2.996E-6	1.050E-7
3-Acetylnitrobenzene	121-89-1(?)	81°C	165.15	2,953	4.792E-7	2.681e-8
4-Acetylnitrobenzene	100-19-6	81.8°C	165.15	3,008	4.279E-7	2.349E-8
2,4,6-Trinitrotoluene	118-96-7	80.1°C	227.13	144.3	1.759E-9	2.769E-9
2,4-Dinitrotoluene	121-14-2	71°C	182.14	376.5	5.431E-7	2.627E-7
2,6-Dinitrotoluene	606-20-2	66°C	182.14	186.5	5.453E-7	5.327E-7
Nitrobenzene	98-95-3	5.7°C	123.11	3,720	4.066E-4	1.346E-5
HMX	2691-41-0	281°C	296.16	4.519	4.235E-17	2.774E-15
RDX	121-82-4	205.5°C	222.12	69.45	1.055E-12	4.249E-12
Common Solvents						
Pentane	109-66-0	−129.7°C	72.15	51.81	0.7383	1.029
Pentadecane	629-62-9	9.9°C	212.42	0.005402	4.345E-6	26.35
Benzene	71-43-2	5.49°C	78.11	2,481	0.1355	0.004266

Toluene	108-88-3	−94.95°C	92.14	775.8	0.04133	0.004909
Styrene	100-42-5	−31°C	104.15	425.5	0.008913	0.002181
Pyridine	110-86-1	−40.70°C	79.10	178,600	0.03064	1.357E-5
Methylene chloride	75-09-2	−94.94°C	84.93	25,240	0.5370	0.001807
Chloroform	67-66-3	−63.41°C	119.38	11,930	0.2088	0.002090
Carbon tetrachloride	56-23-5	−22.73°C	153.82	2,348	0.1671	0.01095
Trichloroethylene	79-01-6	−84.7°C	131.39	914.0	0.1773	0.02549
p-Dichlorobenzene	106-46-7	53.09°C	147.00	122.1	0.001934	0.002329
Acetone	67-64-1	−94.8°C	58.08	246,800	0.3051	7.180E-5
Methyl ethyl ketone	78-93-3	−86.65°C	72.11	129,200	0.1323	7.382E-5
Acetonitrile	75-05-8	−43.82°C	41.05	265,800	0.09967	1.539E-5
Methanol	67-56-1	−97.54°C	32.04	1,144,500	0.1982	5.549E-6
Ethanol	64-17-5	−114.1°C	46.07	406,200	0.08656	9.817E-6
Aniline	62-53-3	−6.02°C	93.13	47,480	7.075E-4	1.388E-6
Nitrobenzene	98-95-3	5.7°C	123.11	3,720	4.066E-4	1.346E-5
Methyl tert-butyl ether	1634-04-4	−108.6°C	88.15	15,290	0.5634	0.003247
Gasoline Components						
Benzene	71-43-2	5.49°C	78.11	2,481	0.1355	0.004266
Ethylbenzene	100-41-4	−94.9°C	106.17	270.4	0.01569	0.006160
methyl tert-butyl ether	1634-04-4	−108.6°C	88.15	15,300	0.5634	0.003247
heptane	142-82-5	−90.6°C	100.20	3,725	0.06942	1.867
2-methylpentane	107-83-5	−153.7°C	86.18	16.69	0.3243	1.674
Toluene	108-88-3	−94.95°C	92.14	775.8	0.04133	0.004909
Toluene	108-88-3	−94.95°C	92.14	775.8	0.04133	0.004909
m-xylene	108-38-3	−47.8°C	106.17	268.3	0.01241	0.004909
o-xylene	95-47-6	−25.2°C	106.17	320.0	NA	NA

a NA, not available.

53

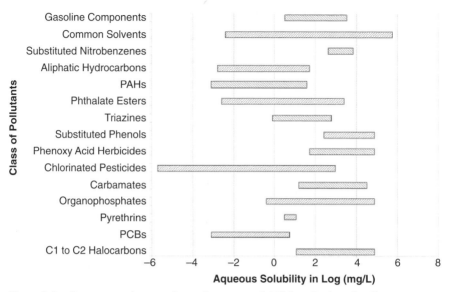

Figure 2.6. Summary and comparison of aqueous solubilities of selected pollutants.

Solution

$$CuBr_{(aq)} \Rightarrow Cu^+_{(aq)} + Br^-_{(aq)}$$

$$K_{sp} = [Cu^+][Br^-]$$

For every mole of CuBr that dissolves, one mole of Cu^+ and one mole of Br^- dissolve, so let $[Cu^+] = [Br^-] = x$. Substituting into the K_{sp} equation yields

$$K_{sp} = 5.3 \times 10^{-9} = x^2, \qquad x = (K_{sp})^{1/2}$$

or

$$x = [Cu^+] = 7.3 \times 10^{-5} \text{ M Cu}^+$$

Finally, to get solubility in ppm,

$$(7.3 \times 10^{-5} \text{ M Cu}^+)(63.6 \text{ g/mol}) = 4.6 \times 10^{-3} \text{ g/L or } 4.6 \text{ mg/L Cu}^+ \text{ or } 4.6 \text{ ppm}$$

Now, let's make the problem a bit more complicated. Say that you have a hazardous waste sludge sample containing $PbCO_3$, $PbCl_2$, $PbCrO_4$, PbF_2, $PbSO_4$, and PbS. Which form of lead waste will determine the maximum concentration of lead in any leachate that may come from the waste? What could be the maximum concentration of Pb^{2+} in the leachate?

TABLE 2.7. Solubility Products for Inorganic Compounds at 25°C (*Handbook of Chemistry and Physics*, 1980)

Substance	Solubility Product at Temperature Noted (°C)	Substance	Solubility Product at Temperature Noted
Aluminum hydroxide	4×10^{-13} (15°)	Lead iodide	7.47×10^{-9} (15°)
Aluminum hydroxide	1.1×10^{-15} (18°)	Lead iodide	1.39×10^{-8} (25°)
Aluminum hydroxide	3.7×10^{-15} (25°)	Lead oxalate	2.74×10^{-11} (18°)
Barium carbonate	7×10^{-9} (16°)	Lead sulfate	1.06×10^{-8} (18°)
Barium carbonate	8.1×10^{-9} (25°)	Lead sulfide	3.4×10^{-28} (18°)
Barium chromate	1.6×10^{-10} (18°)	Lithium carbonate	1.7×10^{-3} (25°)
Barium chromate	2.4×10^{-10} (28°)	Magnesium ammonium phosphate	2.5×10^{-13} (25°)
Barium fluoride	1.6×10^{-6} (9.5°)	Magnesium carbonate	2.6×10^{-5} (12°)
Barium fluoride	1.7×10^{-6} (18°)	Magnesium fluoride	7.1×10^{-9} (18°)
Barium fluoride	1.73×10^{-6} (25.8°)	Magnesium fluoride	6.4×10^{-9} (27°)
Barium iodate, Ba(IO$_3$)$_2$ 2H$_2$O	8.4×10^{-11} (10°)	Magnesium hydroxide	1.2×10^{-11} (18°)
Barium iodate, Ba(IO$_3$)$_2$ 2H$_2$O	6.5×10^{-10} (25°)	Magnesium oxalate	8.57×10^{-5} (18°)
Barium oxalate, BaC$_2$O$_3$ 3.5H$_2$O	1.62×10^{-7} (18°)	Manganese hydroxide	4×10^{-14} (18°)
Barium oxalate, BaC$_2$O$_4$ 2H$_2$O	1.2×10^{-7} (18°)	Manganese sulfide	1.4×10^{-15} (18°)
Barium oxalate, BaC$_2$O$_4$ 0.5H$_2$O	2.18×10^{-7} (18°)	Mercuric sulfide	4×10^{-53} to
Barium sulfate	0.87×10^{-10} (18°)		2×10^{-49} (18°)
Barium sulfate	1.08×10^{-10} (25°)	Mercurous bromide	1.3×10^{-21} (25°)
Barium sulfate	1.98×10^{-10} (50°)	Mercurous chloride	2×10^{-18} (25°)
Cadmium oxalate CdC$_2$O$_4$ 3H$_2$O	1.53×10^{-8} (18°)	Mercurous iodide	1.2×10^{-28} (25°)
Cadmium sulfide	3.6×10^{-29} (18°)	Nickel sulfide	1.4×10^{-24} (18°)
Calcium carbonate (calcite)	0.99×10^{-8} (15°)	Potassium acid tartrate [K$^+$] [HC$_4$H$_4$O$_6^-$]	3.8×10^{-4} (18°)
Calcium carbonate (calcite)	0.87×10^{-8} (25°)		
Calcium fluoride	3.4×10^{-11} (18°)	Silver bromate	3.97×10^{-5} (20°)
Calcium fluoride	3.95×10^{-11} (26°)	Silver bromate	5.77×10^{-5} (25°)
Calcium iodate, (Ca(IO$_3$)$_2$ 6H$_2$O	22.2×10^{-8} (10°)	Silver bromide	4.1×10^{-13} (18°)
Calcium iodate, (Ca(IO$_3$)$_2$ 6H$_2$O	64.4×10^{-8} (18°)	Silver bromide	7.7×10^{-13} (25°)

(*Continued*)

TABLE 2.7. Solubility Products for Inorganic Compounds at 25°C (*Handbook of Chemistry and Physics*, 1980) (continued)

Substance	Solubility Product at Temperature Noted (°C)	Substance	Solubility Product at Temperature Noted
Calcium oxalate, $CaC_2O_4\ H_2O$	1.78×10^{-9} (18°)	Silver carbonate	6.15×10^{-12} (25°)
Calcium oxalate, $CaC_2O_4\ H_2O$	2.57×10^{-9} (25°)	Silver chloride	0.21×10^{-10} (4.7°)
Calcium sulfate	2.45×10^{-5} (25°)	Silver chloride	0.37×10^{-10} (9.7°)
Calcium tartrate, $CaC_4H_4O_6\ 2H_2O$	0.77×10^{-6} (18°)	Silver chloride	1.56×10^{-10} (25°)
Cobalt sulfide	3×10^{-26} (18°)	Silver chloride	13.2×10^{-10} (50°)
Cupric iodate	1.4×10^{-7} (25°)	Silver chloride	215×10^{-10} (100°)
Cupric oxalate	2.87×10^{-8} (25°)	Silver chromate	1.2×10^{-12} (14.8°)
Cupric sulfide	8.5×10^{-45} (18°)	Silver chromate	9×10^{-12} (25°)
Cuprous bromide	4.15×10^{-8} (18–20°)	Silver cyanide $[Ag^+][Ag(CN)^-_2]$	2.2×10^{-12} (20°)
Cuprous chloride	1.02×10^{-6} (18–20°)	Silver dichromate	2×10^{-7} (25°)
Cuprous iodide	5.06×10^{-12} (18–20°)	Silver hydroxide	1.52×10^{-8} (20°)
Cuprous sulfide	2×10^{-47} (16–18°)	Silver iodate	0.92×10^{-8} (9.4°)
Cuprous thiocyanate	1.6×10^{-11} (18°)	Silver iodide	0.32×10^{-16} (13°)
Ferric hydroxide	1.1×10^{-36} (18°)	Silver iodide	1.5×10^{-16} (25°)
Ferrous hydroxide	1.64×10^{-14} (18°)	Silver sulfide	1.6×10^{-49} (18°)
Ferrous oxalate	2.1×10^{-7} (25°)	Silver thiocyanate	0.49×10^{-12} (18°)
Ferrous sulfide	3.7×10^{-19} (18°)	Silver thiocyanate	1.16×10^{-12} (25°)
Lead carbonate	3.3×10^{-14} (18°)	Strontium carbonate	1.6×10^{-9} (25°)
Lead chromate	1.77×10^{-14} (18°)	Strontium fluoride	2.8×10^{-9} (18°)
Lead fluoride	2.7×10^{-8} (9°)	Strontium oxalate	5.61×10^{-8} (18°)
Lead fluoride	3.2×10^{-8} (18°)	Strontium sulfate	2.77×10^{-7} (2.9°)
Lead fluoride	3.7×10^{-8} (26.6°)	Strontium sulfate	3.81×10^{-7} (17.4°)
Lead iodate	5.3×10^{-14} (9.2°)	Zinc hydroxide	1.8×10^{-14} (18–20°)
Lead iodate	1.2×10^{-13} (18°)	Zinc oxalate, $ZnC_2O_4\ 2H_2O$	1.35×10^{-9} (18°)
Lead iodate	2.6×10^{-13} (25.8°)	Zinc sulfide	1.2×10^{-23} (18°)

Pollutant	K_{sp}
$PbCO_3$	7.4×10^{-14}
$PbCl_2$	1.7×10^{-5}
$PbCrO_4$	2.8×10^{-13}
PbF_2	3.6×10^{-8}
$PbSO_4$	6.3×10^{-7}
PbS	3×10^{-28}

Solution. In general, this can be solved by looking at the compound with the highest K_{sp} value, since this will be the one most responsible for the solubility of Pb^{2+} (but for similar K_{sp} values you must also consider stoichiometry). In this case, $PbCl_2$ will determine the overall solubility of Pb^{2+} from the waste.

$$K_{sp} = 1.7 \times 10^{-5} = \left[Pb^{2+}\right]\left[Cl^{-}\right]^2$$

For each mole of $PbCl_2$ that dissolves, one mole of Pb^{2+} ion is released and two moles of Cl^{-}. We will let $x = Pb^{2+}$. For every Pb^{2+} there are two Cl^{-}, so if $x = Pb^{2+}$, $Cl^{-} = 2x$. Substituting into the K_{sp} equation yields

$$K_{sp} = 1.7 \times 10^{-5} = [x][2x]^2 = 4x^3$$
$$x = 0.016 \text{ M } Pb^{2+}$$
$$\text{in ppm: } (0.016 \text{ M } Pb^{2+})(207.2 \text{ g/mol})(1000 \text{ mg/g}) = 3400 \text{ mg/L}$$

There is one complicating variable concerning solubility, both for organic and inorganic pollutants. Numerous field investigations report "dissolved" concentrations of pollutants that exceed their known maximum solubilities. But how can this be? The answer is the presence of a second "dissolved" phase, usually in the form of colloidal inorganic or organic particles. There is no conclusive way to isolate free ion in the dissolved phase from these phases for measurement. Actually, the dissolved phase is commonly defined operationally as the pollutant concentration present in a sample that has been filtered through a 0.20- or 0.45-μm filter. But there are many naturally occurring inorganic and organic particles (which can contain sorbed pollutants) that can pass through these filters. The most common second phase that has been found in field samples is natural organic matter (NOM) present in the dissolved phase (DOM). NOM is the organic matter left over from microbial decay of natural plant and animal material. NOM can vary greatly in molecular weight, and in general no two NOM molecules are the same. Scientists have attempted to characterize NOM using a variety of techniques (IR, NMR, MS) in order to develop a "representative" structure. Two such structures are shown in Figure 2.7. The purpose of these structures is not to have a working molecular model of NOM but to illustrate the presence of important functional groups and hydrophobic centers in the molecule. Every pollutant studied has some affinity for sorbing or binding to NOM molecules. We will discuss NOM in greater detail in Section 2.7.2.

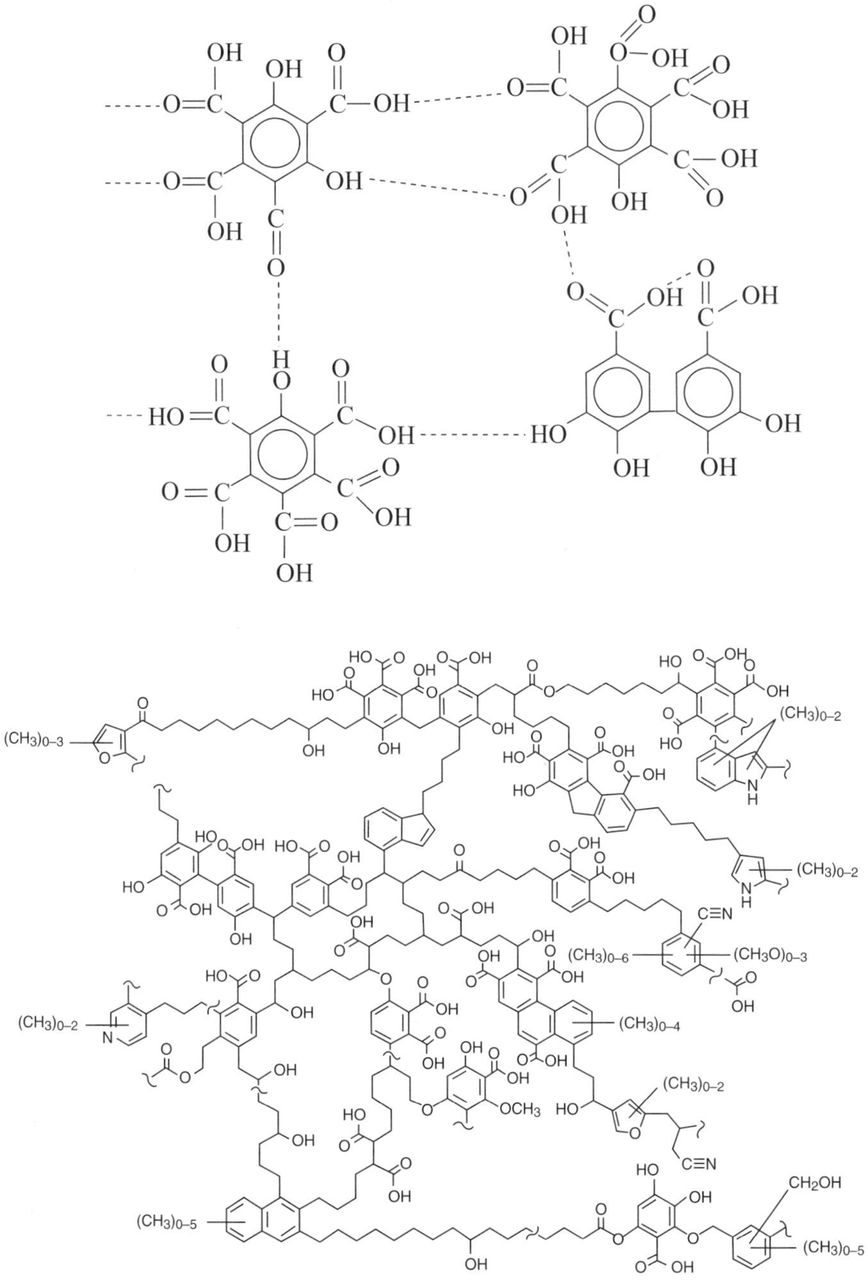

Figure 2.7.   Representative structures of natural organic matter. [Reprinted with permission from Wiley Interscience (Stumm, W. and J. J. Morgan, 1981) and Springer Publishing (Schulten, H. R. and M. Schnitzer, 1993).]

The variety of forms in which a pollutant may be present gives rise to the term speciation. For example, Cd^{2+} may be present as the free hydrated cation (an ion surrounded by water of hydration) or as the bound $NOM–Cd^{2+}$ complex. But note that given the variety of ionic binding sites in the NOM molecule, a variety of $NOM–Cd^{2+}$ complexes can be present, which greatly complicates the transport modeling process. We'll discuss this further in Chapters 3 and 8.

2.4.4 Vapor Pressure

Vapor pressure is the pressure of a compound's vapor phase at equilibrium with the compound's pure phase, which can be a liquid or solid depending on the temperature. As a reference point we use conventional standard conditions (25°C and 1.00 atmosphere of pressure) to describe standard vapor pressures. As with aqueous solubilities, and especially for the broad range of pollutants of interest to this textbook, vapor pressures for many compounds are difficult to measure experimentally. So, again we will rely on the SPARC program to generate a range of values for selected compounds. These are also shown in Table 2.6 and are summarized by type of pollutant in Figure 2.8. Note that the range extends from 0.77 atm to $10^{-11.7}$ atm. Vapor pressure is important in the fate and transport modeling of atmospheric pollutants, since the vapor pressure determines both whether a chemical will volatilize and, in general, how fast it volatilizes in an open system. Pollutants with high vapor pressures volatilize faster than those with low vapor pressure. This parameter will be

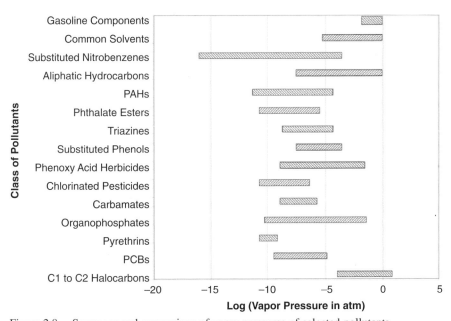

Figure 2.8.   Summary and comparison of vapor pressures of selected pollutants.

important in Chapter 9, where we will study the fate and transport of pollutants in the atmosphere.

2.4.5 Henry's Law Constant

A Henry's law constant (K_H) is the ratio of the equilibrium vapor pressure to aqueous solubility of a substance (at a given temperature), expressed by

$$X(aq) \leftrightarrow X(g)$$

$$K_H = \frac{\text{Pressure in atms } (P_x)}{\text{M solubility (mol/L)}} \tag{2.4}$$

K_H can be calculated by taking the ratio of experimentally determined or predicted vapor pressures and aqueous solubilities, or can be measured using the dynamic purge technique described by Mackay et al. (1979). Since the K_H is a ratio of two experimentally difficult measurements, its experimental determination is subject to the same errors as its principle measurements. Thus, again, we will turn to the SPARC estimation program for K_H values. These are summarized in Table 2.6 and in Figure 2.9 by type of pollutant. Note the large range in values reflecting the large range in vapor pressures and aqueous solubilities.

K_H is important in environmental chemistry, perhaps more important than vapor pressure, for estimating the volatilization of pollutants from water. Although many hydrophobic pollutants have low vapor pressure, they have high K_H values;

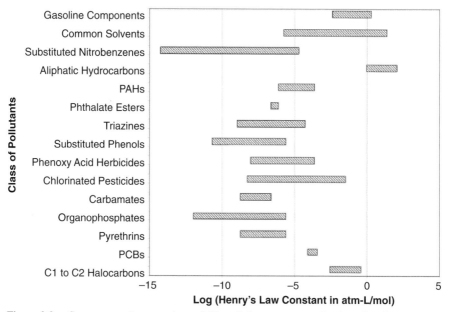

Figure 2.9. Summary and comparison of Henry's law constants of selected pollutants.

when they are placed in water, they readily escape to the overlying gas phase. This equilibrium preference for the gaseous phase gives many pollutants a higher-than-expected volatilization rate. Classic examples are DDT and PCBs. This will be expressed in our river and lake fate and transport models as a first-order removal rate.

2.5 REACTIONS AND EQUILIBRIUM

The three basic types of chemical reactions are precipitation, acid–base, and oxidation–reduction. We covered precipitation to the extent necessary in Section 2.4.2 (and we will cover it again in Section 3.2.2). We will spend a little more time covering acid–base and oxidation–reduction reactions in this section. First, we will extend your knowledge of acid–base chemistry through the concept of buffers and pC–pH diagrams.

2.5.1 Acid–Base Chemistry

By far the most important buffer system in nature is the carbonate system, and we will use it here to introduce you to a new way of looking at pH and acid–base equilibrium and to introduce the concept of chemical speciation. The diagrams we will be constructing are referred to as pC–pH diagrams (also referred to as distribution or alpha plots), where the pC represents the −log of concentration of any chemical species in solution. Note the C in this case does not represent the chemical element carbon, but instead the concentration of any compound.

The concentration of a weak acid or base in a solution (for example, H_2CO_3, HCO_3^-, or CO_3^{2-}) can be calculated using simple equilibrium expressions at any given pH value. In some cases it is useful to look at the equilibrium distribution of each of the protonated and nonprotonated species in solution at the same time. A pC–pH diagram (Figure 2.10) is an excellent tool for viewing these concentrations simultaneously. As the name implies, this diagram shows the concentrations of all chemical species (expressed as the negative log of concentration), with pC values on the vertical axis and pH on the horizontal. To construct a pC–pH diagram, the total concentration of the acid or base is needed along with the corresponding equilibrium equations and acid-dissociation constants (K_a).

We will first construct a pC–pH diagram for a system that is closed to the atmosphere, such that no additional atmospheric CO_2 can enter the water. For calculation purposes we assume that carbonate in the water comes solely from the dissolution of carbonate-containing minerals. Refer to Figure 2.10 during the following discussion.

All pC–pH diagrams have two lines in common: the line describing the concentration of hydroxide (OH^-) as a function of pH and the line describing the concentration of hydronium ion (H_3O^+) as a function of pH. These are based on the equilibrium relationship

$$H_2O \Leftrightarrow H^+ + OH^-$$

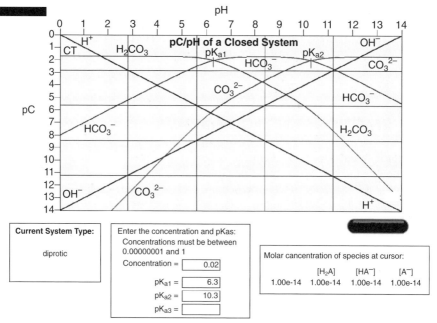

Figure 2.10.   A pC–pH diagram for a closed 0.0200 M carbonate system. (Diagram from Fate®.)

where

$$K_w = \left[H^+\right]\left[OH^-\right] = 1.00 \times 10^{-14} \text{ (at 25°C)}$$

By rearranging and taking the negative log of each side, we obtain

$$-\log K_w = -\log\left[H^+\right] - \log\left[OH^-\right]$$
$$14 = pH + pOH$$
$$pOH = 14.0 - pH$$

Thus, when pH equals zero, pOH equals 14. This results in a line from (pH = 0.0, pC = 14.0) to (pH = 14.0, pC = 0.0). The slope of this diagonal line is

$$\frac{\Delta(-\log[OH])}{\Delta pH} = +1$$

Similarly, a line can be drawn representing the hydronium ion concentration as a function of pH. By definition,

$$-\log\left[H^+\right] = pH, \qquad \text{therefore} \qquad \frac{\Delta(-\log[H^+])}{pH} = -1$$

When the pH equals 0, the $-\log[H^+]$ equals 0. This results in a line from (pH = 0.0, pC = 0.0) to (pH = 14.0, pC = 14.0).

The next line (or set of lines) normally drawn on a pC–pH diagram is the one representing the total concentration of acid or base, C_T. We will use a total carbonate concentration of 0.0200 M and assume it is constant, since we are dealing with a closed system. When pC–pH diagrams are drawn by hand, C_T is drawn as a straight horizontal line at pC_T on the y-axis, as a guide for filling in the curves for the different acid/base species. This line is actually broken into regions of two or more lines for the dominant chemical species, depending on the number of protons present in the acid. Vertical (dotted) lines representing the pK_a values (negative logs of the K_a values) are then added. These vertical lines become the guides for adding the curves representing the different carbonate species (CO_3^{2-}, HCO_3^-, H_2CO_3). At the lowest pK_a, along the C_T line, for example, falls the intersection of the lines representing the most acidic and second most acidic species (HCO_3^- and H_2CO_3 in the case of the carbonate system), one with a negative whole number slope and one with a positive whole number slope. We will limit our discussion of the mathematical derivation of these lines to a diprotic system (the carbonate system), but after reviewing the equations you should be able to derive the governing equations for triprotic and monoprotic systems.

For the diprotic system, the equilibrium equations for H_2A (H_2CO_3), HA^- (HCO_3^-), and A^{2-} (CO_3^{2-}) (where A refers to the anionic species, CO_3^{2-} for the carbonate example) are

$$H_2A \Leftrightarrow HA^- + H^+, \qquad \text{where } K_{a1} = \frac{[HA^-][H^+]}{[H_2A]}$$

$$HA^- \Leftrightarrow A^{2-} + H^+, \qquad \text{where } K_{a2} = \frac{[A^{2-}][H^+]}{[HA^-]}$$

The mass balance equation (a description of all carbonate forms) is

$$C_T = [H_2A] + [HA^-] + [A^{2-}]$$

When the equilibrium equations are solved for H_2A, HA^-, and A^{2-} (in terms of C_T, $[H^+]$ and the equilibrium constants) and combined with the mass balance equation, three equations are obtained:

$$[H_2A] = C_T \left(\frac{1}{1 + (K_1/[H^+]) + (K_1 K_2/[H^+])^2} \right)$$

$$[HA^-] = C_T \left(\frac{1}{([H^+]/K_1) + 1 + (K_2/[H^+])} \right)$$

$$[A^{2-}] = C_T \left(\frac{1}{([H^+]^2/K_1 K_2) + ([H+]/K_2) + 1} \right)$$

If a pH-dependent constant, α_H, is defined as

$$\alpha_H = \frac{[H^+]^2}{K_1 K_2} + \frac{[H^+]}{K_2} + 1$$

then the previous equations for the diprotic system can be simplified to

$$[H_2A] = \frac{C_T[H^+]^2}{K_1 K_2 \alpha_H} \tag{2.5}$$

$$[HA^-] = \frac{C_T[H^+]}{K_2 \alpha_H} \tag{2.6}$$

$$[A^{2-}] = \frac{C_T}{\alpha_H} \tag{2.7}$$

By taking the log transform of each equation, we now have equations describing every line in the pC–pH diagram. The utility of a pC–pH diagram is that all of the ion concentrations can be estimated at the same time for any given pH value. The computer program (the pC–pH Simulator) used to create this diagram, included with your textbook, allows the user to select an acid system, enter the pK_a values, and draw the pC–pH diagram. After the diagram is drawn, the user can point the cursor at a given pH and the concentration of each species will be given.

Now we will develop a pC–pH diagram for an open system, one that is open to a gas that can dissolve and contribute chemical species (dissolved gases and ions) to the solution. The pC–pH diagrams for open systems are similar to those described for the closed systems. The primary difference is that in an open system a component of the system exists as a gas and the system is open to inputs from the atmosphere. In other words, the system can exchange matter with the atmosphere. The most important environmental examples of such systems are lakes, rivers, and oceans, which exchange carbon dioxide with the atmosphere and therefore contain dissolved carbon dioxide ($CO_2(aq)$), carbonic acid (H_2CO_3), bicarbonate ion (HCO_3^-), and carbonate ion (CO_3^{2-}).

The reactions occurring in this system are

1. $CO_{2(aq)} \leftrightarrow CO_{2(g)}$
2. $CO_{2(g)} + H_2O \leftrightarrow H_2CO_3$
3. $H_2CO_3 \leftrightarrow HCO_3^- + H^+$
4. $HCO_3^- \leftrightarrow CO_3^{2-} + H^+$
5. $H_2O \leftrightarrow H^+ + OH^-$

The equilibrium relationships, corresponding to the line number, for this system are, respectively:

$$K_2 = \frac{[H_2CO_3]}{P_{CO_2}} = 10^{-1.47}$$

$$K_3 = \frac{[H^+][HCO_3^-]}{[H_2CO_3]} = 10^{-6.35}$$

$$K_4 = \frac{[H^+][CO_3^{2-}]}{[HCO_3^-]} = 10^{-10.33}$$

$$K_5 = [H^+][OH^-] = 10^{-14}$$

where P_{CO_2} is the partial pressure of CO_2 in the atmosphere.

Open-system pC–pH diagrams, such as the one shown in Figure 2.11, contain lines describing the concentration of hydroxide (OH⁻) and hydronium ion (H⁺) identical to those for closed systems. However, because open systems can exchange matter with the atmosphere, the total inorganic carbon (C_T) concentration is not constant as it is for a closed system, where all carbon present was derived from an initial input of calcium carbonate. Rather, in an open system, C_T varies as a function of pH. Still, the total inorganic carbon concentration is the sum of all inorganic carbon species, as it was for closed systems. In this case,

$$C_T = [H_2CO_3] + [HCO_3^-] + [CO_3^{2-}].$$

The concentration of H_2CO_3, HCO_3^-, and CO_3^{2-}, as a function of pH and partial pressure, P_{CO_2}, can be calculated from the equilibrium relationships given previously. The equations for these lines are

$$[H_2CO_3] = (K_2)(P_{CO_2}) = (P_{CO_2})10^{-1.47} - \log[H_2CO_3]$$
$$= -\log(P_{CO_2}) + 1.47$$

$$[HCO_3^-] = \frac{(K_3)[H_2CO_3]}{(H^+)^2} = \frac{(K_3)(P_{CO_2})(10^{-1.47})}{H^+} = \frac{(10^{-6.35})(P_{CO_2})(10^{-1.47})}{H^+}$$

$$-\log[HCO_3^-] = -\log(P_{CO_2}) + 7.82 - pH$$

$$[CO_3^{2-}] = \frac{(K_4)(10^{-6.35})(P_{CO_2})(10^{-1.47})}{(H^+)^2} = \frac{(10^{-10.33})(10^{-6.35})(P_{CO_2})(10^{-1.47})}{(H^+)^2}$$

$$-\log[CO_3^{2-}] = -\log(P_{CO_2}) + 18.15 - 2pH$$

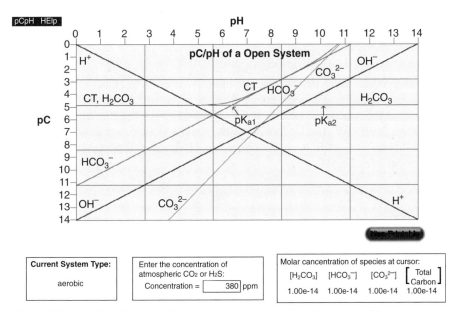

Figure 2.11. A pC–pH diagram for an open carbonate system in contact with an atmosphere of 380 ppm CO_2. (Diagram from Fate®.)

As mentioned previously and demonstrated by the above equations, the concentrations of H_2CO_3, HCO_3^-, and CO_3^{2-} vary as a function of both pH and P_{CO_2}. This means that as P_{CO_2}, the global carbon dioxide concentration in the atmosphere, has varied naturally over the years, often associated with climate changes, the concentration of H_2CO_3, HCO_3^-, and CO_3^{2-} in surface waters has changed. It also means that P_{CO_2} changes caused by global warming will alter the surface water concentrations of these species.

A pC–pH diagram for the open carbonate system ($P_{CO_2} = 380\,ppm$ CO_2 or $3.80 \times 10^{-4}\,atm$) in the atmosphere) is shown in Figure 2.11. The intercept of the $[H^+]$ line and the HCO_3^- line represents the pH of the system, which decreases with increasing atmospheric CO_2 concentration.

As you can see, the pH of the system determines the chemical speciation of protonated and deprotonated species, and pC–pH diagrams can be used to determine the dominant form of protonated–deprotonated compound at a specific pH of interest. So how does this relate to the buffering of pH? Recall from your general chemistry course the concept of buffers and locate the two important buffer points in Figures 2.10 and 2.11. These are located where the concentrations of H_2CO_3–HCO_3^- and HCO_3^-–CO_3^{2-} are close to each other, within 0.5 to 1.0 pH units of the two pK_a values (6.33 and 10.33; refer to any titration in a general chemistry text). At or near these pH values, the pH remains relatively stable with small inputs of acid or base, the effects of which are effectively absorbed by the weak acid–base pairs. We will discuss the effect of pH on metal speciation further in Section 3.2.2.

2.5.2 Oxidation–Reduction Chemistry

Oxidation–reduction (redox) is probably the most common, and the most complicated, type of reaction in the environment. In general chemistry you mostly studied redox reactions from the standpoint of batteries and electrochemical cells, but there is much more to this complicated area of chemistry. Earlier in this chapter, we described pH as a master variable. Another important chemical parameter of an aqueous solution, which determines, and in many cases drives, other chemical reactions, is the apparent electrode potential of the aqueous system. We refer to this potential of the aqueous system as the E_H, measured in volts. But where does this term come from? We start by looking at the activity of electrons in solution (their availability to participate in reduction reactions), designated by $\{e^-\}$, and use the same notation as for pH

$$p\varepsilon = -\log\{e^-\} \tag{2.8}$$

Thus, $p\varepsilon$ is a measure of the electron activity. Under typical conditions in surface waters (pH of 7.0, in the presence of dissolved oxygen), the $p\varepsilon$ is approximately 14. But it decreases to approximately 4 in the presence of reduced iron, and it drops further to approximately −4 when sulfide and methane are present. $p\varepsilon$ is related to E_H by

$$E_H = \frac{2.3RT}{nF}p\varepsilon \tag{2.9}$$

where R is the ideal gas law constant, T is the absolute temperature (in Kelvin), and F is the Faraday constant (96,485 coulombs/mol of electrons). At 25°C and for a one mole transfer of electrons ($n = 1$), the equation reduces to

$$E_H = 0.059 p\varepsilon$$

The oxidation state of most natural waters is controlled by microbial activity; thus, the reactions controlling the chemistry of these waters are said to be biologically mediated. In order to understand these reactions and their implications, we must first understand the concept of terminal electron acceptors (TEAs).

Among biological energy-producing oxidation reactions, we are most familiar with the use of oxygen to oxidize food (as occurs in our bodies); yet there are many other ways of utilizing carbon-based compounds as food. Food, in the general sense, consists of reduced carbon in the form $(CH_2O)_n$. When we eat (oxidize) this food, we remove electrons from the reduced carbon, split off water, and oxidize the carbon to CO_2, which we exhale. But what happens to the electrons we have taken off the carbon? There must be an acceptor for these electrons, since there is no free current running through our bodies—we do not light up bulbs if we hold them in our hands (or mouth). This is where molecular oxygen comes into the equation in human bodies; it takes the electrons freed up from the carbon and is reduced to water (O_2 goes from an oxidation state of 0 to an oxidation state of -2 in water). These paired redox reactions are illustrated for glucose in the equations below, where glucose is our reduced carbon source (food) and oxygen is the terminal electron acceptor (where the released electrons go). The standard free energy change for each half-reaction ($\Delta G°$) under the conditions in natural water is shown to the right (Schwarzenbach et al., 1993).

$$C_6H_{12}O_{6(glucose)} + 6H_2O_{(1)} \Rightarrow 6CO_{2(g)} + 24H^+_{(aq)} + 24e^- \qquad -984.0 \text{ kJ/mol glucose}$$

$$O_{2(g)} + 4H^+_{(aq)} + 4e^- \Rightarrow 2H_2O_{(1)} \qquad -313.2 \text{ kJ/mol } O_2$$

You should recall from general chemistry that you can add half-reactions to obtain an overall energy value for the complete process. In order to add these half-reactions, you must have an equal number of electrons on opposite sides of the two equations and cancel out the electrons when you add the reactions. In order to do this we must multiply the oxygen equation by 6, and when we do this you must also multiply the $\Delta G°$ value by 6. This yields

$$C_6H_{12}O_{6(glucose)} + 6H_2O \Rightarrow 6CO_2 + 24H^+ + 24e^- \qquad -984.0 \text{ kJ/mol glucose}$$

$$6O_2 + 24H^+ + 24e^- \Rightarrow 12H_2O \qquad -1879.2 \text{ kJ/6 mol } O_2$$

Adding the two equations together and reducing the sum to its simplest form yields

$$C_6H_{12}O_{6(glucose)} + 6O_2 \Rightarrow 6H_2O + 6CO_2 \qquad -2863. \text{ kJ/mol glucose}$$

Note that this reaction yields 2863 kJ for each mole of glucose oxidized. Again, in this reaction electrons are taken from the glucose and added to gaseous oxygen, producing water and releasing energy from the chemical bonds. Neither reaction is possible without coupling to the other half-reaction. As mentioned, this is the way people think of food becoming oxidized, but in reality there are many more ways the overall reaction is completed by other organisms, especially in aqueous environments not in contact with our highly oxygen-rich atmosphere.

In the reaction discussed above, oxygen (O_2) was the terminal electron acceptor (TEA), but many other TEAs also exist. Other common TEAs that are present in the deep water of lakes and rivers, in anoxic lake and river sediments, and in anoxic groundwater systems include

$$\mathbf{2NO_3^-} + 12H^+ + 10e^- \Rightarrow N_2 + 6H_2O \qquad\qquad -714\ kJ/mol$$

$$\mathbf{MnO_2(s)} + HCO_3^- + 3H^+ + 2e^- \Rightarrow MnCO_3(s) + 2H_2O \qquad -100.4\ kJ/mol$$

$$\mathbf{FeOOH(s)} + HCO_3^- + 2H^+ + e^- \Rightarrow FeCO_3(s) + 2H_2O \qquad 4.6\ kJ/mol$$

$$\mathbf{SO_4^{2-}} + 9H^+ + 8e^- \Rightarrow HS^- + 4H_2O \qquad\qquad 170.4\ kJ/mol$$

$$\mathbf{CO_2} + 8H^+ + 8e^- \Rightarrow CH_4 + 2H_2O \qquad\qquad 188.0\ kJ/mol$$

The chemicals in bold print represent the TEAs of interest. Each of these half-reactions can be manipulated and added to the glucose half-reaction above to obtain the energy produced by the microbe per mole of glucose. The magnitude of these $\Delta G°$ values determines which terminal electron acceptor will be used first (i.e., the combination yielding the most energy will be favored and used first, since it gives the microbes that utilize it an evolutionary advantage, then the next most energetic, and so on).

The reason for discussing these reactions is to note the oxidation state (E_H) where each reaction occurs in nature and polluted systems. Not all environments contain oxygen, and the chemistry (and fate and transport) of metals and organic pollutants can significantly change depending on the oxidation potential of the surrounding system. Figure 2.12 shows the oxidation potential where each of the reducing reactions can occur. Note the E_H of the systems. Highly oxygenated systems have an E_H value of approximately +0.81 V, and this value decreases as you move through the other terminal electron acceptors (oxidized and reduced species). Extremely anaerobic water can have E_H values of −0.40 V, indicating a highly reducing environment. In such environments, transformation reactions of the pollutants can occur. We will discuss these TEA reactions further in the lake and groundwater chapter (Chapters 5 and 8, respectively). In addition, specific transformation reactions occurring in anaerobic groundwater will be discussed in Section 2.8.

The redox potential of a water is obviously important in the fate and transport of metal pollutants, since the oxidation state of transition metals is strongly influenced by E_H. But what about organic pollutants? As we will see in Section 2.8, organic pollutants are subject to a variety of transformation reactions. Now, we will use one of these reactions, the reduction of nitrobenzene pollutants, to illustrate the effect of the pH and E_H on the concentration of reactants and the resultant speed (rate) of the transformation reaction. This example is taken from experiments conducted using anaerobic groundwaters by Dunnivant et al. (1992). When organic matter is present in groundwater, the water may become anaerobic from high micro-

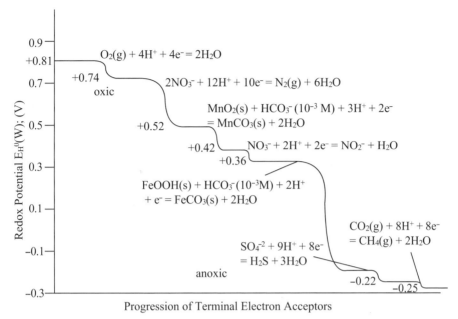

Figure 2.12.   The electrochemical potential of water as a function of the terminal electron acceptor being used.

bial activity, which will lower the E_H of the water to reducing conditions. Many organic chemicals can be transformed (degraded) to their reduced counterparts by biotic (involving microorganisms) and abiotic (not directly involving microorganisms) reactions under these conditions. Figure 2.13 shows such a reduction reaction for 3-chloronitrobenzene at a pH of 7.2 and E_H of −0.207 V. Note the exponential shape of the plot (concentrations are given in natural log units), which follows first-order kinetics (discussed in Chapter 3). Reduction products 3-chlorophenylhydroxylamine (3-ClPhA) and 3-chloroaniline (3-ClAn) are formed and increase in concentration as 3-chloronitrobenzene is removed from the system. This reduction follows the expected reaction path shown in Figure 2.14. 3-Chloronitrosobenzene, the second degradation product in Figure 2.14, is not shown in Figure 2.15 because it is so reactive that no concentration builds up during the experiment (nitrosobenzene is immediately converted to 3-chlorophenylhydroxylamine). Thus, this redox reaction not only changes the chemical form of the pollutant (a nitrobenzene), but also creates a more toxic form of the pollutant (an aniline).

But what are the effects of changing the pH and E_H of the system? These effects are shown in Figures 2.16 and 2.17, respectively. Increasing the pH and decreasing the E_H of the system increases the rate of reaction (k_{NOM}) in an exponential manner. An increasing negative E_H represents increasing reducing conditions (potential). In many of our transport models discussed in later chapters, we will include a degradation term; therefore, it is important to carefully characterize your system with respect to all chemical parameters that will affect the reaction rate. Another important parameter to characterize is the concentration of the chemical agent involved in

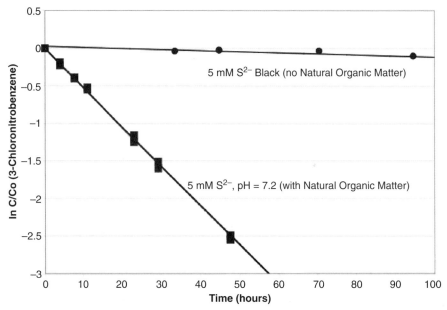

Figure 2.13.   The reduction of 3-chloronitrobenzene at a pH of 7.2 and E_H of -0.207 V. [Reprinted with permission from *Environmental Science and Technology*, Dunnivant et al. (1992), copyright 1988 American Chemical Society.]

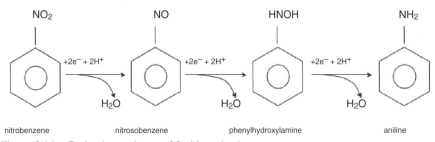

Figure 2.14.   Reduction pathway of 3-chloronitrobenzene.

the reduction reaction. In the experiments discussed in Figures 2.13 through 2.17, the active reducing agent is natural organic matter. But how do the reactivities of different types of organic matters differ? This is illustrated in Figure 2.18. Surprisingly, the reactivities of the 10 natural organic matters shown, as judged by second-order reaction rate, are very similar and vary only over a factor of approximately 15. This indicates that most organic matter will reduce nitrobenzenes under anaerobic conditions, with the reaction rate highly dependent on E_H (Figure 2.17). Figure 2.19 summarizes the reduction of several substituted nitrobenzenes as a function of natural organic matter concentration (in mg/L) and E_H of the system. The E_H value on the x-axis is the E_H associated with the transfer of the first electron to the sub-

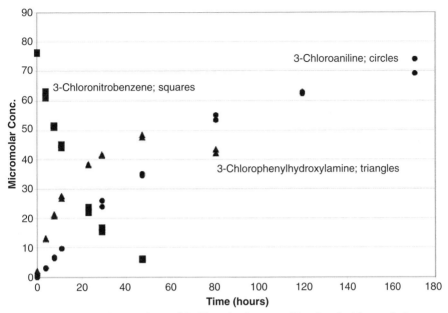

Figure 2.15.   Reduction products of 3-chloronitrobenzene. [Reprinted with permission from *Environmental Science and Technology*, Dunnivant et al. (1992), copyright 1988 American Chemical Society.]

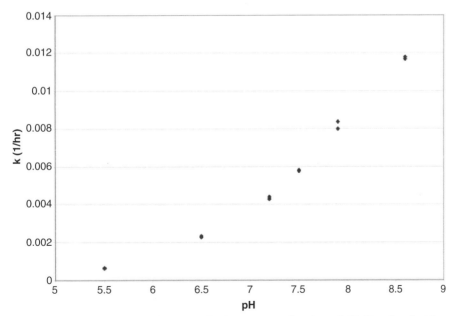

Figure 2.16.   The reduction of 3-chloronitrobenzene as a function of pH. [Reprinted with permission from *Environmental Science and Technology*, Dunnivant et al. (1992), copyright 1988 American Chemical Society.]

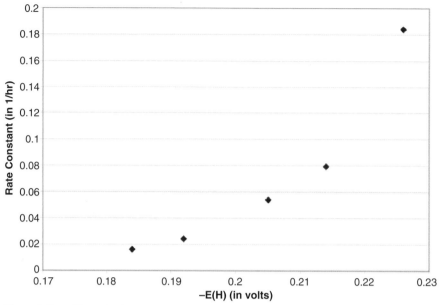

Figure 2.17. The reduction of 3-chloronitrobenzene as a function of E_H. [Reprinted with permission from *Environmental Science and Technology*, Dunnivant et al. (1992), copyright 1988 American Chemical Society.]

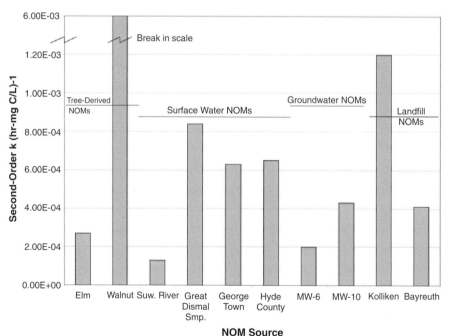

Figure 2.18. Comparison of second-order rate constants for a variety of NOM samples. [Reprinted with permission from *Environmental Science and Technology*, Dunnivant et al. (1992), copyright 1988 American Chemical Society.]

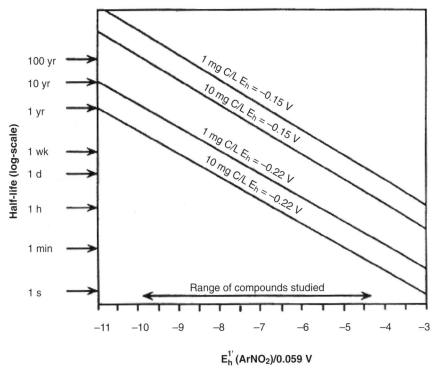

Figure 2.19.   A summary of the effects of NOM concentration and E_H on the reduction of 3-chloronitrobenzene. [Reprinted with permission from *Environmental Science and Technology*, Dunnivant et al. 1992, copyright 1988 American Chemical Society.]

stituted nitrobenzene molecule (there are six electrons required in the complete reduction reaction). This study shows the importance of carefully evaluating each chemical parameter in determining the fate and transport of a reactive pollutant, although rarely can such a complete characterization of a system be conducted.

2.6   COMPLEXATION

Chemical speciation, or the way in which chemicals distribute themselves between different forms and phases, is central to studies of fate and transport. In order to understand the behavior of a chemical in the environment, we need to know its physical and chemical properties, as determined by the form in which it exists. In order to be held in solution, molecules of a dissolved species must be subject to interactions with other molecules of solvent or other solutes. Most free metal ions in aqueous solution exist as hydrated or aquated cations (e.g., $Ca^{2+} \cdot 6H_2O$). The exact number of hydration waters varies for different metals, but four to six waters of hydration are the most common.

A complex, or coordination compound, is a dissolved species formed from two or more simpler species (e.g., Pb^{2+} reacting with Cl^- to form $PbCl^+$). Complexes consist of the central metal cation and a ligand or anion. Depending on the combination of cations and ligands, the complex can have an overall positive, neutral, or negative charge. The number of ligands (or water molecules) that surround a metal cation is referred to as the coordination number. Ligands come in a variety of forms. Basically, a ligand is any chemical species that can complex the metal cation. A few common examples relevant to environmental chemistry include Cl^-, OH^-, CO_3^{2-}, HCO_3^-, HPO_4^{2-}, and $H_2PO_4^-$, as well as (a) naturally occurring organic compounds such as amino, humic, and fulvic acids and (b) synthetic organic compounds such as ethylenediaminetetraacetic acid (EDTA) and nitrilotriacetic acid (NTA). The charge and shape of the ligand determines how it will complex the metal cation. For example, the simple anions only have one binding site and are referred to as unidentate. Other ligands, such as EDTA and NTA, have multiple charges and molecular shapes that allow more than one site on the ligand to bind with the metals. These are referred to as multidentate or chelating agents.

Complexation can occur in two ways, as inner-sphere or outer-sphere complexes. An outer-sphere complex consists of a hydrated metal cation associated with (bonded to) a ligand but retaining its waters of hydration. An inner-sphere complex results from replacement of hydration waters with the ligand. The bonding forces of inner-sphere complexes are more polar-covalent in nature and are stronger than the bonds present in outer-sphere complexes. In general, cations with greater charge and lesser radius form more inner-sphere-type complexes with a given ligand. Another way of phrasing this is to say "the higher the surface charge density, the more inner-sphere the complex." Outer-sphere complexes are frequently formed by metals such as Na^+, K^+, Ca^{2+}, Mg^{2+}, and Sr^{2+} with ligands such as Cl^-, HCO_3^-, SO_4^{2-}, and CO_3^{2-}. Inner-sphere complexes are frequently formed by metals such as Ag^+, Cd^{2+}, Zn^{2+}, and Hg^{2+} with ligands such as S^{2-} and SH^-. We will return to complexation in Chapter 3 (Section 3.2.2) where we will work equilibrium problems associated with complex solutions.

2.7 EQUILIBRIUM SORPTION PHENOMENA

We have already discussed several equilibrium processes, and these are summarized in Table 2.8 along with some new ones for this section. Most chemists, especially those dealing with pollutant fate and transport phenomena, prefer to work with a system at equilibrium. This makes the mathematical expressions much simpler, and it allows us to ignore many poorly understood kinetic processes. In this section, we will look at pollutant sorption (attraction) phenomena between the aqueous phase and other phases present in natural water. We will use several terms to describe these processes, but in general they all mean the same thing. For example, when metals associate with particles in water, it is usually through an ion exchange mechanism on the surface of the particle and it is technically correct to use the term "adsorption" to describe the process. However, when hydrophobic pollutants associate with particles, it is more of a solvation process, since the interaction is not site- or charge-

TABLE 2.8. A Summary of Important Equilibrium Constants Used in Environmental Chemistry

K	Name	Mathematical Expression	Description
K_{eq}	Typical equilibrium constant	$K_a = \dfrac{|A^-||H^+|}{|HA|}$	A common expression used to represent the equilibrium concentration in a chemical reaction
K_H	Henry's law constant	$\dfrac{\text{Partial pressure of } x}{\text{Water con. of } x \text{ (M)}}$	Used to describe the equilibrium partitioning of a pollutant between the gas and liquid phase (units of atm-L/mol)
K_{DOM}	Organic matter–water partition coefficient	$\dfrac{\text{Conc. of } x \text{ in DOM}}{\text{Conc. of } x \text{ in water}}$	Used to describe the partition or binding of pollutants between dissolved organic matter and water (L/kg)
K_{ow}	Octanol–water coefficient	$\dfrac{\text{Conc. of } x \text{ in octanol}}{\text{Conc. of } x \text{ in water}}$	Used to model the partitioning of pollutants in biota (bioconcentration) (unitless)
K_d	Sediment–water distribution coefficient	$\dfrac{\text{Conc. of } x \text{ in sediment (mg/kg)}}{\text{Conc. of } x \text{ in water (mg/L)}}$	Used to show how pollutants adsorb to a solid phase (units of L/kg)
K_p	Sediment–water partition coefficient	$\dfrac{\text{Conc. of } x \text{ in sediment (mg/kg)}}{\text{Conc. of } x \text{ in water (mg/L)}}$	Used to show how pollutants partition onto or into another phase (i.e., organic matter on a sediment particle) (units of L/kg)
K_{oc}	Organic carbon–water partition coefficient corrected for the presence of organic matter	$\dfrac{K_d \quad \text{or} \quad K_p}{\text{Fraction of organic matter}}$	Used to correct the K_d or K_p for the presence of organic matter coatings. (units are the same as K_d or K_p)

specific, and we use the term "partitioning." Some researchers use the term "sorption" to include both processes, since some organic compounds are slightly polar in nature and the associated process can be a mixture of partitioning and adsorption. You should try to keep these terms straight, since some researchers become very agitated when the terms are used incorrectly.

2.7.1 Sorption Surfaces

Clays. There are a variety of surfaces in natural waters, including inorganic and organic colloids (mineral phases and humin—one form of natural organic matter). Colloidal particles are defined as very small particles that do not settle out of the solution during the time scale of interest, which can range from hours to decades in length. Colloidal particles can consist of very small inorganic minerals, natural organic matter (generally classified as dissolved), or a combination of both. In addition, the mineral phases can be coated with precipitants such as iron or manganese oxides and hydroxides. Due to small particles' high ratio of surface area to volume, they can account for a large amount of adsorbed and/or partitioned mass of the pollutant in solution. Natural organic matter was mentioned at the end of Section 2.4.3, where representative structures were given. NOM comes in a variety of sizes and forms, and there seem to be an endless number of ways to characterize and describe it (discussed later). For the moment, we will classify NOM as either (a) dissolved or (b) sorbed to particles (colloidal and large particles) or humin (insoluble chunks of NOM). There is no officially defined size of colloids, but environmental chemists usually filter natural water samples through a 0.20- to 0.45-μm filter and call everything that passes through the filter "dissolved," including colloidal. The reader should be aware that there are other definitions of dissolved, but we will use this most common definition in our discussions.

If you have had a course in geology, you know that many minerals exist in nature. Soil scientists interested in sorption phenomena study many forms of particulate matter (those that do not pass a 0.45-μm filter), especially aluminum oxides, iron oxides and hydroxides, manganese oxides and hydroxides, and clay minerals. Clays and NOM coatings on inorganic particles are important in the adsorption phenomena of all pollutants, including metals and organics. Iron and manganese minerals and precipitants, meanwhile, are important primarily for sorption of metals and polar or ionic organics.

The term *clay* has two meanings: a clay *particle* is any particle smaller than 2 μm in size, regardless of composition, whereas a clay *mineral* is distinguished by its chemical composition and crystallographic structure. These two definitions tend to overlap, since most particles in the <2-μm fraction of most soils and sediments are some form of clay mineral.

We will discuss two clay mineral phases, kaolinite and montmorillonite, shown in Figures 2.20 and 2.21, respectively. We will begin with a summary, and then we elaborate on the terms used. Clay minerals are phyllosilicates, hydrous aluminum-silicate sheet structures. Clays minerals are composed of alternating sheets of (a) silicon in tetrahedral coordination with oxygen and (b) aluminum in octahedral coordination with oxygen. Kaolinite is composed of one-to-one (1 : 1) layers, each com-

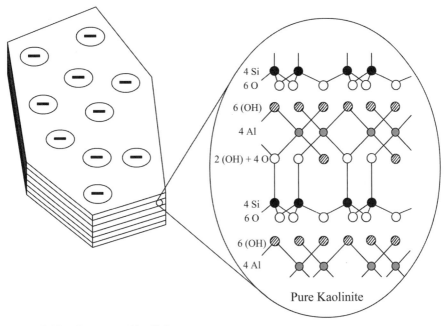

Figure 2.20.   Structure of kaolinite.

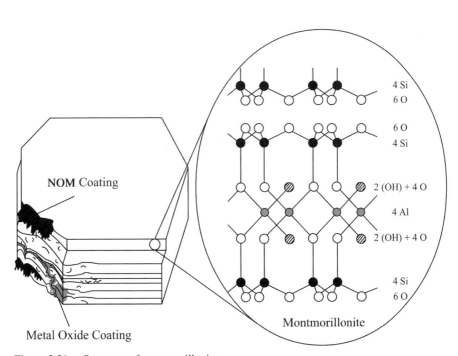

Figure 2.21.   Structure of montmorillonite.

posed of one of each kind of sheet, with a chemical structure of Al_2O_3–$2SiO_2$–$2H_2O$. Montmorillonite is a 2:1 layered clay (each octahedral sheet is bounded by two tetrahedral sheets, to form a layer), with a chemical structure of Na_2O–$7Al_2O_3$–$22SiO_2$–nH_2O or CaO–$7Al_2O_3$–$22SiO_2$–nH_2O.

In order to understand how and why adsorption of metals to clays occurs, we must further expand on clays' chemical structure and three-dimensional shape. Clays have a characteristic structure of layers composed of two alternating types of sheets. One sheet consists of Al^{3+}, O^{2-}, and OH^- ions, where the negative ions form an octahedral structure around the Al^{3+}. The relative numbers of Al^{3+}, O^{2-}, and OH^- must satisfy the valences of the entire continuous structure in two dimensions. This sheet is commonly referred to as the gibbsite sheet or octahedral sheet, since it has the same general chemical formula $[Al_2(OH)_6]$ as the mineral gibbsite. The second type of sheet is composed of Si^{4+}, O^{2-}, and OH^- ions. The Si^{4+} ion forms the center of a tetrahedron of oxygen atoms, while the bases of the tetrahedrons form hexagonal rings. This sheet is referred to as the silica sheet or tetrahedral sheet of the clay structure.

Clay structures consist of layers composed of various combinations of octahedral and tetrahedral sheets. The simplest combination, one of each sheet, forms a kaolinite clay. Each octahedral sheet is linked to one tetrahedral sheet through the sharing of oxygens by the Si and Al atoms. This results in the structures shown in Figure 2.20 for kaolinite. Clays with this structure are referred to as 1:1-type clays, since for every octahedral sheet there is one tetrahedral sheet. Note how combinations of O^{2-} and OH^- are used to satisfy the valence charge and result in a neutral structure. The resulting clay crystal is built up with a succession of the 1:1 (gibbsite–silicate) layers, one on top of another. Successive layers of kaolinite are relatively difficult to separate because of the hydrogen bonding between gibbsite and silicate sheets in adjacent 1:1 layers. The hydrogen in the OH^- from the gibbsite sheet is bound to the O^{2-} in the silicate sheet of the adjacent layer. This rigid structure will become important in the next section, when we discuss isomorphic substitution and surface charge.

Another common form of clay is the 2:1 structure, where an octahedral sheet is sandwiched between two tetrahedral sheets. An idealized clay of this type is montmorillonite, illustrated in Figure 2.21. These 2:1 clays have very interesting properties with respect to absorption phenomena. For example, each 2:1 layer is rigidly held together, but adjacent 2:1 layers may be loosely held together depending on the chemicals or ions that are present in the interstitial area (the space between adjacent 2:1 layers). In the absence of interstitial ions, dry montmorillonite layers are held together by a combination of electrostatic forces (resulting from isomorphic substitution discussed later) and van der Waals dispersion forces (between adjacent O^{2-} groups from each layer) (Sposito, 1984). The interstitial spaces between adjacent layers in 2:1 clays can be expanded in the presence of water (or other solvents), as water molecules commonly occupy these interstitial areas. Water acts to hold the layers loosely together through hydrogen bonding or by the presence of hydrated ions that may be present in the interstitial space. In contrast, adjacent kaolinite layers are held so tightly together that essentially no ions can migrate between the layers. The importance of the expandable nature of 2:1 clays is that it provides more surface area for diffusion and sorption of metal ions or organic pollutants.

There are many other possible configurations of gibbsite and silicate sheets, including the brittle platy minerals called micas. Muscovite is basically a montmorillonite layering structure with potassium between the two 2 : 1 layers. Potassium acts to collapse the 2 : 1 layers and holds the layers tightly together. Chlorite, another common mineral, is composed of an Mg–Al gibbsite-type layer sandwiched between two 2 : 1 layers.

Isomorphic Substitution.   The chemical structure of clays lends itself to imperfections, and these imperfections are referred to as isomorphic substitutions. Essentially any cation with a coordination number of 4 or 6 can be substituted for Si^{4+} and Al^{3+} in individual sheets. Montmorillonite rarely, if ever, is found in the pure form. However, this substitution does not occur in kaolinite, which always has a chemical structure of $Al_4Si_4O_{10}(OH)_8$. For montmorillonite, the most common substitution is in the tetrahedral sheets, where Al^{3+} replaces Si^{4+}. Greater varieties of substitution occur in the octahedral sheet. Most commonly, the substitution of Al^{3+} for Si^{4+} and Mg^{2+} for Al^{3+} leaves a deficiency of positive charges in the montmorillonite layers. These substitutions occur in the crystal lattice when the clay forms, resulting in a permanent charge. The charge deficiency may be compensated for in a variety of ways: (1) by replacement of O^{2-} by OH^-, (2) by introduction of excess cations into the octahedral sheet, which may have some of its cation sites unfilled, and (3) by adsorption of cations onto the surface of individual layers. Although all of these may happen, the last will be our main focus, since it occurs after the clay is formed and can account for removal of metals from solution. Overall, isomorphic substitution results in permanent, nonspecific, diffuse charges that are spread across the clay surface.

The Inorganic Hydroxyl Group.   In our previous discussions, we have presented the clay sheet as a continuous two-dimensional surface. However, clay particles are $<2\,\mu m$ in diameter; therefore, there will also be considerable clay edges present in a soil or sediment sample. The most common and reactive functional group in clays is the hydroxyl group that is exposed on the outer periphery of the clay on the truncated ends of the sheets. Two types of edges occur, with silanol groups originating from the silicon tetrahedral edge and aluminol groups originating from the aluminum octahedral edge. These are similar in reactivity and only differ in the fact that silanol groups do not form inner-sphere complexes (discussed earlier). The charge of silanol and aluminol groups is highly pH-dependent, so that these groups will in general be protonated at low pH values and deprotonated (anionic) at high pH values. This is important later when we discuss the adsorption of metal ions.

Measurement of Surface Charge.   The actual charge of a soil/sediment suspension can be determined by a variety of experiments. The most common measure of charge is the *point of zero charge* (PZC), which is the pH value of a soil suspension at the point when the total net particle charge vanishes (Sposito, 1984). This can be determined by titrating a sample and measuring the mobility of the particles under an applied voltage. Another measure of charge is the *point of zero salt effect*

(PZSE). The PZSE is determined by locating the common point of intersection for several graphs of surface charge (σ_H in Figure 2.22) versus pH, each determined at a fixed ionic strength of the background electrolyte.

Factors Affecting Metal Sorption. Clearly both of these variables (pH and ionic strength) will affect adsorption of metal pollutants. For example, as the surface charge of a particle changes, by a change in either ionic strength or pH, the affinity of the surface for a metal pollutant will change. As the pH is increased, the particle surface becomes more anionic on average and absorbs more and more of the metal from solution. As we will see in Section 2.7.4, as more metal is adsorbed onto the solid, the observed K_d (the ratio of pollutant concentration on the clay to the water phase) will increase.

Other factors that affect the adsorption of metals include the oxidation state of the metal, composition of salts contributing to ionic strength, other surface-complexed cations, and the concentration of suspended solids. Summaries of these

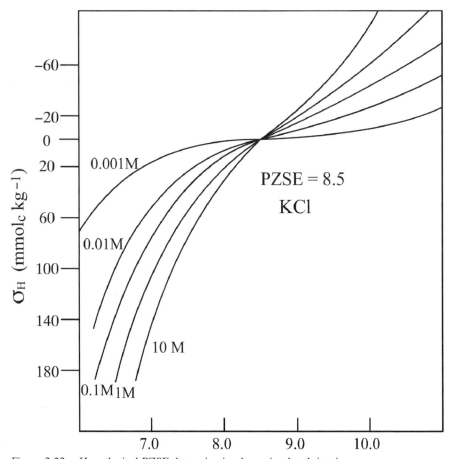

Figure 2.22. Hypothetical PZSE determination by a simulated titration.

effects can be found in Ames and Rai (1978), Bell and Bates (1988), Looney et al. (1987), and Tichnor (1993).

Clay Particles in Nature. Clays in natural environments are rarely free of coatings; they are not "clean." They are normally coated with inorganic precipitates or organic molecules resulting from the degradation of plant and animal material (illustrated in Figure 2.21). Both of these types of coatings will affect surface charge, sometimes imparting a charge of their own. This brings us to the next topic, a discussion of coating on mineral surfaces.

2.7.2 Organic Matter

One of the most important factors influencing sorption phenomena is the presence of organic matter in a sample. Virtually all samples have some organic matter present, but the type and concentration can vary dramatically. In principle, the sources of organic matter are obvious: Any plant, animal, or excrement of these can be incorporated into a water or soil sample. As you can imagine, the chemical variability of the resulting compounds is unlimited. However, upon introduction into a natural system, they undergo complex microbial and abiotic transformations that produce a set of compounds generally referred to as fulvic, humic, and humin materials. For simplicity and consistency, the term "natural organic matter" will be used in this textbook to refer to any organic compound present in the sample. Compounds entering a natural system include proteins (polypeptides and nucleotides), lipids (fats, waxes, oils, and hydrocarbons), carbohydrates (cellulose, starch, hemicellulose, lignin), and porphyrins and plant pigments (chlorophyll, hemin, carotenes, and xantophylls) (Stumm and Morgan, 1996). The products of microbial digestion and degradation of these compounds make up NOM. A typical NOM sample contains a mixture of "fresh" organic matter additions as well as "aged" organic matter. Thus, a single NOM sample will contain thousands to tens of thousands of chemically different structures. Generally, 20–30% of the compounds in a NOM sample can be identified by conventional means as protein-like materials, polysaccharides, fatty acids, and alkanes (Schnitzer, 1986). The remaining 70–80% of the NOM consists of complex, altered residues of plants and animals. Molecular weights of NOM found "dissolved" in water range from 500 to 5000 atomic mass units (Thurman, 1985).

There have been intensive efforts to characterize the structure of NOM. These efforts are summarized in Thurman (1985), Hayes et al. (1989), and Suffet and MacCarthy (1989). However, no one expects to establish a single structural formula to describe NOM. We have simply attempted to identify important functional groups, molecular sizes, and chemical properties for these compounds. Two of these structures were shown earlier in Figure 2.7. The first is an early structure of fulvic acid. The second is a more elaborate conceptualization of NOM structure from Schultten and Schnitzer (1993). The key point of the latter figure is that there are numerous ionic sorption sites and hydrophobic centers in the large NOM molecules. These will be important when we discuss sorption of pollutants by NOM.

Other characterization attempts have concentrated on the chemical functionality of NOM. Researchers have devised chromatographic techniques to separate or fractionate NOM based on chemical properties, such as hydrophilicity and hydrophobicity (Leenheer, 1980; Leenheer and Huffman, 1976). These provide the environmental chemist with a means of characterizing different NOM molecules based on their reactivity, but you must realize that, as with most classifications, these may be rather arbitrary. A simple version of this classification identifies the functional groups observed in different NOM molecules. These groups are summarized in Table 2.9 (partially based on Killops and Killops, 1993).

Another important way of characterizing NOM is by changing the pH of a water sample and observing the behavior of the various NOM components (Hayes et al., 1989). *Fulvic acids* are the fraction of NOM that is soluble under all pH conditions. *Humic acids* are defined as the organic matter that is precipitated from an aqueous solution when the pH is decreased below 2. Given that the pH of most natural waters is between 5.5 and 9, both humic and fulvic acids will be present in most natural water samples. In contrast, *humin* is the fraction of NOM that is not soluble in water at any pH value. Thus, humin will be present in or associated with soil or sediment.

The most important characterization of NOM in water defines whether it is present in the dissolved form or sorbed to a solid. For this characterization, we use another operational definition. Dissolved organic matter (DOM) is defined as the organic matter that will pass though a 0.45-μm filter (Gelman type A/E glass fiber filters are usually used for this distinction). The organic matter retained by the filter is considered to be in the particulate form, usually sorbed to inorganic particles. NOM can be attached to inorganic particles through a variety binding mechanisms, including hydrogen bonding, van der Waals dispersion forces, cationic bridging, and hydrophobic effects.

2.7.3 Organic Sorbates

First, we will consider the adsorption of ionic pollutants, specifically metals. Metals, being cations in aqueous solutions, will be adsorbed to anionic sites (negative charges on a clay or depronated functional groups on the NOM). Solution conditions that favor the formation of negative sites will favor increased adsorption of metals. A good exercise at this point would be to return to Table 2.9 and Figure 2.7 and identify functional groups that may be important in attracting cationic pollutants.

Now consider the binding of organic pollutants, specifically nonpolar, nonionizable pollutants such as polychlorinated biphenyls (PCBs). These are commonly referred to as hydrophobic pollutants. These types of pollutants are not attracted to ionized functional groups—in fact, they are repelled by these groups. Hydrophobic pollutants are attracted to hydrophobic centers in the NOM molecule and hydrophobic mineral surfaces. The intermolecular forces responsible for these attractions are van der Waals dispersion forces, which simply follow the old saying "like dissolves like" (e.g., hydrophobic liquid coatings dissolve hydrophobic pollutants). In this regard, the attraction is not really adsorption but is more like a solution or dissolv-

TABLE 2.9. Functional Groups Observed in Different NOM Molecules

Symbol	Name	Resulting Compound
ROH	Hydroxyl	Alcohol Phenol
	Carbonyl	Aldehyde Ketone Quinone
	Carboxyl	Carboxylic acid
—O—	Oxo	Ether
—NH$_2$	Amino	Amine
	Amido	Amide
—SH	Thio	Thiol
	R = CH$_2$ Indenyl R = O Furanyl R = NH Pyrryl R = S Thiophenyl	Indene Furan Pyrrole Thiophene
	R = CH Phenyl R = N Pyridinyl	Benzene Pyridine
	Pyranyl	Pyran
	LAW	H$_2$LAW
	Quinone	Hydroquinone
	Iron (II) Porphyrin	Iron (III) Porphyrin

ing phenomenon. As noted earlier, environmental chemists regard this type of attraction as partitioning or sorption, where the pollutant dissolves or partitions into the hydrophobic center of the NOM. The difference between adsorption and sorption is the basis for environmental chemists using (a) distribution coefficients (K_d) to describe the adsorption of metals and (b) partition coefficients (K_p) to describe the partitioning of hydrophobic pollutants to environmental particles. As with the adsorption of metals, favored by conditions that create negative sites, sorption of hydrophobic pollutants is favored by conditions that promote the formation of hydrophobic centers in the NOM or coiling of the NOM.

2.7.4 Partition Coefficients, K_d and K_p

One of the most important parameters that can determine the fate of a pollutant in an aqueous system, especially in rivers, lakes, and groundwater, is its distribution coefficient (K_d) or partition coefficient (K_p) between different media. These coefficients are a measure of how a pollutant distributes itself between the water phase and the particulate (or solid) phase. Pollutants on the solid phase are considerably less bioavailable and therefore less toxic. These sorbed pollutants can also settle out of solution in lakes or become immobile in groundwater and be effectively removed, at least temporally, from the system. An example of the buildup of pollutants in sediments is given below. We will present ways of calculating these coefficients in Chapter 3.

Distribution coefficients are concerned with adsorption, defined as the net accumulation of matter (pollutants) at the interface between a solid and a liquid. The matter (pollutant) that accumulates at the surface is referred to as the adsorbate. The solid surface on which the pollutant accumulates is the adsorbent. Partition coefficients, as we discussed earlier, are concerned with the partitioning and induced-dipole interactions between two nonpolar compounds (i.e., PCB and hydrophobic regions of NOM). Even though adsorption is not occurring in this process, the effects on the system are similar, and the terms adsorbate and adsorbent are still used.

As soil and sediments wash into lakes and streams, the particles aggregate and form larger particles that will settle in calm (quiescent) waters. When these particles have accumulated pollutants, the settling of these particles to the sediments can act as a removal mechanism (referred to as a sink) for pollutants. Over time, and when clean water and sediments return to the water body, the pollution will be buried by clean material and removed from interaction with the ecosystem. Such an example is shown in Figure 2.23 for PCBs in Lake Hartwell in South Carolina (United States). Note that as you move down into the sediment from the water–sediment interface, you are moving back in time. Figure 2.23a shows an area of the lake that is subject to considerable mixing and input of pollutants. This is indicated by the variable but high presence of PCBs in each section of the sediment column. In Figure 2.23b, the PCB-contaminated sediments start to be buried by cleaner, more recent deposited sediment. Figure 2.23c shows an even further burial of contaminated sediments. Thus, the accumulation of pollutants on soil-sediment particles is an important factor in fate and transport processes. We will return to distribution

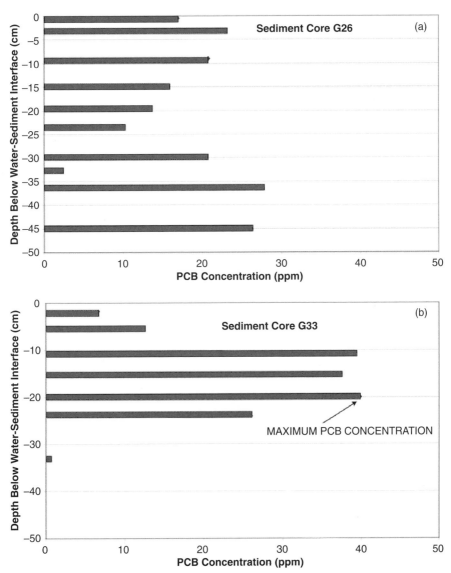

Figure 2.23.   PCB concentration as a function of sediment depth in Lake Hartwell Sediments. [Data from Germann (1988).]

coefficients and partition coefficients in the lake, stream, and groundwater modeling chapters.

2.7.5   Ion Exchange Phenomena for Ionic Pollutants

Another way of representing the adsorption process is as an ion exchange reaction. Here we visualize the negative surface or edge sites of a clay or environmental particle as being saturated, or nearly saturated, by native cations such as H^+, Na^+, and

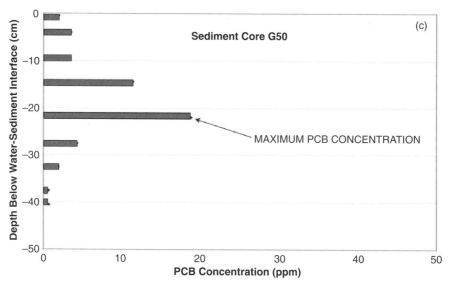

Figure 2.23. continued

K^+. Cationic pollutants, such as heavy metals, generally have a higher charge (or charge density) and therefore a higher affinity for these sites and displace the native, readily exchangeable ions. Thus, we can model the process as ion exchange. There will be a finite number of sites for cations to adsorb to the surface, and the relative abundance of these sites on the particles of a soil or other material is expressed as the *ion exchange capacity*. A more formal definition is the number of moles of adsorbed ions that can be desorbed from a unit mass of solid under a given set of conditions (i.e., temperature, pressure, solution composition, solid–solution ratio, etc.). Soil scientists refer to this measure as the cation exchange capacity (CEC), which is defined as the concentration of sorbed cations that can be readily exchanged for other cations. The CEC is usually reported in units of meq/100 g soil, as an exchangeable charge per mass, and is an indication of a soil's ability to store nutrients or absorb metals.

As a general rule, the affinity of a soil or sediment for a metal cation will increase with the tendency of the cation to form inner-sphere complexes. For a series of uniform valence metals, this tendency is directly related to ionic radius, R, for two reasons:

1. The ionic potential (z/R; charge/radius) decreases with increasing R.

2. A larger radius implies a greater tendency for the metal to polarize (distort) in response to an electric field (the surface charge of the soil particle).

Using these two guidelines, selectivity sequences for metal ions can be established based on ionic radii:

$$Cs^+ > Rb^+ > K^+ > Na^+ > Li^+$$
$$Ba^{2+} > Sr^{2+} > Ca^{2+} > Mg^{2+}$$
$$Hg^{2+} > Cd^{2+} > Zn^{2+}$$

where adsorption of the ion increases from left to right. Unfortunately, ionic radius alone is not sufficient to predict selectivity for transition metals. Extensive experimentation has established the following order of selectivity (the Irving–Williams order) (Stumm and Morgan, 1981):

$$Mn^{2+} < Fe^{2+} < Co^{2+} < Ni^{2+} < Cu^{2+}$$

Outer-sphere complexes, meanwhile, are responsible for the effect of pH on metal adsorption. As pH increases above the pH_{ZPC} (the pH where there is no surface charge), the net surface charge of the particle increases, thus increasing the electrostatic attraction of a mineral surface for the metal. Usually NOM is present to some degree and competes with the surface for the metal. If the NOM is present in the dissolved phase, the adsorption to the surface will decrease if the NOM has a greater affinity for the metal than does the surface. If the affinity of the NOM for the metal is less than that of the surface, little or no change in the surface adsorption will occur due to the presence of NOM. This scenario will be complicated further if the NOM subsequently adsorbs to the mineral surface, a common phenomenon in nature.

2.8 TRANSFORMATION/DEGRADATION REACTIONS

In this section, we will discuss various ways that pollutants are removed from the system through transformation and degradation reactions. In general, these terms can be used interchangeably, but it should be emphasized that just because the original pollutant is removed from a system, the degradation process may not be beneficial to or safe for the environment. For example, many pollutants can be transformed or degraded to other or more toxic pollutants (e.g., the reduction of nitrobenzenes to anilines and the conversion of carbon tetrachloride to vinyl chloride). Transformations can be divided into abiotic (chemical reactions without the aid of microorganisms), photochemical, nuclear, and biological (microbial). Abiotic, photochemical, and biological transformations are generally more important for organic pollutants. A notable exception is the methylation of inorganic mercury.

2.8.1 Abiotic Chemical Transformations/Degradations

Abiotic chemical degradations refer to the removal of a pollutant exclusively through chemical reactions without the direct aid of "active" microorganisms. In other words, live, viable microorganisms do not use their internal enzymes in the process. Abiotic reactions can, however, include indirect effects of microbial activity, which may regulate the pH or E_H of the environment and thus allow or control the presence of abiotic oxidizing or reducing agents. The abiotic nature of these oxidizing and reducing agents is complicated because once a microbial cell, animal material, or plant material dies, intracellular chemicals (such as enzymes) are released and incorporated into natural organic matter. As you can see, this is a gray area and subject to interpretation. There is a general consensus among chemists that if a degradation process can occur in a system that has been sterilized (by filtration, irradiation, autoclaving, etc.), it is abiotic.

TABLE 2.10. Summary of Abiotic Reactions

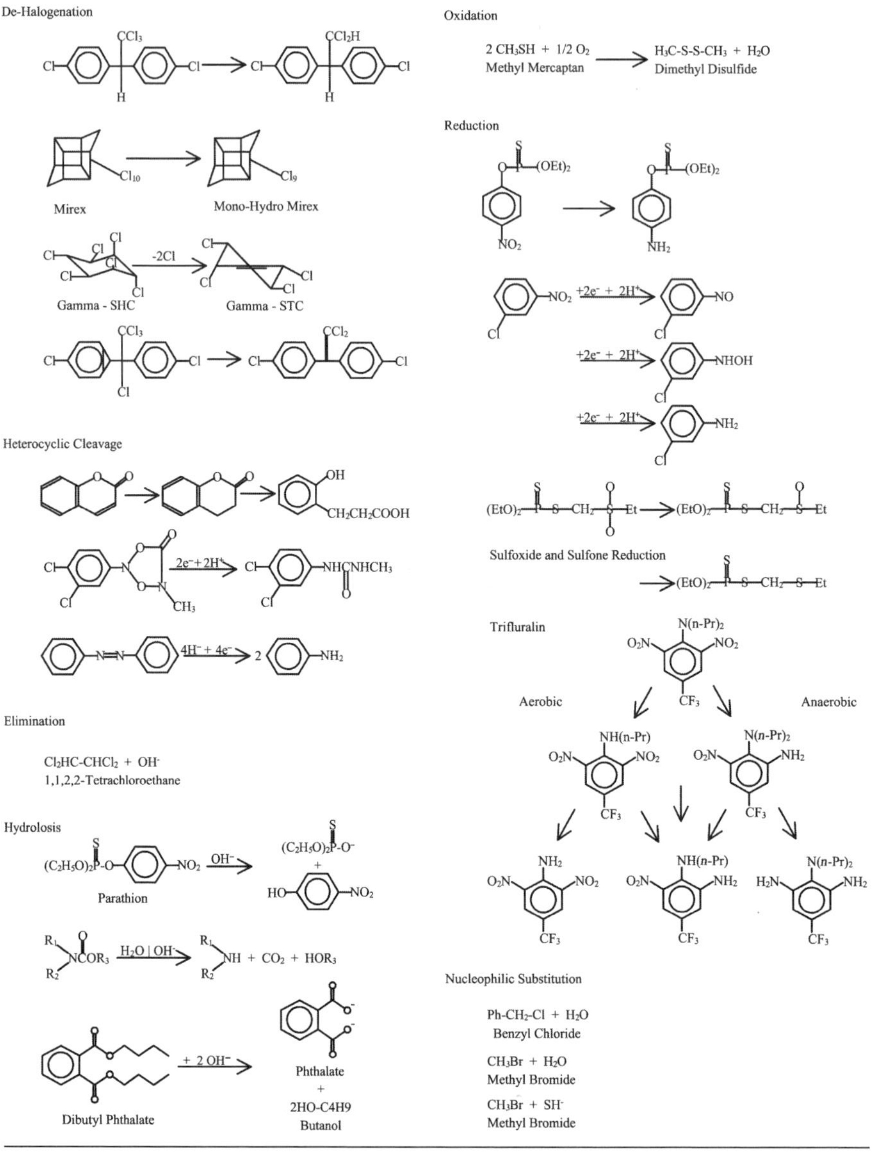

De-Halogenation

Mirex → Mono-Hydro Mirex

Gamma - SHC $\xrightarrow{-2Cl}$ Gamma - STC

Heterocyclic Cleavage

$\xrightarrow{2e^- + 2H^+}$

$\xrightarrow{4H^+ + 4e^-}$

Elimination

$Cl_2HC-CHCl_2 + OH^-$
1,1,2,2-Tetrachloroethane

Hydrolosis

$(C_2H_5O)_2\overset{S}{P}-O-\bigcirc-NO_2 \xrightarrow{OH^-}$
Parathion

$(C_2H_5O)_2P-O^- + HO-\bigcirc-NO_2$

$\overset{R_1}{\underset{R_2}{N}}\overset{O}{C}OR_3 \xrightarrow{H_2O\ |\ OH^-} \overset{R_1}{\underset{R_2}{N}}H + CO_2 + HOR_3$

Dibutyl Phthalate $\xrightarrow{+\ 2\ OH^-}$ Phthalate + 2HO-C4H9 Butanol

Oxidation

$2\ CH_3SH + 1/2\ O_2 \longrightarrow H_3C-S-S-CH_3 + H_2O$
Methyl Mercaptan → Dimethyl Disulfide

Reduction

$\bigcirc-NO_2 \xrightarrow{+2e^- + 2H^+} \bigcirc-NO$

$\xrightarrow{+2e^- + 2H^+} \bigcirc-NHOH$

$\xrightarrow{+2e^- + 2H^+} \bigcirc-NH_2$

$(EtO)_2\overset{S}{P}-S-CH_2-\overset{O}{S}-Et \longrightarrow (EtO)_2\overset{S}{P}-S-CH_2-\overset{O}{S}-Et$

Sulfoxide and Sulfone Reduction

$\longrightarrow (EtO)_2\overset{S}{P}-S-CH_2-S-Et$

Trifluralin

O_2N-...-CF_3 with $N(n-Pr)_2$, NO_2

Aerobic / Anaerobic

Nucleophilic Substitution

$Ph-CH_2-Cl + H_2O$
Benzyl Chloride

$CH_3Br + H_2O$
Methyl Bromide

$CH_3Br + SH^-$
Methyl Bromide

The large variety of abiotic reactions, summarized in Table 2.10, include nucleophilic substitution (dehalogenation), elimination, dealkylation, sulfoxide and sulfone reduction, heterocyclic ring cleavage, hydrolysis, oxidation, nitro reduction, and azo reduction.

Many of these reactions, such as nucleophilic substitition, elimination, hydrolysis, and dealkylation, are detailed in organic chemistry textbooks, and occur in

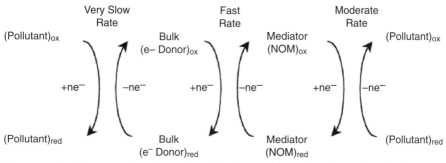

Figure 2.24.   Generalized reaction scheme for the reduction of a pollutant by abiotic reactions.

aqueous environments. The general nature of abiotic reduction reactions is similar to that of reactions performed by microorganisms and is illustrated in Figure 2.24. Here, a bulk electron donor (such as sulfide) is supplied by the environment (abiotic or biotic) and reduces an abiotic electron mediator, which in turn reacts with the pollutant. In most cases, the bulk electron donor is present in excess and reacts slowly or not at all with the pollutant. But what are the abiotic electron mediators? A number have been found to exist, including iron and manganese minerals and a variety of functional groups and centers in NOM, including bound transition metals, quinines, and porphyrin structures.

   In general, abiotic chemical transformations/degradations are easier to study than biochemical reactions involving microorganisms. This is because chemists can eliminate microbiology in their experiments through sterilization, and then they may easily control system-specific pH, E_H, and chemical concentrations in the laboratory. In addition, chemical laboratory studies are more transferable between the laboratory and the environment, since we can measure the chemical conditions of a system and reproduce them in the laboratory. Biologists, on the other hand, have difficulty selectively growing microorganisms in the field and relating rates of biochemical reactions from the laboratory to the field. Abiotic reactions are also thought to be especially important for the degradation of some pollutants by nucleophilic substitution, elimination, and hydrolysis, and reduction reactions can be important in certain environments (in anaerobic lakes, sediments, and aquifer systems).

2.8.2   Photochemical Transformation/Degradation Reactions

No discussion of environmental chemistry or fate and transport would be complete without specifically considering photochemical reactions, a special form of abiotic reactions that break apart organic molecules. These reactions occur both in the atmosphere and in surface waters. In aqueous systems, these enter into fate and transport reactions as first-order or pseudo-first-order reaction rates, which are dependent on the concentration of one reactant or one reaction parameter (reaction kinetics are discussed further in Chapter 3). In atmospheric systems, second-order reaction kinetics are used. In order to understand photochemical reactions, we must first review the availability of wavelengths present in the atmosphere at the Earth's surface and the energy associated with these wavelengths.

Figure 2.25 shows the distribution of the spectrum of electromagnetic radiation at the Earth's surface. The top solid line shows the distribution of wavelengths entering the Earth's atmosphere, while the attenuated (lower) solid line shows the distribution at the Earth's land surface. This lower line is the more important to our discussions. Radiation from ultraviolet (200 nm, 0.2 µm in Figure 2.25) to visible (750 nm or 0.75 µm) wavelengths can be involved in photochemical reactions. An easy calculation from general chemistry relates wavelength of a photon to energy of a mole of these photons. This can be achieved using the following equation:

$$E = h\nu = hc/\lambda$$
$$= 6.022 \times 10^{23} \, hc/\lambda \text{ for a mole of photons}$$

where h is Planck's constant (6.63×10^{-34} J-s), ν is the frequency of light (1/s), c is the speed of light in a vacuum (2.99×10^{8} m/s), and λ is the wavelength in meters.

Table 2.11 relates energy of some single bonds to the corresponding wavelengths of light (i.e., the wavelength needed to break the bond). As seen from these calculations, UV and visible wavelengths possess sufficient energy to break covalent bonds in many molecules, but infrared radiation possesses only enough energy to cause an increase in bond vibration and does not break chemical bonds. A pho-

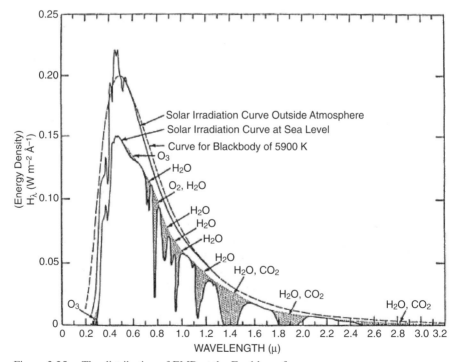

Figure 2.25. The distribution of EMR at the Earth's surface.

TABLE 2.11. Relationship Between Bond Energy and
Minimum Wavelength of Light Required to Break the Bond

Bond	Bond Energy (kJ/mol)	Wavelength, λ (nm)
O–H	465	257
C–H	415	288
N–H	390	307
C–O	360	332
C–C	348	344
C–Cl	339	353

tochemical reaction in which photons are absorbed by a pollutant molecule and the molecule is broken apart is referred to as direct photolysis. But just because a molecule absorbs a photon of sufficient energy does not mean that it will be destroyed, since there are a number of ways the molecule can dissipate the energy. Thus, the quantum yield (the number of photons actually producing a photochemical reaction) of the absorbed photons must also be considered.

Other indirect types of photolysis can also occur. For example, an atmospheric molecule such as O_2 can absorb energy from a photon, become excited, and transfer the energy to a pollutant molecule by collisions. If the transfer of energy is sufficient to break a bond in the pollutant, destruction can occur. Other compounds in the atmosphere and in surface water can be photosensitized (made more likely to photo-degrade a pollutant), which may lead to the formation of free radicals that can subsequently react with pollutant molecules. An important radical formation in the atmosphere is the reaction of an excited oxygen atom with water to form the hydroxyl radical (sometimes called the vacuum cleaner of the atmosphere):

$$O_3 \Rightarrow O_2 + O*$$
$$O* + H_2O \Rightarrow 2OH*$$

Another, less common, reaction also produces these radicals:

$$H_2O + h\nu \Rightarrow HO* + H$$

The OH* radical is very reactive, both in the formation of smog and in the destruction of atmospheric pollutants. An important creator of free radicals in surface water is natural organic matter, which can form singlet oxygen, a strong oxidizer of pollutant molecules.

While photochemical reactions are very important in the formation of smog, and in some cases are important removal mechanisms of atmospheric pollutants, the atmospheric fate and transport models we will use in the textbook do not allow for the inclusion of degradation terms. This is because the atmospheric fate and transport models are only valid on short time and distance scales, before most photochemical degradation reactions have time to become important. Photochemical reactions can, and will, be included as first-order degradation terms in lake and river transport models.

TABLE 2.12. **Properties of Common Forms of Radiation**

Property	Type of Radiation		
	α	β	γ
Charge	2+	1–	0
Mass	6.64×10^{-24} g	9.11×10^{-28} g	0
Relative penetrating power	1	~100	~10,000
Composition	He nucleus	Electron	High-energy photon

2.8.3 Nuclear

Nuclear reactions are often one of the least studied types of reactions by under-graduates, but are perhaps the easiest to understand. The three most common and important types of nuclear decay involve the release of alpha (α), beta (β), or gamma (γ) forms of radiation. These are summarized in Table 2.12. Alpha particles consist of ejected helium nuclei (two protons and two neutrons). These particles have the greatest mass of radioactive emissions, but cannot easily penetrate substances. Beta radiation consists of high-speed electrons emitted from an unstable nucleus and have intermediate penetrating power. Gamma radiation consists of high-energy photons and has the greatest penetrating power.

All known radioactive decay reactions follow simple first-order kinetics. Half-lives range from fractions of a second to billions of years. An extensive list of the important environmental radionuclides is given on your CD in Fate®.

Sources of anthropogenic radionuclides released into the environment include medical research and treatment, nuclear power production, nuclear weapons pro-duction, and nuclear weapons testing, with the latter three accounting for most of the waste generation and environmental contamination.

2.8.4 Biological

Microorganisms offer an amazing diversity of biochemical reactions to degrade organic pollutants. In fact, virtually every organic compound that has been tested can be degraded by microorganisms under laboratory settings and at sufficiently fast rates to be potentially significant. Microorganisms use enzymes (natural catalysts) to degrade pollutants relatively quickly. You will recall from general chemistry that catalysts do not affect the direction of the reaction (ΔG or the equilibrium), but they speed up the reaction (approach to equilibrium) by lowering the activation energy. Yet despite the potential for these reactions, most toxic organic chemicals (pollu-tants) are relatively stable in the environment; that is, they do not rapidly degrade. Of course, in the absence of microbial processes, there are abiotic and photochem-ical degradation reactions, as discussed above, but most of these require special con-ditions. If pollutants did rapidly degrade when released into the environment, after all, pollutants would not be a problem.

There are as many ways that microorganisms can degrade pollutants as there are microbial enzymes. We like to think of microorganisms mineralizing pollutants to stable inorganic forms of carbon, hydrogen, phosphorous, sulfur, and so on, but as in chemical degradations, not all biological degradations are this complete. Complete mineralization usually requires a diversity of microorganisms to complete the process.

When an established (acclimated) colony of microorganisms is present, the rate-limiting step is usually uptake through the cell wall, and the kinetics of the reaction can be simplified to pseudo-first-order rate laws. Of course, the difficult part of this assumption is determining if and when an established community of microorganisms is present in a lake or river, and it is especially difficult to test these conditions in groundwater aquifers. Another highly complicating factor in isolating and documenting biochemical reactions is that abiotic degradations often occur at the same time and under the same conditions. Chemists can eliminate microorganisms from their experiments, but suffice it to say biologists cannot eliminate chemistry from their experiments.

We mentioned in the first paragraph of this section that almost every organic compound can be degraded by some type of microorganism. The point of testing for biodegradability is twofold: (1) to determine whether degradation can occur and (2) to determine the rate of the reaction. The determination of whether a degradation will happen is a bit complicated but easy compared to extrapolating rates. The laboratory rate must be extrapolated to an estimated field rate, since the rate is largely determined by the number and distribution of microbial cells in the system. Documenting this in sediments or soils is virtually impossible. For example, let's look at the PCB contamination in lake sediments. Researchers for PCB manufacturers and users have documented fairly well that microbes can degrade PCBs under laboratory conditions and have proposed that PCB contamination in lake and river sediments should not undergo costly remediation because microbes will take care of the problem given sufficient time. But what is sufficient time? Laboratory removal rates of PCBs from sediment slurries are relatively fast (weeks to months), but river systems such as the Hudson River have been contaminated for decades with little reduction in PCB contamination. Obviously, transferring results from laboratory experiments to the natural environment is difficult.

When considering whether an organic pollutant can be degraded by microorganisms, it is important to mimic the environmental setting as close as possible. This includes pH, E_H, nutrient concentrations, pollutant concentrations, and temperature, just to name a few parameters. Students should note that there are many types of microorganisms in the environment that have a variety of biochemical enzymes to perform these reactions, and a variety of terminal electron acceptors can be used (refer to the oxidation–reduction section). Biologically mediated oxidation reactions of organics in the environment can be especially important, due to their potentially rapid rates of removal of pollutants.

2.9 SUMMARY

In this chapter, we have illustrated that many chemical factors are important in the fate and transport of pollutants in aquatic and atmospheric systems. In fact, the chemistry of a system can dictate the type of fate and transport model that we use. We have covered a lot of information and concepts in this chapter, and a good review of these would be to make an outline of the chapter. As you list the concepts, predict how these will affect inorganic versus organic pollutants. This organization will be very important when we make our summary at the end of the next chapter.

Concepts

1. List the global volumes of freshwater in groundwater, lakes, rivers, and glaciers in increasing order. Which is the most available for human use?

2. List the average residence times of water in increasing order in the following compartments: lakes, oceans, rivers, groundwater, and glaciers.

3. List two world regions that will probably have shortages of fresh water in the near future.

4. List and describe the five most important unique properties of water.

5. What intermolecular force is responsible for water's unique properties?

6. Explain the term *bioconcentration*.

7. Define a colloid.

8. List two important sorption surfaces commonly found in water systems.

9. How does isomorphic substitution affect clay surface charges and sorption phenomena?

10. What is meant by natural organic matter (NOM)? How does it relate to DOM?

11. List five functional groups in NOM.

12. Contrast adsorption and partitioning with respect to the surface concentration of a pollutant.

13. How do sorption reactions affect pollutant fate and transport in aquatic systems? Specifically, will a pollutant be more or less mobile in the sorbed state?

14. List the four major transformation reactions of pollutants and give a two-to-three sentence description of each.

Exercises

1. By how much do water's strong intermolecular forces increase the actual temperature range of liquid water over the theoretical range?

2. Mathematically show how a ppb is approximately equal to µg/L. Why is this equality only approximate?

3. Calculate the water solubility of CO_2 for a water in contact with an atmosphere containing 1000 ppm CO_2 (1000×10^{-6} atm). K_H is given in the text of this chapter.

4. Define what is meant by the term "pollutant water solubility."

5. How can a pollutant's water solubility affect its fate and transport in water systems?

6. Calculate the solubility of BaF_2 at 18°C.

7. Calculate the solubility of silver chloride at 25°C.

8. Calculate the pH of a water solution containing 5.89×10^{-5} M H^+.

9. Calculate the pH of a water solution containing 4.92×10^{-4} M OH^-.

10. Hand-draw a pC–pH diagram for a closed system containing 5.29×10^{-3} M total carbonate. Calculate the concentration of H_2CO_3, HCO_3^-, and CO_3^{2-} at a pH of 5.90. Check your results using the pC–pH Simulator®.

11. Hand-draw a pC–pH diagram for a closed system containing 8.37×10^{-5} M total sulifde. Check your results using the pC–pH Simulator®.

12. Hand-draw a pC–pH diagram for an open system of water in contact with an atmosphere containing 50.5 ppm CO_2 (50×10^{-6} atm). Check your results using the pC–pH Simulator®.

13. The concentration of CO_2 in the atmosphere is predicted to rise to ~700 ppm (partial pressure of 700×10^{-6} atm) by the year 2100. First, use the pC–pH Simulator® to estimate the pH of rainwater falling through an atmosphere of 380 ppm CO_2 (the approximate current concentration). Next, use the simulator to predict the pH of rainwater in the year 2100.

14. Use the $\Delta G°$ values from Section 2.5.2 for the half-reactions of NO_3^-, MnO_2, $FeOOH$, SO_4^{2-}, and CO_2 to calculate the combined energy generated (in kJ/mol) from the oxidation of a mole (1.00) of glucose with each. Which TCE yields the most energy?

15. If you have a water with an E_H of 0.55 V, would you consider it to be an oxidizing or reducing environment?

16. If you have a water with an E_H of −0.48 V, would you consider it to be an oxidizing or reducing environment?

REFERENCES

Ames, L. L. and D. Rai. *Processes Influencing Radionuclide Mobility and Retention. Element Chemistry and Geochemistry, Conclusions and Evaluation*, Vol. 1, EPA 520/6-78-007, 1978, Battelle Pacific Northwest Laboratories, Richland, WA 99352.

Bell, J. and T. H. Bates. Distribution coefficients of radionuclides between soils and groundwater and their dependence on various test parameters. *Sci. Total Environ.* **69**, 297–317 (1988).

Dunnivant, F. M., D. L. Macalady, and R. P. Schwarzenbach. Reduction of substituted nitrobenzenes in aqueous solutions containing natural organic matter. *Environ. Sci. Technol.* **26**, 2133–2141 (1992).

Germann, Geoffrey G. Masters Thesis, *The Distribution and Mass Loading of Polychlorinated Biphenyls in Lake Hartwell Sediments*, Environmental Systems Engineering, Clemson University, 1988.

Handbook of Chemistry and Physics, Weast, R. C. and M. J. Astle, eds. 61st edition, p. B-242, Boca Raton, FL1980.

Harris, D. C. Quantitative Chemical Analysis, 5th Ed. W. H. Freeman & Co., N. Y. 1999.

Hayes, M. H. B., P. MacCarthy, R. L. Malcolm, Jr., S. Swift. Humic Substances II: In Search of Structure, Wiley-Interscience, N. Y. 1989.

Killops, S. D. and V. J. Killops. *An Introduction to Organic Geochemistry*, Longman Scientific & Technical, Harlow, Essex, England; John Wiley & Sons, New York, 1993.

Leenheer, J. A. and E. W. D. Huffman, Jr. *J. Res. U.S. Geol. Surv.* **4**, 737–751 (1976).

Leenheer, J. A. Study of sorption of complex organic solute mixture on sediments by dissolved organic carbon fractionation analysis. In: *Contaminants and Sediments*, Vol. II, Baker, R. A. ed, Ann Arbor Science, Ann Arbor, MI, pp. 267–277, 1980.

Looney, B. B., M. W. Grant, and C. M. King. *Estimation of Geochemical Parameters for Assessing Subsurface Transport at the Savannah River Plant*, DPST-85-904, Savannah River Laboratory, Aiken, SC 29808, 1987.

Mackay, D., W. Y. Shiu, and R. P. Sutherland. Determination of air–water Henry's law constants for hydrophobic pollutants. *Environ. Sci. Technol.* **13**, 333–337 (1979).

Morel, F. M. M. and J. G. Hering. *Principles and Applications of Aquatic Chemistry*, John Wiley & Sons, New York, 1993.

Schnitzer, M. Binding of humic substances by soil mineral colloids. In: *Interactions of Soil Minerals with Natural Organic and Microbes*, Soil Science Society of America Special Publication No. 17, 1986.

Schulten, H. R. and M. Schnitzer. Naturwissenschaften **80**, 29–30 (1993).

Schwarzenbach, R. P., P. M. Gschwend, and D. M. Imboden. *Environmental Organic Chemistry*, John Wiley & Sons, New York, p. 407, 1993.

Sposito, G. *The Surface Chemistry of Soils*, Oxford University Press, New York, 1984.

Stumm, W. and J. J. Morgan. *Aquatic Chemistry: An Introduction Emphasizing Chemical Equilibria in Natural Waters*, 2nd edition, John Wiley & Sons, New York, p. 515, 1981.

Stumm, W. and J. J. Morgan. *Aquatic Chemistry: Chemical Equilibria and Rates in Natural Waters*, 3rd edition, John Wiley & Sons, New York, 1996.

Suffet, I. H. and P. MacCarthy. *Aquatic Humic Substances: Influence of Fate and Treatment of Pollutants.* Advances in Chemistry Series # 219, American Chemical Society, Washington, DC, 1989.

Thurman, E. M. *Organic Geochemistry of Natural Waters*, Martinus Nijhoff/Junk, Boston, MA, 1985.

Tichnor, K. V. actinide sorption by fracture-infilling materials, *Radiochim. Acta* **60**, 33–42 (1993).

Woodwell, G. M., Wurster, C. F., and Isaacson, P. A. DDT residues in an east coast estuary: A case of biological concentration of a persistent insecticide. *Science* **156**, 821–824 (1967).

CHAPTER 3

QUANTITATIVE ASPECTS OF CHEMISTRY TOWARD MODELING

3.1 INTRODUCTION

In the previous chapter, we introduced several important chemical concepts and processes that contribute to pollutant fate and transport. In this chapter, we will consider the mathematical operations that allow us to incorporate these processes into models of environmental processes. We should note that fate and transport models, even when using sophisticated numerical methods of analysis, cannot incorporate all of the chemistry that we will discuss. Some of the chemistry, such as chemical speciation, must be taken into account in the risk assessment portion of the process. As you read each of the following sections, note how each process will affect mobility of a pollutant in an aqueous and in an atmospheric system and how it will affect toxicity (which is generally associated with the free, unbound form of the pollutant). In the closing section of this chapter, we will attempt to "bring it all together" as we relate each chemical process to fate and transport.

3.2 CALCULATION OF THE FREE METAL ION CONCENTRATION IN NATURAL WATERS

3.2.1 Calculating Chemical Equilibria

Students begin their study of chemical equilibrium in general chemistry with basic equilibrium expressions and Le Chatelier's principle. However, these courses rely on many simplifications regarding equilibria. These simplifications were fine when you were working in simple solutions—solutions containing only one or two salts at low concentration—but are rarely acceptable in environmental chemistry (as in groundwater, estuary, and ocean water). Courses in quantitative analysis, a sophomore-level chemistry course, go on to consider more complex equilibrium that utilize activity coefficients. Now we must further develop your concepts of chemical equilibrium in order to study metal speciation in natural and industrial aqueous solutions.

A Basic Introduction to Pollutant Fate and Transport, By Dunnivant and Anders
Copyright © 2006 by John Wiley & Sons, Inc.

First, we will review the basic principles that you learned (and should remember) from general chemistry. Then we will make the chemical system a little more complicated, to a point where we can still do calculations but only with considerable effort. Finally, we will make the problems even more complicated (and reflective of reality) and find that we cannot solve them by hand. This is where we can use programs such as GEOCHEM, MINTECH, and MINEQL+, which contain numerical solving routines to work these problems.

First, let's discuss general chemistry. In your introduction to chemistry, we kept systems and reactions simple, maybe a little too simple, considering that students coming out of this course usually have some common misconceptions concerning ions in solution. For example, we teach simple reactions such as the following precipitation reaction:

$$Pb^{2+} + 2Cl^- \rightarrow PbCl_2(s)$$

However, as we'll see in this section, lead is present in solution with chloride not only as Pb^{2+}, but also as $PbCl^+$, $PbCl_2(aq)$, $PbCl_3^-$, and $PbCl_4^{2-}$. You may wonder why this matters. There are several answers to this question. The most relevant to environmental chemistry are, first, that this provides us with a much deeper understanding of equilibrium processes; second, that speciation determines how a metal will be transported through aquatic systems; and third, that different forms of a specific metal ion will exhibit different toxicities or bioavailabilities. In general, the most toxic form of each metal ion is the hydrated free metal form (e.g., Pb^{2+} or Cd^{2+} instead of $PbCl^+$ or $CdCl^+$). However, there are important exceptions to this rule, such as methylmercury and dimethylmercury, highly toxic forms of mercury. Thus, we now have two important inputs for risk assessment models concerning bioaccumulation and toxicity: pollutant concentration (e.g., ppm, ppb, etc.) reaching the receptor organism, determined by fate and transport models, and chemical speciation of the pollutant in contact with the receptor organism, expressed in activities.

Review of Equilibrium Calculations from Previous Courses. In general chemistry, you were given reactions such as

$$aA + bB \leftrightarrow cC + dD$$

where a, b, c, and d are the respective (stoichiometric) number of atoms or molecules of chemical species A, B, C, and D in the balanced equation. The equilibrium expression for this reaction was expressed as

$$K = \frac{|C|^c \, |D|^d}{|A|^a \, |B|^b} \tag{3.1}$$

where K is the equilibrium constant. You were told (and you probably did an experiment in lab to demonstrate) that K was constant for a system, even as the concentrations of the species A, B, C, and D varied. However, this was an idealized experiment. More typically, the equilibrium constant has been found to vary as the concentration changes, as illustrated in Figure 3.1. You should recall Le Chatelier's principle, which states that if a system is at equilibrium, such as the reaction above,

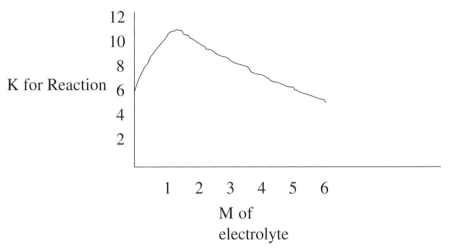

Figure 3.1. Illustration of K as a function of molar concentration of electrolyte.

and you add additional A or B, the system will react to readjust the concentrations of A, B, C, and D to reestablish the same K value.

There is a relatively simple way to correct for this observation. When activities rather than concentrations are used in the equilibrium expression, the equilibrium constant *does* remain constant:

$$K = \frac{(A_C)^c (A_D)^d}{(A_A)^a (A_B)^b} = \frac{[C]^c \gamma_c [D]^d \gamma_D}{[A]^a \gamma_A [B]^b \gamma_B}$$

where A is the activity and γ is the activity coefficient (from Chapter 2) of the species indicated by subscript. Note that in order for K to remain constant, all activities must remain constant. In order for this to occur, if the concentration of a species increases, its activity coefficient, γ, must decrease. The activity coefficient is a function of total concentration of all ions in solution (also known as ionic strength), as seen in Eq. (3.1). Recall that $\gamma * C = A$. For distilled water, the concentration is equal to the activity, which means that the activity coefficient must be equal to 1.00.

General Rule. If you add an inert salt (a salt that dissolves but does not undergo a chemical transformation) to a solution, generally you will increase the solubility of another salt, when the two salts do not share a common ion. (*Note:* This is not true when the anion of the added salt acts as a ligands to form insoluble complexes with the cation of interest.)

Let's work an example to illustrate this rule. Consider the solubility of $CaSO_4$ in distilled water ($K_{sp} = 2.4 \times 10^{-5}$). In distilled water (in the strict sense, K_{sp} values are only accurate for distilled water; these values are compiled from many μ values and are extrapolated to $\mu = 0$), the concentration is equal to the activity, and

$$K_{sp} = A_{Ca^{2+}} A_{SO_4^{2-}} = [Ca^{2+}][SO_4^{2-}]$$

Since the stoichiometry is one-to-one, we can let x equal the Ca^{2+} ion concentration and the SO_4^{2-} ion concentration. Thus, $K_{sp} = x^2$, and $x = [Ca^{2+}] = [SO_4^{2-}] = (2.4 \times 10^{-5})^{1/2} = 4.9 \times 10^{-3}\,M$.

Recall that the solubility of $CaSO_4$ is limited by the attraction between the Ca^{2+} and SO_4^{2-} ions. Anything that we do to decrease these interactions will increase the solubility. One way that we can decrease these interactions is to add another electrolyte. In the distilled water, only the attractions to polar water molecules decreased these attractions. Now, let's look at the solubility of $CaSO_4$ in a 0.02 M solution of KNO_3. As the Ca^{2+} and SO_4^{2-} ions go into solution, the calcium ion will be surrounded by the $\delta-$ (negative dipole) end of H_2O as well as by the negative ions SO_4^{2-} and NO_3^-, while sulfate will be surrounded by the positive end of H_2O and the positive ions K^+ and Ca^{2+}.

Returning to the problem, we also assume that the ionic strength from the $CaSO_4$ is insignificant compared to the ionic strength of KNO_3 (based on their relative concentrations). Thus, KNO_3 determines the ionic strength. By using this ionic strength to calculate the activity coefficients, we can plug these coefficient into the solubility expression to calculate the concentrations of the two ions in solution. This was covered in Section 2.4.2 and will be repeated here.

$$\mu = 0.500(C_1 Z_1^2 + C_2 Z_2^2) = 0.500([K^+](1^2) + [NO_3^-](1^2))$$
$$\mu = 0.500[(0.0200) + (0.0200)] = 0.0200\ M$$
$$\log \gamma = -\frac{0.512 Z^2 \sqrt{\mu}}{1 + \alpha \sqrt{\mu}/305} \qquad \text{at } 25°C$$
$$\log \gamma_{Ca} = -\frac{0.512(2)^2 \sqrt{0.0200}}{1 + 600\sqrt{0.0200}/305} = -0.230$$
$$\gamma_{Ca} = 0.590$$
$$\log \gamma_{SO_4^{2-}} = -\frac{0.512(2)^2 \sqrt{0.0200}}{1 + 450\sqrt{0.0200}/305} = -0.240$$
$$\gamma_{SO_4^{2-}} = 0.580$$
$$K_{sp} = A_{Ca^{2+}} A_{SO_4^{2-}} = [Ca^{2+}]\gamma_{Ca^{2+}}[SO_4^{2-}]\gamma_{SO_4^{2-}} = x^2 \gamma_{Ca^{2+}} \gamma_{SO_4^{2-}} = K_{sp} = 2.4 \times 10^{-5}$$
$$K_{sp} = x^2 \gamma^{Ca^{2+}} \gamma_{SO_4^{2-}} = x^2(0.590)(0.580) = 2.4 \times 10^{-5}$$

So, x, the concentration of Ca^{2+} and SO_4^{2-}, equals $8.37 \times 10^{-3}\,M$.

Note that the concentration of $CaSO_4$ in distilled water (calculated above) was $4.9 \times 10^{-3}\,M$, while in the presence of another ionic salt the concentration increased to $8.37 \times 10^{-3}\,M$. Thus, the general rule was followed. Common ion problems, which you also worked in general chemistry and quantitative analysis, are worked in a similar manner.

Now, we will develop a more general method for solving equilibrium/speciation problems. In the remainder of this discussion, we will ignore activities, in order to make the calculations a little more manageable. However, if you are concerned with obtaining the most accurate estimate of concentration, you must use equations such as the extended Debye–Hückel equation to first calculate the ionic strength of the solution (as was done in the example above).

The approach we will use below uses the algebraic observation that if you have as many equations to describe a system (here, an aqueous solution) as you have unknowns (here, concentrations) then you can mathematically solve the problem. To solve these types of problems, we will use the charge balance equation, the mass balance equation, and the basic equilibrium expressions.

The charge balance is an algebraic statement of the sum of the molar concentrations of cations and anions, and it can be represented by

$$\sum \text{positive charges on the ions} = \sum \text{negative charges on the ions}$$

For example, say that you have K_3PO_4 in aqueous solution. The charge balance equation would be

$$[H^+]+[K^+]=[OH^-]+[H_2PO_4^-]+2[HPO_4^{2-}]+3[PO_4^{3-}]$$

Note that the concentrations for each of the above are in molar units, and we multiply each concentration by the charge of the ion to make this a charge balance. For example, say that PO_4^{3-} is present at $0.300\,M$. We are concerned with the "charge" concentration, not the molar concentration of the ions. Since there are three charges per ion, the value used in the charge balance is $3*[PO_4^{3-}]=3*0.300\,M=0.900\,M$.

The mass balance states that the quantity of a particular element in all species put into a solution must equal the amount of that element delivered to the solution. Thus, although a chemical input into a solution may dissociate into or form a number of different species, the sum of the concentrations of a particular element in the system must be equal to its concentration from the input chemical(s). The following examples illustrate this concept.

For the reaction $CH_3COOH \leftrightarrow CH_3COO^- + H^+$, if the total acetic acid added is equal to $0.0500\,M$, then the mass balance would be

$$\text{Total mass of acid and associated species}$$
$$= 0.0500 \text{ M} = [CH_3COOH]+[CH_3COO^-]$$

For a $0.105\,M$ solution of Na_2S, the reaction would be $Na_2S \rightarrow 2\,Na^+ + S^{2-}$, but S^{2-} reacts with water to form HS^- and H_2S. The mass balance for all sulfur species would be

$$\text{Total S} = 0.105 \text{ M} = [S^{2-}]+[HS^-]+[H_2S]$$

The general steps for this approach to equilibrium are as follows:

Step 1: Write the pertinent reactions.
Step 2: Write the charge balance.
Step 3: Write the mass balance.
Step 4: Write the equilibrium expressions with appropriate constants.
Step 5: Count the equations and unknowns.

Step 6: If the number of equations is equal to the number of unknowns, then solve.

In the following example, we will look at a salt that does not react with water (other than hydration reactions characteristic of all metal ions). The problem statement is this: Calculate the concentration of Hg_2^{2+} in a saturated solution of Hg_2Cl_2. Again, in this example we will ignore activities.

Step 1: Write the pertinent reactions:

$$Hg_2Cl_2 \leftrightarrow Hg_2^{2+} + 2\,Cl^- \qquad K_{sp} = 1.2 \times 10^{-18}$$

Also, you should always include the dissociation of water, since it will affect the charge balance in aqueous solutions:

$$H_2O \leftrightarrow H^+ + OH^- \qquad K_w = 1.00 \times 10^{-14}$$

Step 2: Write the charge balance:

$$[H^+] + 2[Hg_2^{2+}] = [Cl^-] + [OH^-]$$

Step 3: Write the mass balance.

Since neither Hg_2^{2+} nor Cl^- reacts with water, the $[H^+]$ and $[OH^-]$ remain constant. Recall that in pure distilled water (not open to the atmosphere) we have

$$[H^+] = [OH^-]$$
$$[Cl^-] = 2[Hg_2^{2+}]$$

Step 4: Write the equilibrium expressions and constants:

$$K_{sp} = [Hg_2^{2+}][Cl^-]^2 = 1.2 \times 10^{-18}$$
$$K_w = [H^+][OH^-] = 1.00 \times 10^{-14}$$

Step 5: Count the equations and unknowns.

Step 6: Solve the following.

For pure water:

$$[H^+] = [OH^-]$$
$$K_w = [H^+][OH^-] = [H^+]^2 = 1.00 \times 10^{-14}$$
$$[H^+] = [OH^-] = 1.00 \times 10^{-7}$$

For Hg_2Cl_2:

$$[Cl^-] = 2[Hg_2^{2+}]\,(\text{from mass balance})$$
$$K_{sp} = [Hg_2^{2+}][Cl^-]^2 = [Hg_2^{2+}](2 * [Hg_2^{2+}])^2 = 1.2 \times 10^{-18}$$

or

$$[Hg_2^{2+}] = (K_{sp}/4)^{1/3} = 6.7 \times 10^{-7} \text{ M}$$

This is the same answer that you would obtain if you solved the problem as you did in general chemistry. But what if we were considering a soluble species that reacts with water? In this case, the general chemistry approach would not work, and you would have to use the five-step process illustrated above. For example, consider the solubility of HgS, which can be the predominant source of Hg ion in the sediments and hypolimnion of lakes.

$$HgS_{(s)} \leftrightarrow Hg^{2+} + S^{2-} \qquad K_{sp} =\sim 5 \times 10^{-54}$$
$$S^{2-} + H_2O \leftrightarrow HS^- + OH^- \qquad K_{b1} = 0.80$$
$$HS^- + H_2O \leftrightarrow H_2S + OH^- \qquad K_{b2} = 1.1 \times 10^{-7}$$

and, as always,

$$H_2O \leftrightarrow H^+ + OH^- \qquad K_w = 1.00 \times 10^{-14}$$

Recall that S^{2-} is a strong base, so $[H^+]$ will not be equal to $[OH^-]$ as in the earlier example.

Step 1: See above equations for $HgS_{(s)}$.

Step 2:

$$2[Hg^{2+}] + [H^+] = 2[S^{2-}] + [HS^-] + [OH^-]$$

Step 3: Since the composition of HgS is one-to-one for the Hg and S and the Hg and S dissociate completely, the total molar concentration of Hg is equal to the total concentration of S species:

$$[Hg^{2+}] = [S^{2-}] + [HS^-] + [H_2S]$$

Step 4:

$$K_{sp} = [Hg^{2+}][S^{2-}] = 5 \times 10^{-54}$$
$$K_{b1} = \frac{[HS^-][OH^-]}{[S^{2-}]} = 0.80$$
$$K_{b2} = \frac{[H_2S][OH^-]}{[HS^-]} = 1.1 \times 10^{-7}$$
$$K_w = [H^+][OH^-] = 1.00 \times 10^{-14}$$

Step 5 & 6: You have 6 equations and 6 unknowns, so you can solve this by substitution. But we can also make assumptions that make the calculation much easier. For example, most of the time you know what the pH of the system is or you know the pH at which you want to evaluate the metal speciation. Let's evaluate the speciation at a pH value of 8.00. This assumption lowers the number of unknowns to 5 because we know the value for $[H^+]$.

$$pH = 8.00, \qquad [H^+] = 10^{-8.00}$$
$$[OH^-] = 10^{-14}/10^{-8.00} = 1.00 \times 10^{-6}$$

Rearrangement of the $K_{\beta 1}$ equation (above) yields

$$[HS^-] = \frac{K_{b1}[S^{2-}]}{[OH^-]}$$

Rearrangement of the K_{b2} equation (above) yields

$$[H_2S] = \frac{K_{\beta 2}[HS^-]}{[OH^-]}$$

and with substitution of the $[HS^-]$ expression yields

$$[H_2S] = \frac{K_{b1} * K_{b2}[S^{2-}]}{[OH^-]^2}$$

Now, we have the values of $[HS^-]$ and $[H_2S]$ expressed in terms of K_{b1}, K_{b2}, $[OH^-]$, and $[S^{2-}]$. All we need in order to solve the problem is an expression for $[S^{2-}]$ expressed in terms of the K values and $[OH^-]$.

From the mass balance equation, we have

$$[Hg^{2+}] = [S^{2-}] + [HS^-] + [H_2S]$$

Substitution of the $[HS^-]$ and $[H_2S]$ expressions from the previous set of equations yields

$$[Hg^{2+}] = [S^{2-}] + \frac{K_{\beta 1}[S^{2-}]}{[OH^-]} + \frac{K_{\beta 1}K_{\beta 2}[S^{2-}]}{[OH^-]^2}$$

Upon rearrangement, this yields

$$[Hg^{2+}] = [S^{2-}] + \left[1 + \frac{K_{\beta 1}}{[OH^-]} + \frac{K_{\beta 1}K_{\beta 2}}{[OH^-]^2} \right]$$

or

$$[S^{2-}] = \frac{[Hg^{2+}]}{1 + \dfrac{K_{\beta 1}}{[OH^-]} + \dfrac{K_{\beta 1}K_{\beta 2}}{[OH^-]^2}}$$

Substitution of the $[S^{2-}]$ expression into the K_{sp} expression yields

$$K_{sp} = [Hg^{2+}][S^{2-}]$$

$$K_{sp} = [Hg^{2+}] \left[\frac{[Hg^{2+}]}{1 + \dfrac{K_{\beta 1}}{[OH^-]} + \dfrac{K_{\beta 1}K_{\beta 2}}{[OH^-]^2}} \right]$$

Rearrangement yields an expression that allows you to calculate the free metal ion concentration.

$$[Hg^{2+}] = \sqrt{\left(K_{sp} \left(1 + \frac{K_{\beta 1}}{[OH^-]} + \frac{K_{\beta 1} K_{\beta 2}}{[OH^-]^2} \right) \right)}$$

As we stated in the original problem, we made the derivation easier by selecting a single pH value for which to calculate the Hg^{2+} concentration. At a pH value of 8.00, $[OH^-]$ is equal to 1.00×10^{-6} M, and, using the equation above, the $[Hg^{2+}]$ is equal to 2.10×10^{-24} M. This equation can be used at other pH values or put into a spreadsheet to generate a plot of $[Hg^{2+}]$ versus pH value.

3.2.2 Equilibrium Applied to More Complex Speciation Problems

Although the problems and derivations above have been a bit tedious, they have been quite manageable. Now, let's make the problem a little more difficult (and a little more true to reality). Three cases are relevant. Case I illustrates the speciation of a metal in the presence of excess metal salt. This situation could occur in mining operations or in industrial treatment processes and plating operations, as well as in laboratory experiments. Case II is more illustrative of situations occurring in the natural environment. Here, we know the total aqueous concentration of the metal, and it is always below the maximum solubility limit. We also know the predominant ligand in solution. From these data (and the equilibrium constants), we can calculate the species distribution for the solution. Case III looks at the combined effects of pH and E_H.

Case I.   Complexation in the Presence of Excess Solid

Problem Statement.   You are interested in studying the speciation of Pb^{2+} in the presence of iodide ion, and you want to determine the predominant chemical form (species) of Pb^{2+}–I^- complexes in solution as a function of I^- concentration. Excess PbI_2 salt is placed in a beaker containing distilled water. Thus, the concentration of hydrated Pb^{2+} ion in solution is controlled by the solubility of the PbI_2 salt ($K_{sp} = 7.9 \times 10^{-9}$, from Table 2.7). We will treat iodide as independent of the solubility of this salt, with other sources available. Lead can form four complexes with iodide, according to the expressions shown below. Complexation data are from Table 3.1.

$$Pb^{2+} + I^- \leftrightarrow PbI^+ \qquad K_1 = [PbI^+]/[Pb^{2+}][I^-] \qquad = \quad 1.00E+02$$
$$Pb^{2+} + 2I^- \leftrightarrow PbI_{2(aq)} \quad \beta_2 = [PbI_{2(aq)}]/[Pb^{2+}][I^-]^2 \quad = \quad 1.40E+03$$
$$Pb^{2+} + 3I^- \leftrightarrow PbI_3^- \qquad \beta_3 [PbI_3^-]/Pb^{2+}[I^-]^3 \qquad = \quad 8.30E+03$$
$$Pb^{2+} + 4I^- \leftrightarrow PbI_4^{2-} \qquad \beta_4 [PbI_4^{2-}]/[Pb^{2+}][I^-]^4 \quad = \quad 3.00E+04$$

To determine the major species, draw a complexation diagram (pPb–pI), much like the pC–pH diagram described above and illustrated in the pC–pH diagrams in Chapter 2, for the system at equilibrium. Use I^- concentrations ranging from 0.0001 to 10 M (pI = −2 to 4). Using your diagram, estimate the concentrations of Pb^{2+}, PbI^+, $PbI_{2(aq)}$, PbI_3^-, and PbI_4^{2-} at an I^- concentration of 0.3162 M (arbitrarily chosen).

TABLE 3.1. Stability Constants (log K or β) for Formation of Complexes and Solids from Metals and Ligands (Selected Data from Morel and Herring, 1993)

	OH^-	CO_3^{2-}	SO_4^{2-}	Cl^-	Br^-	F^-	NH_3	$B(OH)_4$
Sr^{2+}		SrL 9.0, SrL(s)	SrL 2.6, SrL(s) 6.5			$SrL_2(s)$ 8.5		
Ba^{2+}		BaL 2.8, BaL(s) 8.3	BaL 2.7, BaL(s) 10.0			$BaL_2(s)$ 5.8		
Cr^{3+}	CrL 10.0, CrL_2 18.3, CrL_3 24.0, CrL_4 28.6, Cr_3L_4 47.8, $CrL_3(s)$ 30.0		CrL 3.0	CrL 0.23		CrL 5.2, CrL_2 9.2, CrL_3 12.0		
Al^{3+}	AlL 9.0, AlL_2 18.7, AlL_3 27.0, AlL_4 33.0, Al_3L_4 42.1, $AlL_3(s)$ 33.5					AlL 7.0, AlL_2 12.6, AlL_3 16.7, AlL_4 19.1		
Fe^{3+}	FeL 11.8, FeL_2 22.3, FeL_4 34.4, Fe_2L_2 25.0, $FeL_3(s)$ 42.7, $FeL_3(s)$ 38.8		FeL 4.0, FeL_2 5.4	FeL 1.5, 2.1	FeL 0.6	FeL 6.0, FeL_2 10.6, FeL_3 13.7		

Ion												
Fe^{2+}	FeL	4.5	$FeL(s)$	10.7	FeL	2.2			FeL	1.4		
	FeL_2	7.4										
	FeL_3	11.0										
	$FeL_2(s)$	15.1										
Co^{2+}	CoL	4.3	$CoL(s)$	10.0	CoL	2.4	CoL	0.5	CoL	1.0	CoL	2.0
	CoL_2	9.2									CoL_2	3.5
	CoL_3	10.5									CoL_3	4.4
	$CoL_2(s)$	15.7									CoL_4	5.0
Ni^{2+}	NiL	4.1	$NiL(s)$	6.9	NiL	2.3	NiL	0.6	NiL	1.1	NiL	2.7
	NiL_2	9.0									NiL_2	4.9
	NiL_3	12.0									NiL_3	6.6
	$NiL_2(s)$	17.2									NiL_4	7.7
											NiL_5	8.3
Cu^{2+}	CuL	6.3	CuL	6.7	CuL	2.4	CuL	0.5	CuL	1.5	CuL	4.0
	CuL_2	11.8	$Cu_4(OH)_6L(s)$	10.2	$Cu_4(OH)_6L(s)$	68.6					CuL_2	7.5
	CuL_4	16.4	$CuL(s)$	9.6							CuL_3	10.3
	Cu_2L_2	17.7	$Cu_2(OH)_2L(s)$	33.8							CuL_4	11.8
	$CuL_2(s)$	19.3	$Cu_3(OH)_2L_2(s)$	46.0								
	$CuL_2(s)$	20.4										
Zn^{2+}	ZnL	5.0	$ZnL(s)$	10.0	ZnL	2.1	ZnL	0.4	ZnL	1.2	ZnL	2.2
	ZnL_2	11.1			ZnL_2	3.1	ZnL_2	0.2			ZnL_2	4.5
	ZnL_3	13.6					ZnL_3	0.5			ZnL_3	6.9
	ZnL_4	14.8					$Zn_2(OH)_3L(s)$	26.8			ZnL_4	8.9
	$ZnL_2(s)$	15.5										
	$ZnL_2(s)$	16.8										

(Continued)

TABLE 3.1. Stability Constants (log$K$ or β) for Formation of Complexes and Solids from Metals and Ligands (Selected Data from Morel and Herring, 1993) (continued)

	OH⁻	CO₃²⁻	SO₄²⁻	Cl⁻	Br⁻	F⁻	NH₃	B(OH)₄
Pb²⁺	PbL 6.3 PbL₂ 10.9 PbL₃ 13.9 PbL₂(s) 15.3	PbL(s) 13.1	PbL 2.8 PbL(s) 7.8	PbL 1.6 PbL₂ 1.8 PbL₃ 1.7 PbL₄ 1.4 PbL₂(s) 4.8	PbL 1.8 PbL₂ 2.6 PbL₃ 3.0 PbL₂(s) 5.7	PbL 2.0 PbL₂ 3.4 PbL₂(s) 7.4		
Hg²⁺	HgL 10.6 HgL₂ 21.8 HgL₃ 20.9 HgL₂(s) 25.4	HgL(s) 16.1	HgL 2.5 HgL₂ 3.6	HgL 7.2 HgL₂ 14.0 HgL₃ 15.1 HgL₄ 15.4 HgOHL 18.1	HgL 9.6 HgL₂ 18.0 HgL₃ 20.3 HgL₄ 21.6 HgL₂(s) 19.8	HgL 1.6	HgL 8.8 HgL₂ 17.4 HgL₃ 18.4 HgL₄ 19.1	
Cd²⁺	CdL 3.9 CdL₂ 7.6 CdL₂(s) 14.3	CdL(s) 13.7	CdL 2.3 CdL₂ 3.2 CdL₃ 2.7	CdL 2.0 CdL₂ 2.6 CdL₃ 2.4 CdL₄ 1.7	CdL 2.1 CdL₂ 3.0	CdL 1.0 CdL₂ 1.4	CdL 2.6 CdL₂ 4.6 CdL₃ 5.9 CdL₄ 6.7	
Ag⁺	AgL 2.0 AgL₂ 4.0 AgL(s) 7.7	Ag₂L(s) 11.1	AgL 1.3 AgL₂ 4.8 Ag₂L(s)	AgL 3.3 AgL₂ 5.3 AgL₃ 6.4 AgL₄ 9.7 AgL(s)	AgL 4.7 AgL₂ 6.9 AgL₃ 8.7 AgL₄ 9.0 AgL(s) 12.3	AgL 0.4	AgL 3.3 AgL₂ 7.2	AgL 0.6 AgHL₂(s) 22.9

(Continued)

	SiO_3^{2-}	S^{2-}	$S_2O_3^{2-}$	PO_4^{3-}	$P_2O_7^{4-}$	$P_3O_{10}^{5-}$	CN^-
Sr^{2+}			SrL 2.0	SrL 5.5 SrHL 14.5 SrH$_2$L 20.3 SrHL(s) 19.3	SrL 5.4 SrOHL 7.7 Sr$_2$L(s) 12.9	SrL 7.2 SrHL 13.6 SrOHL 9.3	
Ba^{2+}			BaL 2.3 BaL(s) 4.8	BaHL(s) 19.8		BaL 6.3 BaHL 12.9 Ba$_2$L(s) 16.1	
Cr^{3+}							
Al^{3+}							
Fe^{3+}	FeHL 22.7		FeL 3.3	FeHL 22.5 FeH$_2$L 23.9 FeL(s) 26.4			FeL$_6$ 43.6
Fe^{2+}		FeL(s) 18.1		FeHL 16.0 FeH$_2$L 22.3 Fe$_3$L$_2$(s) 36.0			FeL$_6$ 35.4
Co^{2+}		CoL(s) 21.3 CoL(s) 25.6	CoL 2.1	CoHL 15.5	CoL 7.9 CoHL 14.1	CoL 9.7 CoHL 14.8	
Ni^{2+}		NiL(s) 19.4 NiL(s) 24.9 NiL(s) 26.6	NiL 2.1	NiHL 15.4	NiL 7.7 NiHL 14.4	NiL 9.5 NiHL 14.7	NiL 7.3 NiL$_4$ 30.2 NiH$_2$L$_4$ 40.8 NiHL$_4$ 36.1
Cu^{2+}		CuL(s) 36.1		CuHL 16.5 CuH$_2$L 21.3	CuL 9.8 CuHL 15.5 CuL$_2$ 12.5 CuH$_2$L 19.2	CuL 11.1 CuHL 15.5	CuL$_2$ 16.3 CuL$_3$ 21.6 CuL$_4$ 23.1

TABLE 3.1. Stability Constants ($\log K$ or β) for Formation of Complexes and Solids from Metals and Ligands (Selected Data from Morel and Herring, 1993) (continued)

	SiO_3^{2-}	S^{2-}	$S_2O_3^{2-}$	PO_4^{3-}	$P_2O_7^{4-}$	$P_3O_{10}^{5-}$	CN^-
Zn^{2+}		ZnL 16.6; $ZnL(s)$ 24.7	ZnL 2.4; ZnL_2 2.5; ZnL_3 3.3; Zn_2L_2 7.0	$ZnHL$ 15.7; ZnH_2L 21.2; Zn_3L_2 35.3	ZnL 8.7; ZnL_2 11.0; $ZnOHL$ 13.1	ZnL 10.3; $ZnHL$ 14.9; $ZnOHL$ 13.6	ZnL 5.7; ZnL_2 11.1; ZnL_3 16.1; ZnL_4 19.6; $ZnL_2(s)$ 15.9
Pb^{2+}		$PbL(s)$ 27.5	PbL 3.0; PbL_2 5.5; PbL_3 6.2; PbL_4 7.3	$PbHL$ 15.5; PbH_2L 21.1; $Pb_3L_2(s)$ 43.5; $PbHL(s)$ 23.8	PbL 9.5; PbL_2 10.2		
Hg^{2+}		HgL 7.9; HgL_2 14.3; $HgOHL$ 18.5; $HgL(s)$ 52.7; $HgL(s)$ 53.3	HgL_2 29.2; HgL_3 30.6		$HgOHL$ 18.6		HgL 17.0; HgL_2 32.8; HgL_3 36.3; HgL_4 39.0; $HgOHL$ 29.6
Cd^{2+}		CdL 19.5; $CdHL$ 22.1; CdH_2L_2 43.2; CdH_3L_3 59.0; CdH_4L_4 75.1; $CdL(s)$ 27.0	CdL 3.9; CdL_2 6.3; CdL_3 6.4; CdL_4 8.2; Cd_2L_2 12.3		CdL 8.7; $CdOHL$ 11.8	CdL 9.8; $CdHL$ 14.6; $CdOHL$ 12.6	CdL 6.0; CdL_2 11.1; CdL_3 15.7; CdL_4 17.9
Ag^+		AgL 19.2; $AgHL$ 27.7; $AgHL_2$ 35.8; AgH_2L_2 45.7; $AgH_2L(s)$ 50.1	AgL 8.8; AgL_2 13.7; AgL_3 14.2; Ag_2L_4 26.3; Ag_3L_5 39.8; Ag_6L_8 78.6	$Ag_3L(s)$ 17.6			AgL_2 20.5; AgL_3 21.4; $AgOHL$ 13.2; $AgL(s)$ 15.7

	Ethylene-diamine	NTA	EDTA	CDTA	IDA	Picolinate	Cysteine	Desferri-ferrioxamine B
Sr²⁺	SrL 6.3	SrL	SrL 10.5 SrHL 14.9	SrL 12.4	SrL 3.1	SrL 1.8 SrL₂ 3.0		SrL 3.1
Ba²⁺	BaL 5.9	BaL	BaL 9.6 BaHL 14.6	BaL 10.5 BaHL 17.8	Ba 2.5	BaL 1.6		
Cr³⁺			CrL 26.0 CrHL 28.2 CrOHL 32.2		CrL 12.2 CrL 23.2			
Al³⁺	AlL 13.4 AlOHL 22.1		AlL 18.9 AlHL 21.6 AlOHL 26.6 Al(OH)₂L 30.0	AlL 22.1 AlHL 24.3 AlOHL 28.1	AlL 9.9 AlL₂ 17.5			
Fe³⁺	FeL 17.9 FeL₂ 26.3	FeL FeL₂	FeL 27.7 FeHL 29.2 FeOHL 33.8 Fe(OH)₂L 37.7	FeL 32.6 FeOHL 36.5	FeL 12.5	FeL₂ 13.9 FeOHL₂ 24.9		FeL 31.9 FeHL 32.6
Fe²⁺	FeL 4.3 FeL₂ 7.7 FeL₃ 9.7	FeL 9.6 FeL₂ 13.6 FeOHL 12.6	FeL 16.1 FeHL 19.3 FeOHL 20.4 Fe(OH)₂L 23.7	FeL 20.8 FeHL 23.9	FeL 6.7 FeL₂ 11.0	FeL 5.3 FeL₂ 9.7 FeL₃ 13.0		FeHL 18.7 FeH₂L 21.0
Co²⁺	CoL 6.0 CoL₂ 10.8 CoL₃ 14.1	CoL 11.7 CoL₂ 15.0 CoOHL 14.5	CoL 18.1 CoHL 21.5	CoL 21.4 CoL₂ 24.7	CoL 7.9 CoL₂ 13.2	CoL 6.4 CoL₂ 11.3 CoL₃ 14.8		CoL 11.2 CoHL 18.0 CoHL 23.6

(Continued)

TABLE 3.1. Stability Constants (log K or β) for Formation of Complexes and Solids from Metals and Ligands (Selected Data from Morel and Herring, 1993) (continued)

	Ethylene-diamine	NTA	EDTA	CDTA	IDA	Picolinate	Cysteine	Desferri-ferrioxamine B
Ni^{2+}	NiL 7.4 NiL$_2$ 13.6 NiL$_3$ 17.9	NiL 12.8 NiHL 17.0 NiOHL 15.5	NiL 20.4 NiHL 24.0 NiOHL 21.8	NiL 22.1 NiHL 25.4	NiL 9.1 NiL$_2$ 15.7	NiL 7.2 NiL$_2$ 12.5 NiL$_3$ 17.9	NiL 10.7 NiL$_2$ 20.9	NiL 11.8 NiHL 18.3 NiH$_2$L 23.8
Cu^{2+}	CuL 10.5 CuL$_2$ 19.6 CuOHL 11.8	CuL 14.2 CuHL 18.1 CuOHL 18.6	CuL 20.5 CuHL 23.9 CuOHL 22.6	CuL 23.7 CuHL 27.3	CuL 11.5 CuL$_2$ 17.6	CuL 8.4 CuL$_2$ 15.6	Cu(II)-Cu(I)CuL CuHL CuH$_2$L	CuL 15.0 CuHL 24.1 CuH$_2$L 27.0
Zn^{2+}	ZnL 5.7 ZnL$_2$ 10.6 ZnL$_3$ 13.9	ZnL 12.0 ZnL$_2$ 14.9 ZnOHL 15.5	ZnL 18.3 ZnHL 21.7 ZnOHL 19.9	ZnL 21.1 ZnHL 24.4	ZnL 8.2 ZnL$_2$ 13.5	ZnL 5.7 ZnL$_2$ 10.3 ZnL$_3$ 13.6	ZnL 10.1 ZnL 19.1 ZnHL 16.4	ZnL 11.0 ZnHL 17.5 ZnH$_2$L 22.9
Pb^{2+}	PbL 7.0 PbL$_2$ 8.5	PbL 12.6	PbL 19.8 PbHL 23.0	PbL 22.1 PbHL 25.3	PbL 8.3	PbL 5.0 PbL$_2$ 8.6	PbL 12.5	
Hg^{2+}	HgL 14.3 HgL$_2$ 23.2 HgOHL 24.2 HgHL$_2$ 28.0	HgL 15.9	HgL 23.5 HgHL 27.0 HgOHL 27.7	HgL 26.8 HgHL 30.3 HgOHL 29.7	HgL 11.7	HgL 8.1 HgL$_2$ 16.2	HgL 15.3	
Cd^{2+}	CdL 5.4 CdL$_2$ 9.9 CdL$_3$ 11.7	CdL 11.1 CdL$_2$ 15.1 CdOHL 13.4	CdL 18.2 CdHL 21.5	CdL 21.7 CdHL 25.1	CdL 6.6 CdL$_2$ 11.1	CdL 5.0 CdL$_2$ 8.3 CdL$_3$ 11.4		CdL 8.8 CdHL 16.2 CdH$_2$L 22.7
Ag^{+}	AgL 4.7 AgL$_2$ 7.7 AgHL 11.9	AgL 5.8	AgL 8.2 AgHL 14.9	AgL 9.9		AgL 3.6 AgL$_2$ 6.1		

112

Then, using the equations shown above and your understanding of equilibrium, calculate the exact concentration of each species at an I^- concentration of 0.3162 M.

Explanation. In the presence of excess PbI_2, the concentration of Pb^{2+} in solution will be controlled by the K_{sp} for PbI_2. Thus, since we know the K_{sp}, we can calculate the negative log of the concentrations of Pb^{2+} at I^- concentrations of 0.0001 M and 10 M:

$$K_{sp} = [Pb^{2+}][I^-]^2 = 7.9 \times 10^{-9}$$

$$[Pb^{2+}] = \frac{7.9 \times 10^{-9}}{[I^-]^2}$$

$$\text{For } [I^-] = 0.0001 \text{ M}, [Pb^{2+}] = 0.790 \text{ M}$$

$$\log(0.790) = -0.102$$

$$\text{For } [I^-] = 10 \text{ M}, [Pb^{2+}] = 7.9 \times 10^{-11}$$

$$\log(7.9 \times 10^{-11}) = -10.102$$

Refer to Figure 3.2 for the following discussion of the construction of a complexation diagram. The Pb^{2+} line on the pPb–pI diagram should be drawn from $(-4, -0.102)$ to $(1, -10.102)$ (where the first number is the log molar I^- concentration, and the second is the log molar Pb^{2+} concentration.) This line is the basis for the remaining calculations.

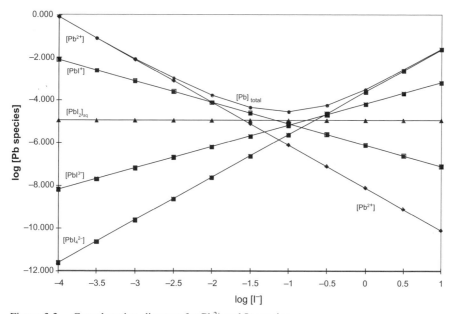

Figure 3.2. Complexation diagram for Pb^{2+} and I^- species.

Note the equilibrium constants shown in the problem statement. The first expression uses K to represent the equilibrium between Pb^{2+} and one I^- ion. The letter "K" is used to represent stepwise formation constants. Notice that the remaining constants are represented by β, which represents the cumulative formation constants (the addition of more than one ligand treated as a single reaction). For example, β_2 represents the formation of PbI_2 from the reaction of one Pb^{2+} and two I^- ions. Since the $[Pb^{2+}]$ concentration is determined by the I^- concentration, the concentration of PbI^+, $PbI_2(aq)$, PbI_3^-, and PbI_4^{2-} can be calculated for each Pb^{2+} concentration. Thus, we can add lines to the diagram representing these species.

For [PbI⁺]:

- An I^- concentration of 0.0001 results in a Pb^{2+} concentration of 0.790 M.
- Substitution into $K_1 = \dfrac{[PbI^+]}{[Pb^{2+}][I^-]} = 100$ yields $[PbI^+] = K_1[Pb^{2+}][I^-] = 7.90 \times 10^{-3}$.
- An I^- concentration of 10 M results in a Pb^{2+} concentration of 7.9×10^{-11}.
- This results in a $[PbI^+]$ concentration of 7.90×10^{-8}.

Thus, a line should be drawn for the $\log I^-$, $\log PbI^+$ pairs from $(-4, -2.10)$ to $(1, -7.10)$.

For [PbI₂(aq)]:

- Again, an I^- concentration of 0.0001 results in a Pb^{2-} concentration of 0.790 M.
- Substitution into $\beta_2 = \dfrac{[PbI_{2(aq)}]}{[Pb^{2+}][I^-]^2} = 1.4 \times 10^3$ yields $[PbI_{2(aq)}] = \beta_2[Pb^{2+}][I^-]^2 = 1.11 \times 10^{-5}$.
- An I^- concentration of 10 M results in a Pb^{2+} concentration of 7.9×10^{-11}.
- This results in a $[PbI_{2(aq)}]$ concentration of 1.11×10^{-5}.

Thus, a line should be drawn for the $\log I^-$, $\log PbI_{2(aq)}$ pairs or from $(-4, -4.956)$ to $(1, -4.956)$.

For [PbI₃⁻]:

- An I^- concentration of 0.0001 results in a Pb^{2-} concentration of 0.790 M.
- Substitution into $\beta_3 = \dfrac{[PbI_3^-]}{[Pb^{2+}][I^-]^3} = 8.3 \times 10^3$ yields $[PbI_3^-] = \beta_3[Pb^{2+}][I^-]^3 = 6.56 \times 10^{-9}$.
- An I^- concentration of 10 M results in a Pb^{2+} concentration of 7.90×10^{-11}.
- This results in a $[PbI_3^-]$ concentration of 6.56×10^{-4}.

Thus, a line should be drawn for the $\log I^-$, $\log PbI_3^-$ pairs or from $(-4, -8.183)$ to $(1, -3.183)$.

For $[PbI_4^{2-}]$:

- An I^- concentration of 0.0001 results in a Pb^{2-} concentration of 0.790 M.

- Substitution into $\beta_4 = \dfrac{[PbI_3^{2-}]}{[Pb^{2+}][I^-]^4} = 3.00 \times 10^4$ yields $[PbI_4^{2-}] = \beta_4[Pb^{2+}][I^-]^4 = 2.37 \times 10^{-12}$.

- An I^- concentration of 10 M results in a Pb^{2+} concentration of 7.90×10^{-11}.

- This results in a $[PbI_4^{2-}]$ concentration of 2.37×10^{-2}.

Thus, a line should be drawn for the $\log I^-$, $\log PbI_3^-$ pairs or from $(-4, -11.63)$ to $(1, -1.63)$.

Using this approach, the pPb–pI diagram in Figure 3.2 can be drawn. Note that the total [Pb] in solution is the sum of all of the Pb species. Now you can use this figure and the equations given above to solve the original problem.

Figure 3.2 also illustrates another very important point concerning the solubility of lead and complex ion formation. In general chemistry, you learned that the solubility of Pb^{2+} was governed by the K_{sp} and, in this case, the concentration of I^- ion. Because of the common ion effect—another concept from general chemistry (Le Chatelier's principle)—as $[I^-]$ increased, you calculated that $[Pb^{2+}]$ would decrease. If you based your calculation solely on this assumption, the solubility would follow the $[Pb^{2+}]$ line in Figure 3.2. However, note the total concentration of Pb in solution. We see that the total Pb decreases to a point but after the formation of complex ions becomes important, at approximately 0.1 M, the total Pb in solution actually increases. Most metals that form complex ions follow this trend, which is important in toxicity and risk assessment.

Case II. Complexation in the Absence of a Solid

Problem Statement. Nitrilotetraacetic (NTA) acid was a common component in detergents and was a chemical of concern (COC) in sewage effluent in the past. It is known for its high metal complexing power; thus it keeps metals in solution, increasing their mobility, and it may reduce their toxicity. Draw a pCd–pNTA diagram for a cadmium system (total cadmium concentration = 0.015 M) using NTA concentrations ranging from 1.00×10^{-1} to 1.00×10^{-14} M. Use the following data:

$$Cd^{2+} + NTA^- \leftrightarrow CdNTA^+ \qquad K_1 = [CdNTA^+]/[Cd^{2+}][NTA^-] = 6.31 \times 10^9$$
$$Cd^{2+} + 2NTA^- \leftrightarrow Cd(NTA)_{2(aq)} \qquad \beta_2 = [CdNTA_{(aq)}]/[Cd^{2+}][NTA^-]^2 = 1.58 \times 10^{15}$$

Explanation. This problem is easier than in the previous section because we only have one metal concentration to be concerned with, namely, $[Cd^{2+}] = 0.015$ M. The total Cd concentration (C_T) is the result of three chemical species:

$$C_T = [Cd^{2+}] + [CdNTA^+] + [CdNTA_2]$$

First, we will develop an expression to calculate the free-ion $[Cd^{2+}]$. The equilibrium expressions K_1 and β_2 are next rearranged to solve for $[CdNTA^+]$ and

[CdNTA$_2$]. These expressions are substituted into the C_T equation, and the equation is rearranged to solve for [Cd^{2+}].

$$C_T = [Cd^{2+}] + [CdNTA^+] + [CdNTA_2]$$

$$C_T = [Cd^{2+}] + K_1[Cd^{2+}][NTA^-] + \beta_2[Cd^{2+}][NTA^-]^2$$

$$C_T = [Cd^{2+}]\left[1 + K_1[NTA^-] + \beta_2[NTA]^2\right]$$

$$[Cd^{2+}] = \frac{C_T}{\left[1 + K_1[NTA^-] + \beta_2[NTA]^2\right]}$$

Similar approaches can be used to derive expressions for [CdNTA$^+$] and [CdNTA$_2$], shown below. In deriving these equations, you must first express the mass balance as the sum of all metal species. Next, you will rearrange the K expressions in terms of the each metal species and substitute the resulting expressions into the mass balance (using the K expressions containing each complexed metal species to solve the equation for that species). Finally, solve the mass balance equation for each complexed metal species as shown below.

$$[CdNTA^+] = \frac{C_T}{\dfrac{1}{K_1[NTA^-]} + 1 + \dfrac{\beta_2[NTA^-]}{K_1}}$$

$$[CdNTA_2] = \frac{C_T}{\dfrac{1}{\beta_2[NTA^-]^2} + \dfrac{K_1}{\beta_2[NTA^-]} + 1}$$

Lines for each chemical species can be drawn by varying the concentration of NTA and calculating the concentration of Cd^{2+}, CdNTA$^-$, and CdNTA$_2$. Such an approach results in the diagram shown in Figure 3.3.

NTA, suggested by the diagram, is an excellent surrogate to illustrate the binding of metal pollutants by NOM and can also be used to explain how metal and organic pollutants' aqueous concentrations can exceed those predicted by solubilities calculated from pure water calculations or experiments. NOM can bind to pollutants, pulling more into solution than would be predicted. Still, NOM-bonded pollutants are less bioavailable than free aqueous pollutants. This will be discussed in the next section.

Case III. The Combined Effects of pH and E_H

Finally, we will look at the combined effects of pH and E_H on speciation. We have discussed that pH and E_H are important chemical parameters in aqueous solutions, and they directly affect the chemical state of metal and organic pollutants, as well as the reactive state of surfaces (sorption and reactive surfaces such as metal oxide coatings and NOM functional groups) in the solution.

Figure 3.4 is referred to as a stability diagram and shows the equilibrium form of Fe as a function of pH and E_H. As you see, these two chemical parameters strongly influence the chemical nature of Fe in solution and as a solid phase. Diagrams such

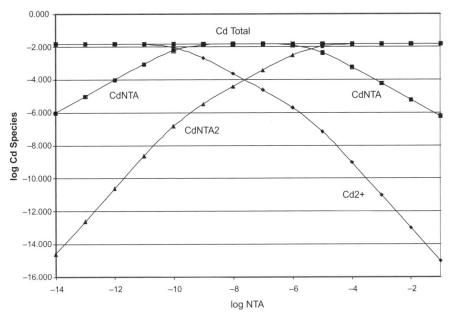

Figure 3.3.   Speciation of Cd^{2+} in solution with NTA.

as this are important in determining which form of Fe (Fe^{2+}, Fe^{3+}, or solids) will be present under the pH and E_H of a natural water. In addition, if Fe^{2+}, present in anaerobic systems, is mobilized and transported to an aerobic region of lake or groundwater and is subsequently oxidized, it will precipitate and form a reactive sorption surface. During the oxidation process, Fe may reduce a pollutant through an abiotic transformation reaction, or after precipitation the Fe^{3+} surface may adsorb a metal pollutant from the water. Thus, the fate of another metal can be greatly affected by changing pH and E_H conditions of a water containing Fe. Stability diagrams can be constructed for all redox-active metal pollutants.

The three cases presented above are instructional, but are still rather simplistic in design. What if you have a common ion or other salts present that also complex with the cation of interest? Or say that you have chloride, nitrate, sulfate, and carbonate present at similar concentrations, and you also have Ca^{2+} and Mg^{2+} present? Also imagine what would happen if you were concerned with the speciation as a function of solution pH and E_H. In these cases, you can throw the calculator away. There are simply too many variables (unknowns) to solve the problem using normal, exact analytical solutions; however, this is exactly what we encounter in aquatic systems. Problems such as this one led to the development of numerical solution models in the late 1960s and early 1970s. There are several of these programs, some designed for freshwater systems and others for higher ionic strength systems such as these found in marine waters or groundwaters. One of the most common, and user friendly, computer programs to solve equilibrium problems such as the ones described above is MINEQL+. This program allows the user to estimate the equi-

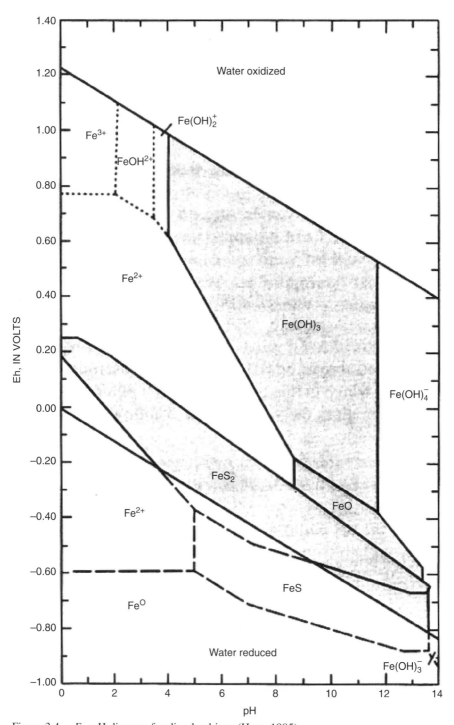

Figure 3.4. E_H–pH diagram for dissolved iron (Hem, 1985).

librium speciation of all cations and anions as a function of pH and E_H. Your instructor may choose to have you use this program or may use a demonstration of how it is used to solve more complicated equilibrium problems. In any case, you should be aware that there is a way to solve these complicated problems.

3.3   METHODS FOR DETERMINING K_d AND K_p

As discussed in Chapter 2, distribution coefficients are central to our modeling of equilibrium sorption processes in environmental media. First, we will mathematically define several terms (refer to Table 2.6 for a description of their use):

For partitioning between the aqueous phase and dissolved organic matter:

$$\text{Pollutant}_{(aq)} \leftrightarrow \text{Pollutant}_{DOM}$$

$$K_{DOM} = \frac{\text{Conc. of pollutant in DOM (mg/kg)}}{\text{Conc. of pollutant in water (mg/L)}} \tag{3.2}$$

For distribution between the aqueous phase and particles in solution:

$$\text{Pollutant}_{(aq)} \leftrightarrow \text{Pollutant}_{particle}$$

$$K_d = \frac{\text{Conc. of pollutant on particle (mg/kg)}}{\text{Conc. of pollutant in water (mg/L)}} \tag{3.3}$$

or

$$K_p = \frac{\text{Conc. of pollutant on organic particle (mg/kg)}}{\text{Conc. of pollutant in water (mg/L)}} \tag{3.4}$$

To describe partitioning as it relates to the concentration of organic matter:

$$K_{Organic\ Carbon} = \frac{K_d \text{ or } K_p}{\text{Fraction of organic matter in sample}} \tag{3.5}$$

As you can see, to calculate the K value from any of these expressions, you need the concentration of pollutant in each of two phases, each with its own set of concentration units. Generally, in order to collect these data in a laboratory, a solution of known pollutant mass and water, with soil, sediment, or DOM, is mixed together for three days and the phases (solid and aqueous) are then separated. Determination of K_{DOM} and concentrations of pollutants in DOM phases is complicated, and the approach of all three K expressions is basically the same, so we will concentrate on K_d and K_p as an example here. After mixing, separation of the soil–water or sediment–water suspension is completed by filtration of the sample through a 0.45-μm filter. The concentration of pollutant in the aqueous phase (or all phases) is then determined. The determination of aqueous concentrations is relatively simple, but solid-phase concentrations require highly involved extraction procedures. In our example, since we know the total mass of pollutant added to each system (each sample vial), we can measure the concentration (mass) of pollutant in the dissolved phase, subtract this mass from the total mass, and thus calculate the concentration of pollutant in the solid-phase. An example of this calculation is shown in Table 3.2 and in the two "A Closer Look" examples at the end of this chapter.

By definition, equilibrium coefficients are constants, but these are considerably more complicated than the typical K expressions for equilibrium reactions in general chemistry. As we have seen, K_d for a metal in a sediment–water suspension is dependent, at a minimum, on pH, the type of other cations present, ionic strength, surface charge, and solids-to-water ratio. Thus, many variables must be controlled in an experiment or must be known to be constant in the environmental medium of interest. Given the ever-changing conditions in the environment, K_d measured in the laboratory can only be an approximation. Furthermore, we are assuming reversible sorption, which is not always the case, especially for metals. Laboratory K_d measurements for metals tend to increase for decreasing total metal concentrations in the aqueous phase, probably due to adsorption competition or saturation of adsorption sites. This is referred to as the Donnon effect.

The usefulness of K_p values, however, stems from the fact that they are usually constant over a wide range of pollutant concentrations in the water phase. This is illustrated in Figure 3.5, illustrating equilibrium partitioning of aqueous methoxychlor (a pesticide) with clay (Karickhoff et al., 1979). The slope of the linear plot is K_p and is independent of water-phase concentration. It is also important to note that Figure 3.5 uses the term K_p, not K_d. This is because methoxychlor is a hydrophobic pollutant and does not undergo site-specific adsorption. Methoxychlor partitions to surfaces, particularly to NOM sorbed onto the clay. Although it is independent of pollutant water-phase concentration, the magnitude of the sorption of hydrophobic pollutants is highly dependent on the amount of organic matter present; the higher the organic matter content, the more pollutant sorbed. This is illustrated in Figure 3.6 (Karickhoff et al., 1979).

TABLE 3.2. Experimental Data for the Determination of a Distribution Coefficient for Cd on EPA Sediment B-2[a]

Total mass (g) of pollutant added to flask	0.00725	
Mass of pollutant recovered in blank (mg)	0.00720	
Mass of pollutant measured in water phase (mg)	0.00542	
Volume of water (L)	0.0300	
Concentration of pollutant in water phase (mg/L)		0.181
Mass of pollutant on solid phase (mg)	0.00178	
Mass of solid phase (kg)	3.58×10^{-5}	
Concentration of pollutant on solid phase (mg/kg)		49.7
K_d		275

[a] Sediment B2 is a highly characterized sediment with respect to chemical and physical composition.

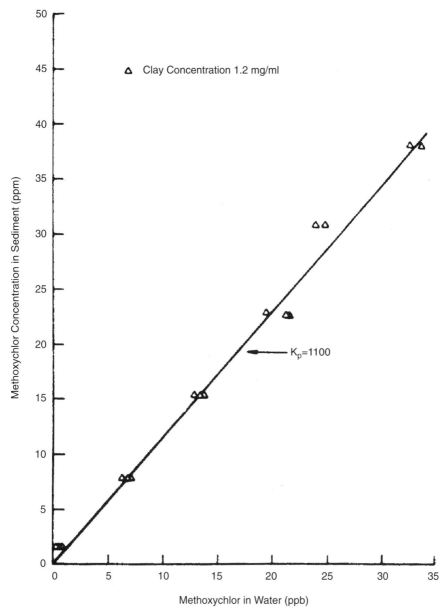

Figure 3.5.   Adsorption isotherm for methoxychlor on clay.'[Taken from Karickhoff et al. (1979), Figure 2, p. 244.]

Also note the slope of the two lines in Figure 3.6, which are defined by Eq. (3.7) given earlier ($K_{oc} = K_p/f_{oc}$). When the partition coefficient is divided by the fraction of organic carbon present in a sample, we refer to the resulting equilibrium constant as K_{oc}. Thus, the partitioning of a pollutant can be directly related to the organic

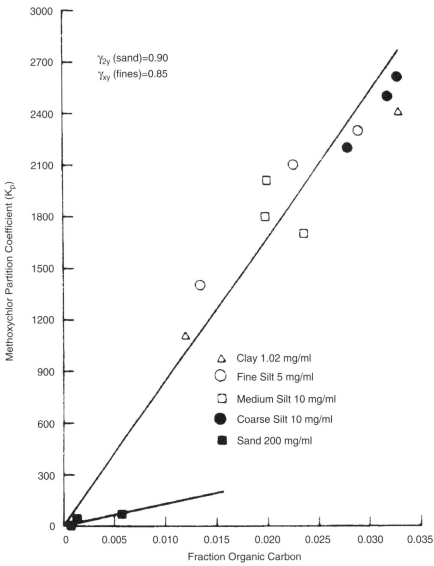

Figure 3.6.   Methoxychlor K_p as a function of sediment organic carbon. [Taken from Karickhoff et al. (1979), Figure 4, p. 245.]

content of a sediment; in general, if you know K_p, K_{oc}, and the organic content of a sediment, you can calculate the equilibrium water concentration of a pollutant. You should mathematically work though this approach. Unfortunately, measuring K_p or K_{oc} for every pollutant would require a very labor-intensive effort. Therefore, we estimate K_{oc} and therefore K_p based on an observation made by Karickhoff et al. (1979). They observed that K_{oc} values are directly correlated to octanol–water equilibrium coefficients (K_{ow}) for the same pollutants, which can be easily and quickly measured in the laboratory by placing a known mass of pollutant in a vial containing a known volume of octanol and water, allowing it to equilibrate, and measuring the pollutant concentration in each liquid phase. Thus, by measuring K_{ow} for a compound in the lab and knowing the general relationship between K_{oc} and K_{ow}, you can calculate its corresponding K_{oc} and K_p. An example of such a K_{ow}-K_{oc} relationship observed for one particular set of hydrophobic compounds is $\log K_{oc} = 1.00 K_{ow} - 0.21$. This equation is illustrated in Figure 3.7 (Karickhoff et al., 1979).

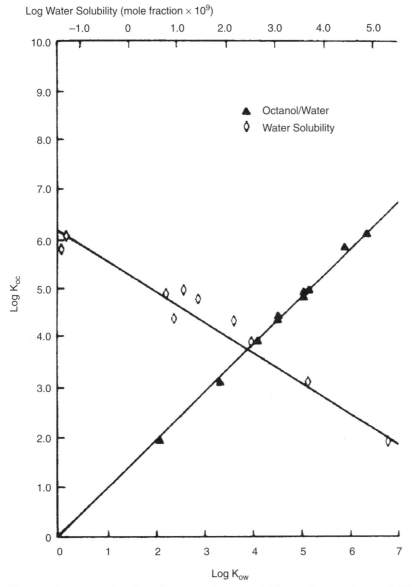

Figure 3.7. K_{oc} as a function of compound water solubility and octanol/water distribution coefficient. [Taken from Karickhoff et al. (1979), Figure 5, p. 247.]

One final note concerning K_d and K_p should be added. In theory, equilibrium coefficients should be independent of suspended solids concentrations. However, numerous investigations for metals and organic pollutants have indicated a strong dependence with respect to solids concentration. In almost every case, K and the concentration of pollutant on the sediment decreased with increasing solids concentration. A summary of these findings can be found in O'Connor and Connolly (1980).

3.4 KINETICS OF THE SORPTION PROCESS

Another factor that must be considered in determining equilibrium coefficients is the time scale required to reach equilibrium. Most studies of inorganic and organic pollutants find that a three-day equilibrium period is sufficient to obtain a constant K_d or K_p. However, this does not mean that true equilibrium concentrations within particle aggregates have been reached. For example, thus far we have limited our discussion to individual particles, but many particles in nature are large aggregates of many smaller inorganic and organic particles. Thus, pollutants must diffuse to the center of the aggregate until a constant concentration is reached (Wu and Gschwend, 1986). Time scales for these diffusion processes range from days to months (Karickhoff, 1984; Karickhoff and Morris, 1985; Coates, 1984). Thus, most environmental particles are in an ever-changing state of sorption equilibrium.

Furthermore, while the sorption process, in terms of measured K_p values, occurs over a three-day period, the desorption process takes much longer. An example is the desorption of a PCB from a sediment suspension shown in Figure 3.8. Three sediment suspensions are shown, and statistical analysis using two kinetic terms, rather than only one, clearly shows that the desorption process of pollutants can be divided into two steps with distinct rates, one rapid and adjacent to the y-axis in Figure 3.8 and one slow that spreads across the entire figure. These distinct steps have been extensively studied and have been observed for a variety of pollutants. Karickhoff and Morris (1985) describe this two-step release in terms of a labile (rapid) and a nonlabile (slow) component. At present, the mechanisms for these two steps are unclear. The labile component may be due to release of the pollutant from the surface of an aggregate or natural organic matter, while the nonlabile component could be accounted for by the slow diffusion of pollutant from the interior of aggregates. This explanation is partially supported by Wu and Gschwend (1986), who proposed a radial diffusion model for the desorption from a sediment aggregate. Also note that the labile desorption rates shown in Figure 3.8 are a function of the suspended solids concentration while the nonlabile rates are not.

The kinetic release of pollutants from sediment suspension can be important sources of pollutants to a system when a sediment is suspended into clear water. This will be important in the lake and stream modeling chapters. Further reading on sorption phenomena and desorption kinetic processes can be found in Coates and Elzerman (1986) and Elzerman and Coates (1987).

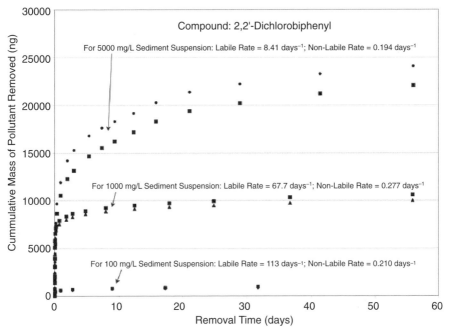

Figure 3.8.   The release of a PCB from sediment suspensions. [Data from Dunnivant (1988).]

3.5   SORPTION ISOTHERMS

3.5.1   A General Approach

No discussion on sorption phenomena would be complete without the derivation of isotherm equations (sorption studies conducted at constant temperature). So, now that we understand the nature of environmental particles and what a distribution/partition coefficient is, we will look a little closer at the mathematical theory of sorption phenomena at surfaces. The processes that promote adsorption/sorption can be divided into three types:

1. Chemical reactions at surfaces such as
   - Surface hydrolysis
   - Surface complexation
   - Surface-ligand exchange
   - Hydrogen bond formation
2. Electrostatic interactions
3. Hydrophobic expulsion

Adsorption/sorption isotherms are determined by equilibrating a system with varying pollutant concentrations (or solids concentrations) and determining the equilibrium solid and liquid phase concentrations for each, as described above. As discussed earlier, K_d and K_p can be calculated from lab experiments; however, a measurement of K at a single dissolved phase concentration is not sufficient to determine whether the K_d or K_p applies at other concentrations. The effect of the dissolved phase concentration on K values is crucial, because pollutant concentrations constantly vary in nature. Therefore, we need to determine adsorption/sorption isotherms in order to establish the general shape of the sorption relationship and to determine the sorbed-to-dissolved pollutant relationship.

Four types of isotherms have been observed and are shown in Figure 3.9. Each of these corresponds to a certain type of adsorption process or set of conditions. Note that three of these are not linear over the entire range of pollutant concentrations tested, indicating that K is not constant at with varying pollutant concentration. The most common type of isotherm encountered for soil and sediment suspensions is the L-type (upper right-hand plot), and an equation of this curve can be easily derived.

The L-type isotherm is referred to as the Langmuir isotherm. Derivation of the equation for this curve requires an assumption that unoccupied adsorption sites (S) on the surface of the adsorbent become occupied by adsorbate (A) in solution, such that

$$S + A \leftrightarrow SA$$

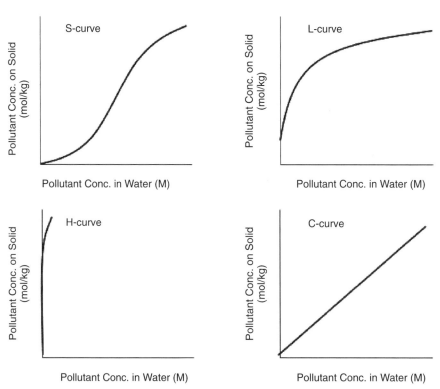

Figure 3.9. General classes of adsorption isotherms.

where SA is the adsorbate on the surface sites. At equilibrium,

$$K_d \text{ or } K_p = \frac{\text{Products}}{\text{Reactants}} = \frac{[SA]}{[S][A]}$$

Therefore, at equilibrium, the maximum concentration of surface sites, S_T, is $[S_T] = [S] + [SA]$.

$$[S] = \frac{[SA]}{K[A]}$$

Solving the equation above for [S] yields and substitution into the previous S_T equation yields

$$[S_T] = [SA] + \frac{[SA]}{K[A]} = [SA]\left(1 + \frac{1}{K[A]}\right)$$

Substituting $[A]K/[A]K$ for 1 and solving for [SA] yields

$$[SA] = [S_T]\frac{K[A]}{1 + K[A]}$$

Now, we need to express the equation in terms of the mass of adsorbent. If we normalize [SA] and $[S_T]$ to the mass of adsorbent where

$$\Gamma_A = \frac{[SA]}{\text{Mass adsorbent}}$$

$$\Gamma_{max} = \frac{[S_T]}{\text{Mass adsorbent}}$$

the above equation becomes

$$\Gamma_A = \Gamma_{max}\frac{K[A]}{1 + K[A]}$$

This equation is presented graphically in Figure 3.10. The first figure shows the general shape of the isotherm (L-type) while the second plots $1/\Gamma_A$ versus $1/[A]$, where A is the notation for the pollutant. Note the slope of the line and Γ_{max} located at the y-intercept.

There are several assumptions in the Langmuir model, including the following:

- Thermodynamic equilibrium exists until monolayer coverage is achieved.
- Adsorption energy is independent of the degree of site coverage (i.e., all surface site have equal energies).
- If more than one adsorbate (pollutant) is present, they do not interact on the surface (as in the case of hydrophobic pollutant partitioning).
- If more than one adsorbate (pollutant) is present and they do interact, they both sorb by the same mechanism.

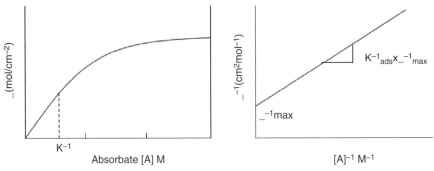

Figure 3.10.   Langmuir adsorption isotherm.

Thus, the following equations allow the Langmuir equation to be applied to systems having multiple adsorbates and adsorbents.

For two adsorbates, A and B, the equation would be

$$\Gamma_A = \frac{\Gamma_{max} K_A [A]}{1 + K_A [A] + K_B [B]}$$

For two adsorbents, 1 and 2, the equation would be

$$\Gamma_A = \frac{\Gamma_1 K_1 [A]}{1 + K_1 [A]} + \frac{\Gamma_2 K_2 [A]}{1 + K_2 [A]}$$

Another common isotherm, which applies to solids with heterogeneous surface properties (i.e., surfaces where all sites do not have equal energies), is the Freundlich isotherm. The Freundlich equation can be represented by

$$\Gamma_A = K [A]^n$$

where $\Gamma_A = [SA]$/mass adsorbent, $[A]$ = concentration of adsorbate in solution, K = the equilibrium constant, and n is a constant. The Freundlich isotherm is represented graphically in Figure 3.11. Unlike the Langmuir isotherm, the Freundlich does not assume monolayer coverage and thus does not limit the maximum amount of pollutant adsorbed or sorbed. This type of isotherm is important in describing partitioning reactions of hydrophobic compounds with NOM. In general, determination of surface coating (sorption or adsorption in moles of pollutant per cm^2 of particle surface) is difficult to measure, so concentration terms (moles of pollutant per mass of sediment) are used in practice. Thus, the experimental results would have the solid-phase concentration on the y-axis and the liquid phase concentration on the x-axis (as seen in Figure 3.9).

While the theoretical approach to Freundlich and Langmuir is attractive, in practice, scientists use experimentally measured K_d and K_p values and assume a linear isotherm such as the ones represented by the linear portion of the L isotherm and the complete plot of the C-type Freundlich isotherm. This is usually appropriate, since linear isotherms hold for relatively low concentrations of pollutants in the environment.

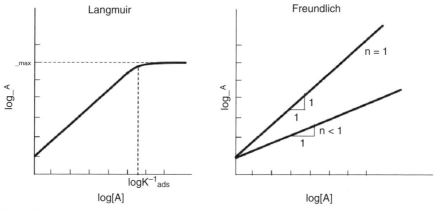

Figure 3.11.   Comparison of Freundlich and Langmuir isotherms.

3.6   KINETICS OF TRANSFORMATION REACTIONS

In Chapters 2 and 3, we have looked at various chemical processes that are important in understanding how chemistry affects fate and transport. These have included acid–base, precipitation, and oxidation–reduction processes and sorption equilibrium. These processes affect the transport but do not necessarily account for any removal of pollutants from the system under study. As noted earlier, most sorption equilibrium reactions are reversible; if they were not, we would not be concerned with these reactions and would treat them as removal reactions. This section deals with (a) transformation and degradation reactions that are responsible for the removal of the original pollutant from the system and (b) the rates of these reactions.

Fortunately for the modelers of environmental systems, most of these transformation/degradation reactions follow first-order kinetics, and when they do not, we can usually make reasonable assumptions that allow us to use pseudo-first-order kinetics. First-order kinetics follows the rate expression

$$\text{Rate} = -\frac{\Delta C}{\Delta t} = kC \tag{3.6}$$

where $\Delta C/\Delta t$ is the change in concentration with change in time, k is the first-order rate constant with units of reciprocal time, and C is concentration (molar, ppm, or any consistent unit).

Upon integration, this equation yields

$$\ln\frac{C_t}{C_0} = -kt \tag{3.7}$$

where ln is the natural log, C_t is the pollutant concentration at any time (t), and C_0 is the initial concentration of pollutant. One of the useful features of first-order reactions is the concept of half-life. The half-life ($t_{1/2}$) is defined as the time when one-half of the initial pollutant concentration has been degraded or removed, and it is related to the rate constant k by

$$\ln 0.5 = -kt_{1/2}$$

or

$$\frac{0.693}{k} = t_{1/2} \tag{3.8}$$

After one half-life, one-half of the pollutant will remain; after two half-lifes, one-fourth of the pollutant will remain; after three half-lifes, one-eighth of the pollutant will remain, and so on. If you do not fully understand or remember first-order reactions, you should review your general chemistry textbook.

As we mentioned, most environmental transformation/degradation reactions are first-order or pseudo-first-order. Let's look a little more closely at pseudo-first-order examples. Many chemical reactions are of the form

$$\text{Rate} = k[\text{pollutant}][\text{some oxidizing or reducing reactant}]$$

The oxidizing or reducing reactant may be NOM, the surface of a mineral phase, a photon of light, or a microbial cell. In general, the concentrations of these oxidizing/reducing reactants are high and relatively constant compared to the low concentrations of pollutant. Thus, we can assume that the concentration of this reactant does not significantly change with time and we can reduce the rate expression to

$$\text{Rate} = k'[\text{pollutant}], \qquad \text{where } k' = k[\text{some oxidizing or reducing reactant}]$$

This approach greatly simplifies the kinetics of the reactions and easily allows the inclusion of kinetics in the fate and transport equations, as shown in the subsequent chapters.

It should be noted that some fate and transport equations rely on other forms of rate constants, including zero- and second-order reactions, but these require different derivations from those given for equations used in this textbook. However, generally there are several transformations/degradations happening at one time, and unless all of the rate expressions describing these degradations are of the same order (zero, first, etc.), they cannot be added together into one rate expression in the fate and transport equation (as discussed in Section 2.8). This gives the use of first-order rate expressions an added advantage, given that most reactions can be adequately described by first-order kinetics.

3.7 PUTTING IT ALL TOGETHER: WHERE CHEMISTRY ENTERS INTO THE MODELING EFFORT

We have covered many important and complicated chemical concepts in Chapters 2 and 3. But where do they all fit into the pollutant fate and transport modeling approach, and when is each important? It depends on the pollutant and the environmental system under study, but we will attempt to summarize the role of chemistry in the modeling effort. We will divide our discussions into a metal pollutant and a hydrophobic (organic) pollutant, and important chemical processes are summarized in Table 3.3. Note that ionizable organic pollutants, such as phenols, will fall in between these two extremes.

Case I: A Metal Pollutant

In our models for aqueous systems, the source of a metal pollutant is largely controlled by K_{sp}, since the dissolved (mobile) phase of the metal cannot exceed its thermodynamically determined solubility (if it does, the metal will precipitate). Vapor pressure and Henry's law constants are of little consequence for metals, which are usually not volatile.

Recall that the purpose of fate and transport modeling is to determine the concentration of pollutant reaching a receptor (human), which in turn serves as the input to risk assessment models (Chapter 10). Although chemical speciation, discussed in Section 3.2.2, is not included in even the most sophisticated fate and transport models, it is considered in risk assessment. Many of the parameters discussed in Chapters 2 and 3 influence the speciation of metals. For example, pH can determine a metal's solubility, speciation, and sorption to mineral surfaces. Metals are more readily available in their hydrated free metal form at low pH values. The adsorption of all metals to particles increases with pH of the solution, due to surface charge availability (thus more dissolved metal is present at low pH values). E_H can greatly affect oxidation states of transition, lanthanide, and actinide metals and therefore change their solubility, speciation, and degree of sorption. Thus, E_H is another important parameter affecting mobility in aquatic systems. As we discussed, increasing ionic strength (salt content) can also have profound effects on metal speciation. A good generalization is that high ionic strength waters exhibit lower toxicity than low ionic strength waters. This is due to the complexation of the toxic free metal ion, generally the most toxic form of the metal, by anions in solution.

Sorption phenomena can greatly influence transport in aquatic systems. Metals adsorbed to mineral surfaces or natural organic matter are less bioavailable than dissolved metal species. Also, adsorbed metals generally share the fate of the particle. In lakes and streams, most particles and the metal ions sorbed to them settle to the bottom of the system and are incorporated into the sediments. Metals adsorbed to dissolved natural organic matter usually stay in the moving water and are transported out of the system under study. Transformation reactions (biological, chemical, and photochemical) are of little consequence to metals, with the exception of radioactive decay for radionuclides. One other rare exception is the methylation of mercury (a biological reaction) that creates a very toxic form of the pollutant.

Case II: Hydrophobic Pollutants

Hydrophobic pollutants represent the other extreme of types of pollutants because these pollutants do not "like" being dissolved in a polar fluid (water). Here, the concentration of pollutants in an aquatic system is controlled by the aqueous solubility and/or Henry's law constant. For atmospheric systems, vapor pressure determines the mass input.

TABLE 3.3. A Summary of Chemical Factors Affecting Fate and Transport for Each Category of Pollutants

Chemical Factor (by section)	Metals	Radionuclides (a Class of Metals)	Ionizable Organics	Hydrophobic Organics
Section 2.4				
pH	Very important	Very important	Potentially important	Usually not important
Solubility	Very important	Very important	Very important	Very important
Vapor pressure	Not important	Not important	Can be important	Can be important
HLC	Not important	Not important	Can be important for uncharged form	Usually important
For inorganics				
Acid–base	Very important	Very important		
Redox	Very important	Very important		
Precipitation	Very important	Very important		
Section 2.5				
Sorption	Very important	Very important	Very important	Very important
Section 2.6 (organic redox transformations)				
Abiotic			Can be important	Can be important
Photochemical			Can be important	Can be important
Biological			Can be important	Can be important

The pH and ionic strength of the system have little to no effect on the fate of hydrophobic pollutants. (Of course, pH does have a large effect on ionizable organics, and the presence of a salt would serve to decrease the solubility of these compounds.) E_H, however, can greatly influence biotic and abiotic degradation reactions. Some pollutants are easily degraded in aerobic environments by microbes, while other pollutants are more easily degraded by biotic and abiotic processes under anaerobic or reducing conditions. Thus, biological and abiotic processes can be very important removal mechanisms for organic pollutants. In the atmosphere and in surface waters, photochemical degradations can also be important.

Sorption phenomena are very important for hydrophobic pollutants, since these pollutants would rather be on any surface than dissolved in a polar solvent like water. Thus, sorption to natural organic matter and mineral surfaces is important. In general, K_p values for hydrophobic pollutants are orders of magnitude greater than K_d values for metal pollutants. As with metal pollutants, sorption phenomena will be important in lake, stream, and groundwater systems.

So, how do we put chemistry into the fate and transport modeling approach? To understand this, we must introduce the concepts of box models and mass balance. In environmental modeling, it is important to define your system, and we do this using boxes. For example, if we are studying the transport of a pollutant in the atmosphere or in a groundwater system, we define the section of the system we are interested in with a box of physical dimensions equal to that of the system under study. Next, we account for all of the pollutant mass entering, reacting in, being retained in, or exiting the system (a mass balance). Of course, this means we use a lot of mathematics, and the second section of this text will deal with the development and use of models to describe the fate and transport of pollutants in lakes, rivers, groundwater, and atmospheric systems.

The basic approach for our mass balance in each of the following chapters will be represented by

Change of mass = sum of	+	sum of internal	− sum of all	− sum of all
in system with	all inputs	sources	outputs	internal sinks
time				
dC/dt	mass of	any source or	mass of	removal from
	pollutant	generation	pollutant	the system
	input	of the pollutant	exiting the	by sorption
		from within	system	or
		the system		degradation
				reactions (3.9)

We have already mentioned most of these terms in this book. For example, the "mass of pollutant input" can be controlled by point and non-point sources, pulse and step inputs, and solubility, vapor pressure, and Henry's law constants. The term "any source or generation of the pollutant from within the system" can be illustrated by desorption from the sediments (K_d or K_p) or by the generation of an atmospheric pollutant by photochemical reactions. The "mass of pollutant exiting the system" can be represented by the outflow from a lake or river or by the specific section of an aquifer or the atmosphere. We will show detailed examples of each of these in the fate and transport chapters.

Finally, let's concentrate on the last term in the mass balance equation, the sum of internal sinks. This is where kinetic transformation/degradation reactions come into the equation. Recall that many of our reactions were found to be or were simplified to be first-order reactions. This makes life much easier for the modelers, since if all of the reactions are of the same rate order (first, in our case), we can add the individual rate constants together and have one overall first-order rate constant. Thus, say we have a pollutant that is biodegraded with a rate constant of 0.05 days^{-1} and is photochemically degraded such that k equals 0.005 days^{-1}. Instead of deriving a much more complicated equation with two kinetic variables, we can simply add the two k values together to obtain an overall rate constant of 0.055 days^{-1} and have one kinetic term in the equation. Any number of first-order rate constants can be added together. This will become clearer in the fate and transport chapters when we give each general transport equation.

A Closer Look: Calculation of a Partition Coefficient from Experimental Data

A K_{oc} is to be determined for the sorption of 2,2'4,4',6,6'-hexachlorobiphenyl (a PCB congener) on a sediment sample. The organic content of the sediment is 2.05%. A 2.05 mg/L solution of 2,2'4,4',6,6'-hexachlorobiphenyl (HCB) is prepared, and ~40 mL of the solution is placed in a vial containing 0.102 g of dry sediment. The final solution volume is 40.0 mL. The sample is mixed for 3 days, the aqueous and sediment phases are separated with a 1.0-μm glass fiber filter, and the aqueous phase is measured for 2,2'4,4',6,6'-hexachlorobiphenyl. A concentration of 0.506 mg/L is measured in the aqueous phase. Analysis of a blank vial (containing water and HCB but no sediment) measures a concentration of 1.88 mg/L of HCB in the dissolved phase. What are the K_p and K_{oc} for the sample?

1	Total mass (mg) of HCB added each flask	0.0820	
2	Mass recovered in blank (mg)	0.0752	
3	Mass of HCB in water phase (mg) of mixture	0.0202	
4	Volume of water (L)	0.0400	
5	Concentration of HCB in water phase (mg/L)		0.505
6	Mass of HCB on solid phase (mg) (determined by difference) (1–3)	0.0550	
7	Mass of solid phase (kg)	1.02×10^{-4}	
8	Concentration of HCB on solid phase (mg/kg)		539
9	K_p		1068
10	K_{oc} (K_p/f_{oc}) (L/kg)		51350

Box 1: First we should determine the mass of HCB in each flask. This is done by multiplying the HCB concentration by the volume of solution: $(2.05 \, \text{mg/L}) \times (0.0400 \, \text{L}) = 0.0820 \, \text{mg}$.

Box 2: Note that not all of the HCB added to each flask was recovered during the analysis of the blank ($1.88 \, \text{mg/L} \times 0.0400 = 0.0752 \, \text{mg}$). This observation is common in laboratory experiments. Thus, we must assume that all of the flasks only have $0.0752 \, \text{mg}$ of recoverable HCB in them. Sources of loss of HCB could include volatilization from the aqueous phase during solution preparation or sorption to the vial walls or vial top.

Box 3: Next the concentration of HCB in the vial containing sediment is calculated: ($0.506 \, \text{mg/L} \times 0.040 \, \text{L} = 0.202 \, \text{mg}$). We will use the concentration in the water to calculate K_p.

Box 4: Next the mass of HCB associated with the sediment is determined by subtracting the mass measured in the dissolved phase from the mass in the blank: $0.0752 - 0.0202 = 0.0550 \, \text{mg}$.

Box 8: Next the concentration of HCB on the sediment is determined: $0.0550 \, \text{mg} \, / \, 1.02 \times 10^{-4} \, \text{kg} = 539 \, \text{mg/kg}$.

Box 9: The K_{oc} is calculated from the ratio of sediment-phase concentration to dissolved-phase concentration: $539 \, \text{mg/kg} \, / \, 0.506 \, \text{mg/L} = 1068 \, \text{L/kg}$.

Box 10: K_p is calculated by multiplying the K_{oc} by the fraction of organic carbon present in the sample: $1068 \, \text{L/kg} \times 0.205 = 219 \, \text{L/kg}$.

A Closer Look: Determination of an Average K_d Based on a Set of K_d Measurements

A more accurate determination of K_d or K_p can be made when the parameter is measured over a range of pollutant concentrations. For example, a set of K_d experiments were conducted to investigate the adsorption of Pb on a soil. It is important to note that all of the vials contained the same mass of sediment. This is important because K_d can be a function of suspended solids concentration (sediment). The experiments were conducted in a manner similar to that described in our other examples. The following results were then compiled:

Aqueous-Phase Concentration (mg/L)	Sediment-Phase Concentration (mg/kg)
0.0500	0.725
0.103	1.42
0.698	10.5
1.50	21.4
3.78	35.5

To solve this problem, we must first plot the data and determine the slope of the line (which equals the K_d). The results are shown in Figure 3.12. Note that if all of the data are used, the plot is a straight line, until it levels off at the highest concentration. This is not uncommon in laboratory experiments when excessively high

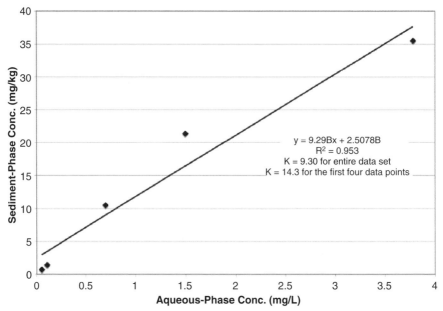

Figure 3.12. Determination of Kd from experimental data.

dissolved metal phase concentrations are used. This phenomenon results primarily from complete coverage of the sorption sites on the sediment and results in an excess concentration in the dissolved phase. When this occurs, it is more accurate to esti-mate K_d from the lower data points. A linear regression of the first four data points in the table yields a slope, and K_d, of 14.3 L/kg.

Exercises

1. Which is a more accurate representation of how toxic a pollutant is, activity or concentration?

2. What is the numeric range of the activity coefficient?

3. Calculate the ionic strength for the following:
   (a) a mixture of 0.050 M $CaCl_2$, 0.025 M NaCl, and 0.045 M KNO_3
   (b) a mixture of 0.097 M $CaCl_2$, 0.015 M KCl, and 0.405 M $NaNO_3$

4. Using your results from exercise 3 and the extended Debye–Hückel equa-tion, calculate the activity of each cation and anion in the mixtures.

5. Acid rain is known to be harmful to the environment for multiple reasons, but few people realize that free Al^{3+} ions are one of its most harmful prod-ucts. Ionic aluminum is toxic to fish at levels as low as 6.2 ppm. Aluminum forms five ligand complexes with OH^- (conplexation constants are given in Table 2.6), but the formation of these is understandably minimized at lower pHs. Draw a speciation plot using a spreadsheet showing the concentrations of each species at pH values ranging from 0–14. Assume that the solution is

in equilibrium with the solid phase, $Al(OH)_3$. K_{sp} values are given in Table 2.5.

6. Mining operations have often been major contributors to groundwater and surface pollution. They introduce toxic metal concentrations as well as increasing the total ion concentration of an aquatic system. An especially troublesome phenomena encountered in mining operations is *acid mine drainage* (AMD). This occurs when groundwater runs over old mining sites or through old mining tunnels. At these sites, many minerals and metals have been exposed from the surrounding rock and are free to react with the water and the atmosphere. Pyrite (FeS_2) is a common mineral found at mine sites that can cause damage to aquatic ecosystems. When pyrite is exposed to air and water, it reacts to form sulfuric acid (H_2SO_4) and iron hydroxide $Fe(OH)_3$. This not only raises the pH of the stream, but also introduces a solid metal hydroxide which can be toxic to fish and aquatic life. However, the pyrite reaction does not always result in solid iron hydroxide. Consider the following complexation reactions, and produce a speciation diagram (using a spreadsheet) for the various iron hydroxide complexes. Determine what the dominant species will be at a pH of 4.5, a common pH found in streams that suffer from AMD.

$$Fe^{3+} + OH^- \Rightarrow FeOH^{2+} \qquad K_1 = 6.31 \times 10^{11}$$
$$Fe^{3+} + 2(OH^-) \Rightarrow Fe(OH)_2^+ \qquad \beta_2 = 2.00 \times 10^{22}$$
$$Fe(OH)_{3(S)} \Rightarrow Fe^{3+} + 3(OH^-) \quad K_{sp} = 1.10 \times 10^{-36}$$
$$Fe^{3+} + 4(OH^-) \Rightarrow Fe(OH)_4^- \qquad \beta_4 = 2.51 \times 10^{34}$$

7. You are curious about the speciation of Ni^{2+} in a solution with OH^- ions. Excess $Ni(OH)_2$ salt is added to a beaker with distilled water. Your goal is to determine the predominant species of Ni^{2+} and OH^- complexes based on the OH^- concentration. Since there is excess salt in the solution, $Ni(OH)_2$ controls the main species present and its concentration by the K_{sp} of the salt (5.48×10^{-16}). Draw a speciation plot using OH^- concentrations from 1.00×10^{-18} to $1.00 \times 10^{-8}\,M$ and the following log K values for binding between the Ni^{2+} and OH^- ions:

For $NiOH^-$:   4.1
For $Ni(OH)_2$:   9.0
For $Ni(OH)_3^-$: 12.0
For $Ni(OH)_4^{2-}$: 17.2

8. Silver bromide is used in photography as a developing emulsion. Given the K_{sp} values (Table 2.5) and stability constants (Table 2.6) for Br^- ligands, determine how the concentration of each Ag species in solution varies as the Br^- concentration changes. Construct a graph of the $\log[Br^-]$ from -1.0 to 6.0 versus $\log[\text{Ag species}]$ from 0.0 to -35. Include a line for total [Ag] as $[Br^-]$ varies.

9. Cobalt is usually emitted during the production of steel and other alloys, and specifically in the production of airline engines and gas turbines. Ionic cobalt can be carcinogenic, but the amount of consumption must be very high in order for cancer to occur. Create a speciation diagram using a spreadsheet for a cobalt hydroxide system without any solid phase present (Case II in this chapter). The total concentration of Co^{2+} species is 20 ppm. The hydroxide concentrations should range from 1×10^{-1} to 1×10^{-14}.

Helpful equations

$$Co^{2+} + OH \Rightarrow CoOH \qquad K_1 = 5.01187 \times 10^{-5}$$
$$Co^{2+} + 2OH \Rightarrow Co(OH)_2 \qquad \beta_2 = 6.30957 \times 10^{-10}$$
$$Co^{2+} + 3OH \Rightarrow Co(OH)_3 \qquad \beta_3 = 3.16228 \times 10^{-11}$$

10. Mercury sulfate was tested in a renegade agricultural operation as a mildew inhibitor. A few acres were dusted with $HgSO_4$, and then irrigation water and rainfall washed the compound into a nearby holding basin. The EPA caught wind of the illegal activity and sent you in to investigate. Since there is no solid phase present, you will be given the K and beta values for Hg and SO_4^{2-} ligands. Your assignment is to show how the concentration of each Hg species changes as the SO_4^{2-} concentration is varied. Create a labeled graph of the log[Hg] from 0.0 to −30.0 versus log[SO_4^{2-}] from −10.0 to 10.0. Be sure to include how the total [Hg] changes overall.

11. An aspiring restaurateur purchases a cheap plot of land to open a restaurant, complete with a less than pristine pond. In an effort to make the restaurant more scenic, she valiantly undertakes the task of cleaning up garbage in the pond. In doing so, she discovers a 55-gallon drum labeled "Barium Waste." She then hires you, a contractor, to determine what has been chemically disposed of in the pond. You find that the concentration of dissolved barium in the pond is 10.5 ppm, with no solid phase present in the sediments. Using the equations in this chapter, determine the concentration of Ba^{2+} in the water as a function of CO_3^{2-} concentration. Use CO_3^{2-} concentrations of 0.000100 M, 0.001 M, 0.100 M, 1.00 M, and 5.00 M.

12. Ethyenediaminetetraacetic acid (EDTA) is a widely used chemical in industry and is very good at complexing metal ions. We can use EDTA as a complexing agent for mercury, which lowers the mercury's toxicity but at the same time increases its mobility by keeping it in solution. Draw a speciation diagram for a closed system (Case II) of 10 ppm Hg as a function of EDTA concentration. Use a log[EDTA] concentration range from 5 to −35.

$$Hg^{2+} + EDTA^{2+} \Rightarrow HgEDTA \qquad K_1 = 3.16 \times 10^{23}$$
$$Hg^{2+} + EDTA^- \Rightarrow HgEDTA^+ \qquad \beta_1 = 1.00 \times 10^{27}$$

13. Cadmium metal ions form three complexes with SO_4^{2-} in the absence of a solid. Draw a speciation plot of Cd^{2+} in the presence of SO_4^{2-} ions using a Cd^{2+} concentration of 40.5 ppm and a concentration of SO_4^{2-} ranging from 1.00×10^{-6} to 1.0 M. The following binding log K values will be useful:

For $CdSO_4$:   2.3
For $Cd(SO_4)_2^{2-}$:   3.2
For $Cd(SO_4)_3^{4-}$:   2.7

14. The leachate from a local landfill has been suspected of containing high cadmium concentrations, and the city downstream from the watershed has begun monitoring the streams for cadmium. In order to better predict transport, they need to determine the distribution coefficient for cadmium between the stream water and the local soil. Two liters of sample water were collected, filtered, and dried. The dried filtered particulate matter weighed 10.0 g and was used to determine the total suspended solids (5000 mg/L). Sediment samples of the local soil (250 mg) were prepared in ~50-mL sample bottles to create the same TSS as the local streams. Cadmium (0.375 mg) was added to each of the sample vials, including the blanks, and water was added for final solution volumes of 50.0 mL. The solutions were mixed for three days, and then they were filtered with 0.20-μm filters and analyzed by flame atomic absorption spectrometry. The equilibrium cadmium concentration in of the aqueous phase was 5.00 mg/L. Calculate the K_d for Cd^{2+} on the soil.

15. Aldrin is a chlorinated pesticide that was used to regulate termite populations until the 1970s. After application, this non-biodegradable pesticide found its way into freshwater systems, poisoning organisms. A sample of lake water sediment was taken to determine how well Aldrin adsorbs onto solid sediment. For this experiment, 0.1147 mg of Aldrin was added to a flask containing 100.0 mL of water and 4.26×10^{-4} kg of lake sediment. Through gas chromatographic–electron capture detection (GC-ECD) the equilibrium concentration of Aldrin in the aqueous phase was determined to be 0.0180 ppm. Determine the mass of Aldrin in the aqueous and solid phases and calculate the K_p value for the pollutant Aldrin.

Total mass (mg) of pollutant added to flask	0.1147
Mass of pollutant in aqueous phase (mg)	
Volume of water (mL)	100.0
Concentration of pollutant measured in	0.0180
aqueous phase (mg/L)	
Mass of pollutant on solid phase (mg)	
Mass of solid phase (kg)	4.26×10^{-4}
Concentration of pollutant on solid phase (mg/kg)	
K_p	

16. Kepone (chlordecone), a carcinogenic, tan to white crystalline solid or powder that is insoluble in water, was used as an insecticide, fungicide, and

larvacide on bananas, tobacco, and other domestic plants. The U.S. EPA banned the use of Kepone in 1975, but the chemical is still used in some countries. Before the termination of its production in the United States, large amounts of Kepone were dumped into the upper James River. This Kepone poses a threat to the fish and other marine animals, as well as to the ground-water supply. Most of the Kepone in the James River has settled into the sand and sediments at the bottom. A K_p needs to be determined for the sorption of Kepone on a sediment sample. A solution of 0.02453 mg/L Kepone is prepared. 100.0 mL of the solution is placed in a vial with 0.000100 kg of dry sediment. The sample is mixed for three days, and the aqueous and solid phases are separated using a 0.20-μm glass fiber filter. The mass of Kepone in the aqueous phase is measured with gas chromatography. From this, the concentration of Kepone in the aqueous phase is determined to be 0.02108 mg/L. Analysis of the blank shows no loss of kepone from absorption onto the vial wall. Find the K_p for the sample using the following chart.

1	Total mass (mg) of kepone added to each flask	
2	Mass of kepone in water phase (mg) of mixture	
3	Volume (L) of water	
4	Concentration of kepone in water phase (mg/L)	
5	Mass of kepone on solid phase (mg)	
6	Mass of sediment in vial (kg)	
7	Concentration of kepone on solid phase (mg/kg)	
8	K_p	

17. The EPA just received a report of a small-scale dumping site of lead-acid batteries in a city under your jurisdiction of investigation. The site contains various car and tractor batteries located on a downhill slope near a stream. It is your job to determine the extent of the contamination of lead pollution in the stream, accounting for the concentration in the aqueous phase as well as the sorption to the solid particles in the stream—in other words, you need to determine the K_d (distribution coefficient) of lead. To determine the K_d, you add 0.100 g clay to 0.1000 L water and then add 0.04976 mg Pb^{2+} to the sample. After allowing the samples to equilibrate for a minimum of 3 days, you analyze them on a Flame Atomic Absorption Spectroscopy system and find that there is 0.500 mg/L Pb in the aqueous phase of each sample. You also run a blank and find that the amount of Pb lost to sorption in the test tubes is 0.01000 mg/L. Use these values to complete the table and determine the K_d of Pb in the samples.

Total Pb in sample (mg)	0.04976
Volume water in sample (L)	0.1000
[Pb] aqueous phase of sample (mg/L)	0.5000
[Pb] lost in blank (mg/L)	0.01000
Mass Pb in aqueous phase (mg)	
Solid-phase mass (mass of clay) (g)	0.1000
Mass of Pb in solid phase (mg)	
[Pb] solid phase (mg/kg)	
K_d	

18. We have discussed equilibrium in terms of K_d and K_p, and we have discussed kinetics in terms of first-order sorption and desorption rates. Explain how each can be important in determining the fate and transport of pollutants in natural water systems.

19. Select a metal pollutant. Using the information from Chapters 2 and 3, explain *all* chemical processes that can influence its fate and transport in an aqueous system. Which processes will increase the transport? Which processes will decrease the transport? Limit your discussion to three typed pages. A good place to start is to outline the important processes in each chapter.

20. Select a hydrophobic pollutant. Using the information from Chapters 2 and 3, explain *all* chemical processes that can influence its fate and transport in an aqueous system. Which processes will increase the transport? Which processes will decrease the transport? Limit your discussion to three typed pages. A good place to start is to outline the important processes in each chapter.

REFERENCES

Bell, J. and T. H. Bates. Distribution coefficients of radionuclides between soils and groundwater and their dependence on various test parameters. *Sci. Total Environ.* **69**, 297–317 (1988).

Coates, J. T. *Sorption Equilibria and Kinetics for Selected Polychlorinated Biphenyls on River Sediments*, Ph.D. Dissertation, Environmental Systems Engineering, Clemson University, Clemson, SC, 1984.

Coates, J. T. and A. W. Elzerman. Desorption kinetics for selected PCB congeners from river sediments. *J. Contam. Hydrol.* **1**, 191–210 (1986).

DeLaune, R. D., C. N. Reddy, and W. H. Patrick, Jr. Effects of pH and redox potential on concentration of dissolved nutrients in an estuarine sediment. *J. Environ. Qual.* **10**(3), 276–279 (1981).

Dunnivant, F. M. *Congener-Specific PCB Chemical and Physical Parameters for Evaluation of Environmental Weathering of Aroclors*, Ph.D. Dissertation, Environmental Systems Engineering, Clemson University, Clemson, SC, 1988.

Elzerman, A. W. and J. T. Coates. Hydrophobic organic compounds on sediments: equilibria and kinetics of sorption. In: *Sources and Fates of Aquatic Pollutants*, Hites, R. A. and S. J. Eisenreich, eds., ACS Advances in Chemistry Series #216, American Chemical Society, Washington, DC, 1987.

Handbook of Chemistry and Physics, 61st edition, Weast, R. C. and M. J. Astle, eds., p. B-242, 1980. Boca Raton, FL.

Harris, D. C. *Quantitative Chemical Analysis*, 5th edition, W. H. Freeman, Table 8–1, p. 180, 1999. Selected data [original source of data: J. Keilland, *J. Am. Chem. Soc.* **59**, 1675 (1939)].

Hem, J. D. Study and interpretation of the chemical characteristics of natural waters. U.S. Geological Survey Water Supply Paper 2254, 1985.

Karickhoff, S. W., D. S. Brown, and T. A. Scott. Sorption of hydrophobic pollutants on natural sediments, *Water Res.* **13**, 241–248 (1979).

Karickhoff, S. W. Organic pollutant sorption in aquatic systems. *J. Hydraulic Eng.* **110**(6), 707–735 (1984).

Karickhoff, S. W. and K. W. Morris. Sorption dynamics of hydrophobic pollutants in sediment suspensions. *Environ. Tox. Chem.* **4**, 467–479 (1985).

Lindsay, W. L. *Chemical Equilibria in Soils*, John Wiley & Sons, New York, p. 107, 1979.

Morel, F. M. M. and J. G. Hering. *Principles and Applications of Aquatic Chemistry*, John Wiley & Sons, New York, 1993.

O'Connor, D. J. and J. P. Connolly. The effect of concentration of absorbing solids on the partition coefficient. *Water Res.* **14**, 1517–1523 (1980).

Sposito, G. *The Surface Chemistry of Soils*, Oxford University Press, New York, 1984.

Stumm, W. and J. J. Morgan. *Aquatic Chemistry: An Introduction Emphasizing Chemical Equilibria in Natural Waters*, 2nd edition, John Wiley & Sons, New York, 1981.

Stumm, W. and J. J. Morgan. *Aquatic Chemistry: Chemical Equilibria and Rates in Natural Waters*, 3rd edition, John Wiley & Sons, New York, 1996.

Wu, S. and P. M. Gschwend. Sorption kinetics of hydrophobic organic compounds to natural sediments and soils. *Environ. Sci. Technol.* **20**, 1213–1217 (1986).

PART III

MODELING

"[Mathematics] The Handmaiden of the Sciences."

—Eric Temple Bell

"For every problem, there is one solution which is simple, neat and wrong."

—Henry Louis Mencken

A Basic Introduction to Pollutant Fate and Transport, By Dunnivant and Anders
Copyright © 2006 by John Wiley & Sons, Inc.

CHAPTER 4

AN OVERVIEW OF POLLUTANT FATE AND TRANSPORT MODELING

Models are often used to represent more complicated or larger systems. Perhaps the most common model we use is a road map. Scientists and engineers use models to help understand how something in the past happened or to predict what will happen in the future. These models can be very simple or very complicated, depending on the system being imitated or the accuracy desired in the calculations. In this chapter, we will develop some simple mathematical models to introduce the scientific modeling approach. Then we will attempt to explain how fate and transport models are developed, starting with the relatively simple models presented in later chapters and ending with the modeling approaches used by professionals to study very complicated environmental systems. Finally we will look at how good our modeling approaches are and what we do with the final numbers produced by these models.

4.1 MODELING APPROACHES

4.1.1 Algebraic Solutions

A linear model is the simplest and easiest to understand, since we tend to think in a linear manner. Figure 4.1 shows a common plot used in chemistry to calibrate an instrument. Instruments that measure pollutant concentrations need daily calibration, accomplished by analyzing solutions with known pollutant concentrations. The instrument responds to these different known concentration levels with a proportional signal. These outputs are created by the instrument's interaction with the sample, and they are expressed in units such as milllivolts, transmission (a function of absorbance), peak height, and peak area. We use these data to make a calibration plot such as the one shown in Figure 4.1. Thus we derive a linear relationship ($y = 10.0x + 0.05$) between instrument response and concentration, and we can then use the instrument to measure a solution containing a pollutant. If the instrument produces a response of 65 units (as shown on the y-axis) for a sample, we can trace

A Basic Introduction to Pollutant Fate and Transport, By Dunnivant and Anders
Copyright © 2006 by John Wiley & Sons, Inc.

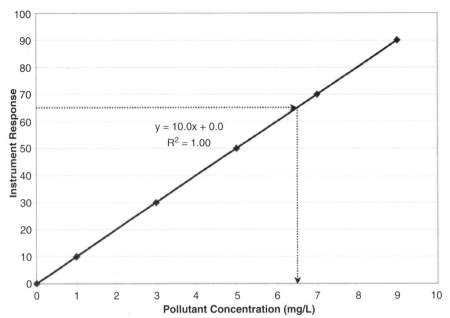

Figure 4.1. A linear calibration model.

the response over to the calibration line and then down to the concentration line and determine the concentration of pollutant in the sample (6.49 mg/L in this example). This is typically how instrumental measurements are made.

This concept can be extended to environmental systems, where we look at flow rate and travel time in a system. If the water or wind flow rate is 2.00 m/sec, then we can calculate that it will take 10 sec to travel 20.0 m. Linear relationships are easy to understand, but unfortunately many environmental processes are not linear. As we discussed in the chemical kinetics section in Chapter 2 and 3, many processes are first-order with respect to reaction rates (a nonlinear relationship). Figures 4.2*a* and 4.2*b* show two common exponential plots that follow the first-order kinetic model. When the concentration values are transformed using a natural log function, a linear plot of ln(C) versus time is obtained. However, there is no simple way to take the log transformation in our relatively more complicated modeling equations and linearize our results. We must learn to identify and interpret log functions and plots; these functions operate much like the linear model from Figure 4.1, by providing an equation that can be used to predict concentrations for the nonlinear models.

4.1.2 Modeling Using Differential Equations

The models used to predict the concentrations of pollutants in environmental media fall into a class of equations referred to as differential equations, which fall into a

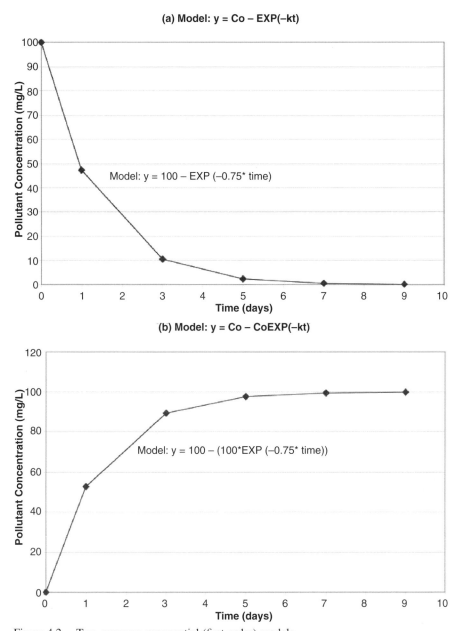

Figure 4.2. Two common exponential (first-order) models.

special class of calculus that we will discuss in a moment. First, it is important that we show the complete development of one fate and transport equation, so that you understand the entire process. To do this we will use the lake pulse and step scenarios, which are the simplest of the systems that we will cover in this textbook. It is not critical that you understand every step, especially if you have not taken calculus, or specifically differential equations. Our goal is to show the overall process in the development of a governing fate and transport equation (like the equations used in Fate®).

The first step in developing the governing equations for the fate of a pollutant in a lake system is to set up a mass balance for the system. First, quantify all of the mass inputs of pollutant to the system. This can be expressed as

$$W = Q_w C_w + Q_i C_i + Q_{\text{trib}} C_{\text{trib}} + P A_s C_p + V C_s$$

where W is the mass input of pollutant to the lake per unit time (kg/time), Q_w is the inflow rate of the wastewater (m³/time), C_w is the pollutant concentration in the wastewater (kg/m³), Q_i is the inflow rate of the main river (m³/time), C_i is the pollutant concentration in the main inlet river (kg/m³), Q_{trib} is the net inflow rate from all other tributaries (m³/time), C_{trib} is the net pollutant concentration of in the tributaries (kg/m³), P is annual precipitation (m/time), A_s is mean lake surface area (m²), C_p is the net pollutant concentration in precipitation (kg/m³), V is the average lake volume (m³), and C_s is the average pollutant release from suspended lake sediments (kg/m³ · time). In most situations, the mass inputs from the smaller tributaries and precipitation are minor compared to the major input source, and the terms C_{trib} and C_p are ignored. We will further simplify the mass input expression here by assuming that the contribution from contaminated sediments is negligible, although this may not always be the case. These assumptions simplify the input expression to

$$W = Q_w C_w + Q_i C_i$$

Next, in order to set up a mass balance for the pollutant across the entire system, we need to incorporate outflow into the expression. The equation for the mass outflow from the system is analogous to the input equation, but with only one outlet considered generally. We also assume that there are no additional sources of pollutant and that chemical degradation yields a constant chemical removal rate for the pollutant. Thus the mass balance becomes

$$\text{Change in mass} = \text{Inflow} - \text{Outflow} + \text{Sources} - \text{Sinks} \tag{4.1}$$

$$V\,dC = (Q_w C_w\,dt + Q_i C_i\,dt) - Q_e C\,dt + 0 - VCk\,dt, \quad \text{or}$$
$$V\,dC = W\,dt - Q_e C\,dt - VCk\,dt \tag{4.2}$$

where dC or ΔC = the change in pollutant concentration in the lake, dt or Δt is the incremental change in time, Q_e is the outlet or effluent flow from the lake (m³/time), Q_w is the waste flow to the lake (m³/time), Q_i is the inlet or influent flow from the lake (m³/time), C is the average lake concentration (kg/m³), and k is the first-order removal rate for the pollutant, 1/time. Note that since we are looking at changes in concentration and time, we must use an operator to indicate this. The operator is the d or Δ symbol.

Equation (4.2), upon substitution and rearrangement, yields

$$Q_e C - W(t) + V\,dC = -VCk\,dt \tag{4.3}$$

Upon rearrangement, this reduces to

$$V\frac{dC}{dt}+(Q_e+kV)C=W(t) \tag{4.4}$$

where Q_e, k, and V of the lake, and thus the quantity (Q_e+kV), are assumed to be constant (usually, they are reasonably constant).

The average detention time (t_0) of water (and thus the pollutant) in the lake, from the time of input to outlet, is defined as

$$t_0=\frac{V}{Q_e} \tag{4.5}$$

Substitution and further rearrangement into the previous equation yields

$$V\frac{dC}{dt}+CV\left(\frac{1}{t_0}+k\right)=W \tag{4.6}$$

Equation (4.6) is a first-order linear differential equation (indicated by the dC/dt, called the derivative of concentration C with respect to time t, and representing the rate of change of concentration with respect to time). It expresses how concentration of pollutant in the lake changes with time in response to the lake volume and the rates of flow and degradation of the pollutant. The technique used for solving this equation, for the function $C(t)$, depends on the nature of the input source (instantaneous or pulse versus continuous or step).

Integration for the Instantaneous (Pulse) Pollutant Input Model.

When the mass input with time from all sources, $W(t)$, is zero, we approach what is referred to as an instantaneous input. In this case, an instantaneous input is characterized as a one-time, finite (known) addition of pollutant mass to the lake. Thus, there is no pollutant added over time, and $W(t)$ is zero. For example, the release of a pollutant by a marine shipping accident would be an instantaneous input, as would a short release from an industry located on the lake. In order to solve this equation for $C(t)$, we must integrate Eq. (4.6). Under these conditions, integration, using a Laplace transformation technique with $W=0$, yields

$$C(t)=C_0 e^{-\left(\frac{Q_c}{V}+k\right)t}, \quad \text{or} \quad C(t)=C_0 e^{-\left(\frac{1}{t_0}+k\right)t} \tag{4.7}$$

where C_0 is the initial pollutant concentration. This equation, specifically in its second form above, would be used to simulate the pollutant concentration versus time in a lake where an instantaneous release occurred. Before we attempt to explain the Laplace integration technique that got us to this equation, we will first look at the purpose of integration.

But what is the purpose of deriving this equation? When you take a mathematical equation involving a nonnegative function and integrate it, you are often developing a way of calculating the area under the curve or function that the equation represents. A plot of the integrated Eq. (4.7) is shown in Figure 4.3. The trace of the curve is an exact solution to the governing equation and represents the pollutant concentration at a given time. The area under the curve is the mass balance

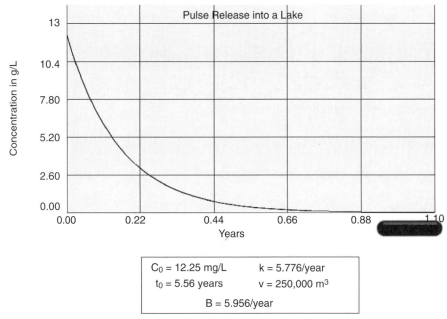

Figure 4.3. Output from Fate for an instantaneous input of pollution to a lake system.

of the system and represents the total mass of pollutant leaving the lake system as a function of time.

So what is a Laplace transformation? As we noted, this is a form of integration, allowing us to take Eq. (4.6) and find a source function for it (an equation that defines the curve or line for a plot of concentration as a function of time). Differential equations used in fate and transport modeling are too complicated to solve using normal integration techniques learned in calculus. In integrations using Laplace transformations, the original equation is first simplified by algebraic substitutions to make it simpler—hence the term *transformation*. Next the resulting simplified equation is integrated using normal calculus techniques. Finally, a reverse transformation is taken on the equation, in which the simplification is reversed and a more complicated form of the equation results, but the entire equation has now been integrated. This technique yielded the governing equation [Eq. (4.7)] used to make Figure 4.3.

Integration for the Continuous (Step) Pollutant Input Model. We now will return to Eq. (4.6) to derive an equation describing the constant release of a pollutant into a lake. This type of release is known as a step input, and an example would be the constant release from an industrial source (constant mass per time). Under these conditions, $W(t)$ is not zero (as assumed in the previous derivation) and normally there is some background concentration of pollutant in the lake system (such that C_0 in the lake cannot be considered to be zero). Here, the net pollutant concentration in the lake (and the water leaving the lake in the effluent river) is the result

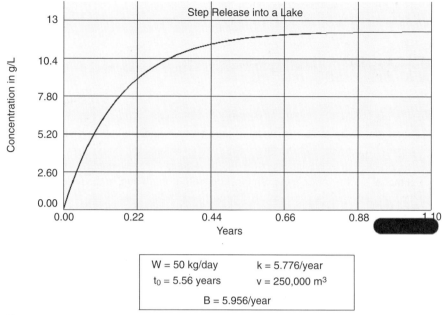

Figure 4.4.   Output from Fate for a step input of pollution to a lake system.

of two opposing forces: (1) the concentration decreases caused by "flushing" of the lake via the effluent river and by first-order pollutant decay and (2) the pollutant concentration increases due to the constant input from the source. If the waste load is constant (which we will assume), integration of Eq. (4.6) (again using a Laplace transformation) yields

$$C(t)=\frac{W}{\beta V}\left(1-e^{-\beta_t}\right)+C_0 e^{-\beta_t} \tag{4.8}$$

where $\beta = 1/(t_0 + k)$ and C_0 is the background concentration of pollutant in the lake. If the background concentration in the lake is negligible, the equation reduces to

$$C(t)=\frac{W}{\beta V}\left(1-e^{\beta_t}\right) \tag{4.9}$$

Equations (4.8) and (4.9) can be used to estimate the concentration of a pollutant in a lake that receives a constant input of pollutant. A plot of this last equation is shown in Figure 4.4, where the line represents the pollutant concentration at a given time in the lake and the area under the plot represents the total mass of pollutant leaving in the outflow.

Equation (4.7) is distinguished from Eq. (4.8) and (4.9) by a distinct boundary condition, which we place on the system to better define it. In the first example, the pulse input, the boundary condition was that the input pollutant concentration from the inlet river, as well as the overall input of pollution with time, was zero.

This simplifies the equation as noted earlier. In the second example, pollution entered the lake at a constant input.

4.1.3 The General Approach for the Models Used in this Text

Differential equation techniques were used to derive all of the governing equations given in the subsequent modeling chapters, which will express changing pollutant concentration with time in rivers, lakes, and groundwater and atmospheric systems. It is beyond the scope and intention of this introductory textbook to show the derivation of these equations. Our goal is to show the use of these equations and learn which model parameters affect pollutant concentrations. Thus, in the following chapters we will only give the governing equation, and this section was intended to give you a brief insight into how these equations are obtained.

But how realistic is our differential equation approach to modeling? For simple systems where you are only looking at a basic understanding, these approaches are fine. However, industry, the government, and the public demand a much more involved (and sometimes more accurate) representation of the environmental system being modeled. Thus, more complicated methods have been developed, and these are the ones used by professionals today. However, you should note that these are based on the same chemical and physical processes and related equations that we used in the examples given above. The difference is that whereas we will hold parameters such as water or air velocity, partition or distribution coefficients, and kinetic degradation rates constant in the governing equations used in the following chapters, professionals use modeling efforts that can allow these parameters to vary with location and time in the system. The technique used by professionals, numerical methods of analysis, is the subject of the next section.

4.1.4 Numerical Methods of Analysis

Unlike in the differential equations approach, where we found a solution to Eq. (4.6), in numerical methods no governing equation is sought. Here we obtain our solution to the problem by simply calculating concentrations across the system subject to our same boundary conditions and/or based on a few known concentrations at defined points in the system. Two common numerical methods of analysis are finite element and finite difference approaches. We will discuss each separately.

First we will discuss the finite difference approach. We begin by placing a grid over the system under study, such as the one shown in Figure 4.5. Each corner point of the squares in the grid, called a nodal point, is represented by an x and y coordinate, and can have a unique equation for calculation of pollutant concentration. The surface area represented by the square is defined to have uniform properties such as water or air flow, water or air velocity, mixing, pollutant distribution coefficient, and so on, but different nodal points can have different values of these parameters. This allows a more realistic mathematical representation of real environmental systems. The goal of the numerical analysis is to calculate pollutant concentrations for each node, located at the center of four nodal points (shown in the upper right-hand nodal

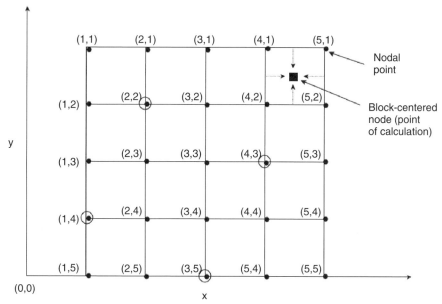

Figure 4.5.   Illustration of a finite difference approach for estimating pollutant concentrations.

cell). It should be noted that the representation in Figure 4.5 is a gross simplification and that the modeling of an actual environmental system could have hundreds to thousands of nodal points composing the grid.

We will return to our lake system example for the following discussion, but now we do not assume that mixing is constant across the lake, as was necessary in the differential equations solution. Water can enter the lake at point (1, 1) and leave at point (5, 5). The grid is overlaid onto a map of the lake, and the appropriate boundary conditions are applied. For simplification purposes, we will assume that our lake is a square corresponding to our nodal points.

Say we want to predict the pollutant concentration at the circled points, nodal points (1, 4), (2, 2), (3, 5), and (4, 3). The approach used in differential equations, in which we integrated and obtained a general solution equation, will not work here, since mixing is different at each node. To solve the equation using finite differences, we use Eq. (4.6) directly. Since each nodal cell can have a unique version of Eq. (4.6) to account for a different magnitude of mixing, we will calculate the concentration of each node (middle of each nodal cell). This is an iterative process governed by one or more boundary conditions. In this case, our boundary condition is the known total mass of pollutant that entered the system. After the computer running the finite difference method completes one set of calculations of each node, it can add up the mass in the system and see if it matches the known total input mass. If it does not, the method makes adjustments to the equations (in our case mixing) and recalculates the entire grid. It repeats this process until an acceptable mass balance

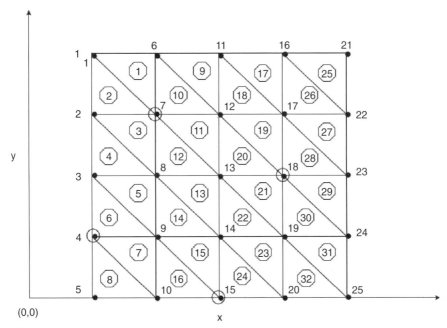

Figure 4.6.   Illustration of a finite element approach for estimating pollutant concentrations.

is achieved. Completion of this process will yield pollutant concentrations for each node in the system.

The finite element approach differs from finite difference in that, in finite element, interpolation functions are used to define the concentration throughout the domain of each triangle instead of at one point, as in the finite difference method. Figure 4.6 represents a finite element grid, similar to the finite difference grid used in Figure 4.5. A triangular grid is used in Figure 4.6, but square or rectangular grids are also possible. Again, Eq. (4.6) is iterated over the grid until an acceptable mass balance is obtained. Such finite element techniques are more commonly used than the finite difference approach in pollutant fate and transport modeling.

It should be noted that all modeling techniques should be calibrated with field-measured pollutant concentrations. A conceptual comparison of the differential equations and numerical methods approaches is shown in Figure 4.7. As noted, if the model results are inconsistent with the field measurements, adjustments to the underlying model may be necessary to successfully model a system.

4.2   THE QUALITY OF MODELING RESULTS

So, how good are typical modeling results? As usual, this depends on a number of factors, such as how realistically your model mimics the system under study, how

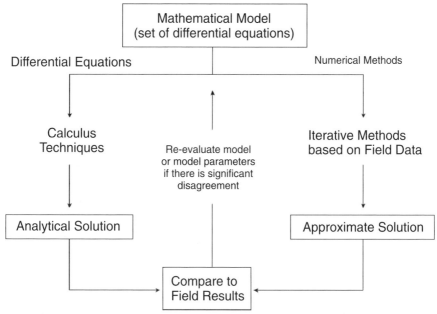

Figure 4.7. A conceptual comparison of the differential equations and numerical methods approaches.

far past your calibration you are attempting to predict, and how long into the future you are extrapolating. Most importantly, it depends on who is asking and answering the question. For example, some modelers will defend their predictions to the death, since they can defend the physics behind the mathematical models. However, it is important to distinguish reality (the environmental system) from the model (a mathematical equation). It is not uncommon to hear a modeler state that the field results are wrong because "they do not match my model results!" One should be careful when dealing with avid modelers.

One of the authors of this textbook (Dunnivant) once gave a presentation of a major field-monitoring project to a group of modelers. The system under study was an uncharacterized, highly fractured, unsaturated groundwater system where we were attempting to gather data to estimate site-specific dispersion (mixing) estimates for a large hazardous waste site located nearby. During the question and answer session that followed the presentation, one modeler asked why we wasted so much money conducting an experiment and measuring pollutant concentrations and mixing since an accomplished modeler could have easily written a model to predict the results. The answer, my colleague, is model validation/calibration. No model is of any value unless based on or verified with experimental data.

Occam's razor is a principle that states that the most simple answer is usually the best. Unfortunately, this rarely applies to equations that model environmental systems. By their very nature, environmental systems are highly complicated, and the accuracy of our experimental and modeling efforts to characterize dispersion in these systems is weak at best. Thus, models need to be as complicated as the system being modeled. Model predictions (explanatory modeling from Chapter 1) should always be compared to experimental data, and the overlay of these two data sets should show consistency. For predictive modeling, we unfortunately have no way to judge accuracy, except to wait and let time evaluate our efforts. In each of the closing sections of the following modeling chapters, we will discuss the limitations of our modeling equations. This will hopefully prepare the reader to know where to ask questions of the modeler and how to judge their modeling results.

4.3 WHAT DO YOU DO WITH YOUR MODELING RESULTS?

As noted in the Chapter 1, the goal of pollutant identification, laboratory analysis of field samples, and fate and transport modeling is to provide data to be used in risk assessment and exposure analysis. Otherwise why would we do all of this work? Risk assessment (and economics) drive virtually every hazardous waste remediation (Superfund) effort today. The predictions used in these efforts rely on models to apply observed chemical and physical principles to the sites under evaluation, in order to predict future concentrations of pollutants at various sites in the system. Chapter 10 of this textbook will focus on risk assessment.

REFERENCES

Burden, R. L. and J. D. Faires. *Numerical Analysis*, 5th edition, PWS-Kent Publishing Company, Boston, 1993.

Ross, S. L. *Introduction to Ordinary Differential Equations*, 3rd edition, John Wiley & Sons, New York, 1980.

Van Nostrand's Scientific Encyclopedia, 5th edition, D. M. Considine, ed; Van Nostrand, New York, 1976.

Wang, H. F. and M. P. Anderson. *Introduction to Groundwater Modeling: Finite Difference and Finite Element Methods*, W. H. Freeman, San Francisco, 1982.

CHAPTER 5

FATE AND TRANSPORT CONCEPTS FOR LAKE SYSTEMS

CASE STUDY: LAKE ONONDAGA

Lake Onondaga is located north of the city of Syracuse in New York State. The lake has a surface area of 11.9 square kilometers (4.6 square miles) and receives water from a drainage basin of 648 square kilometers (248 square miles). In the late 1880s and early 1900s, Lake Onondaga was a prized recreation resort for the citizens of Syracuse, but as the city grew and became more industrialized, the lake became more and more polluted. From the time Syracuse was founded, the city directly dumped raw sewage into the lake. In 1884, the Solvay Process Company began the production of soda ash and released high concentrations of salt (chloride, sodium, and calcium) directly into the lake. Swimming was banned in the lake in 1940, primarily due to health issues relating to the sewage. In 1946, Allied-Signal Corp. began chlorine production using the mercury cell process and directly discharged the mercuric waste into the lake. In 1970, fishing was banned due to mercury pollution. As a result of these uncontrolled pollutions events, steps were taken to slowly improve the quality of water in the lake. The city of Syracuse slowly upgraded its sewage treatment facilities by installing primary treatment in 1925, constructing the METRO wastewater facility in 1960, and upgrading the METRO to secondary and tertiary treatment in 1979. In 1977, Allied-Signal closed a chlorinated benzene plant and one chlorine production facility. In 1986, the soda ash manufacturing operation was closed. In 1995, Lake Onondaga was added to the Federal Superfund National Priority List. In more recent years, remediation efforts have been planned but little direct action has been taken, other than limiting current pollutant inputs to the lake, conducting studies, and planning for the future. For example, the 1990 Onondaga Lake Management Conference initiated research and planned remediation projects. In 1992 the Corps of Engineers completed the Lake Onondaga Water Technical Report, outlining possible lake remediation alternatives. In 1994, aquatic habitat restoration projects began. In 1996, Allied-Signal started a Remedial Investigation and Feasibility Study for the lake. Even with all of these remediation-planning efforts, Lake Onondaga is still considered by most environmentalists the most contaminated lake in the United States.

What are the results of the previous pollution emission, and what can be done to remediate the lake? The primary pollutants include (1) sewage waste consisting of nutrients such as phosphorus, ammonia, nitrite, and many harmful bacteria, (2) turbidity in the lake resulting from excess microbial growth due to the abundant nutrients present in the lake, (3) high salinity from the soda ash production, and (4) mercury from the chlorine production facilities.

Given adequate time and sufficient flow through the lake, sewage emissions can be removed from a lake system. In low flow systems such as Lake Onondaga, nutrients are mostly recycled in the lake during summer stratification and fall over-turn (something we will learn about in this chapter). Some nutrients are removed very slowly by burial in the lake sediments. However, more soluble nutrients such as ammonia, nitrite, and nitrate are difficult to remove by any process.

The largest health hazard comes from the mercury released into the lake. It has been estimated that 165,000 pounds of mercury were released to the lake between 1946 and 1970. Since metals do not degrade, all of the mercury is still present in the lake and will remain in the lake unless direct and expensive remedial actions are taken (such as dredging or natural burial by sedimentation, which will be discussed in this chapter). You should recall our discussions on bioconcentration from Chapter 2. This is a common process in lake systems, and mercury that was buried in the lake sediments of Lake Onondaga has undergone methylation by microbes and has been bioconcentrated in many fish species. This led to a ban on fishing in and eating fish from the lake.

A variety of cleanup efforts are underway for Lake Onondaga, and water quality has significantly improved since the 1970s. Catch and release fishing was reinstated in 1986. A consortium of agencies and public interest groups have agreed on eight goals for Lake Onondaga. These include

1. Development of a eutrophication model for the Seneca River
2. Development of a lake productivity model
3. Development of a hydrodynamic model for the lake outlet
4. Funding of studies on the release of nutrients and toxic substances from lake sediments under changing dissolved oxygen levels
5. Establishment of a long-term baseline water quality program
6. Drafting of an urban/suburban nonpoint source pollution plan
7. Drafting of a fish and wildlife management plan
8. Development of a demonstration project of manipulated littoral zone habitat structures; the project indicated that fencing and wave breaks could significantly increase plant survival, growth, and diversity and that these habitats also increase survival of young fish.

In this chapter we will learn physical and chemical processes that explain the fate and transport of pollutants in lake systems such as Lake Onondaga. First, we will look at the formation, geological history, and seasonal history of lakes. Then we will focus on transport, by conceptually and mathematically describing mixing processes and chemical reactions specific to lakes. We will look at two basic models for predicting the transport of pollutants in lakes, based on pulse and step inputs. Finally, we will look at some of the ways to remediate a lake after a contamination event has occurred, and most importantly we will learn the limitations of what we can actually do to return the lake to pristine conditions.

Information on Lake Onondaga was obtained from the Onondaga Lake Partnership website (http://www.onlakepartners.org).

5.1 INTRODUCTION

Inland surface waters cover slightly less than 2% of the Earth's surface and only account for less than 1% of the terrestrial fresh water in the world. While this may seem a small percentage, lakes are common features in temperate and subarctic regions of the Northern Hemisphere. Lakes are a highly used source of recreation and food, and the quality of water in these systems is extremely important to local populations and economies. As we saw in the bioconcentration example in Table 2.5, an extremely small concentration of pollutant in water can result in significant and health-threatening pollutant concentrations in the biota living in, on, or near a polluted water body.

In this chapter we will look at types of lakes and how they are formed, pollutant input sources specific to surface water bodies, the chemical nature of lakes as a function of seasons and how this affects pollutant transformation reactions, the pulse and step fate and transport models for lakes, the limitations of our relatively simple models, and ways to remediate contaminated lakes.

5.2 TYPES OF LAKES AND LAKE-FORMING EVENTS

Lakes come in a large variety of shapes, sizes, and depths. Of course the most prominent ones are those that appear as large blue shapes on national and global maps, and although these lakes contain the largest percentage of fresh surface water on Earth, they account for only a small percentage of total number of lakes. For example, Lake Baikal in Siberia contains approximately 20% of the Earth's fresh surface water, while another 12% is contained in Lake Superior in North America. An illustration and comparison of the surface area of the largest lakes of the world is shown in Figure 5.1. A summary of the major lakes of the world, with respect to location, surface area, length, and depth, is given in Table 5.1. Table 5.2 contains volume data of the Great Lakes of North America.

While the large lakes in Figure 5.1 are impressive, they are not representative of lakes in general and due to their size require special fate and transport modeling approaches (numerical methods of analysis in Chapter 4). We are more concerned with smaller lakes, which are more common and more widely distributed. To put the size of lakes into perspective, refer to Figure 5.2a, which shows the global distribution of lakes versus surface areas. Lake Nicaragua in North America (one of the smaller lakes located on the lower left-hand side of Figure 5.1) has a surface area of 8030 km². This area in the lower right-hand corner of Figure 5.2a is repre-

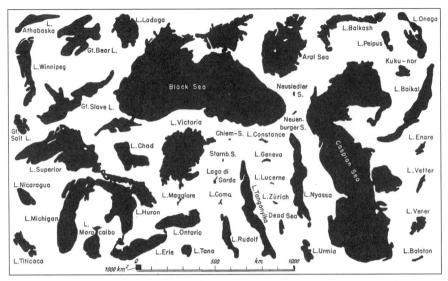

Figure 5.1. Approximate comparison of the surface area of the world's largest lakes. (*Note*: The Black Sea is basically an ocean body, since it is directly connected to the ocean.) [From Ruttner (1963). Reprinted with permission from the University of Toronto Press.]

sented by fewer than 50 lakes worldwide. In contrast, look at more common lake sizes, for example, $0.1–1.0 km^2$. Figure 5.2 indicates that there are between one million and ten million lakes in this size range. Similar observations can be made using Figure 5.2*b* for the depth of lakes.

Lakes are formed through a variety of geologic events, but mostly through glacial, volcanic, and tectonic activity. Other lake-forming events include landslides, dissolution of limestone bedrock, and human-made reservoirs. A summary and description of these formation events is given in Table 5.3. From a fate and transport modeling standpoint, land features around a lake and physical features of the lake are important. For example, lakes located in flat regions are subject to frequent mixing by the wind. Important physical features include surface area and depth. With the exception of glacial lakes, lakes in flat regions tend to be shallow and are easily mixed. Moderate and deep lakes thermally stratify very easily, which inhibits mixing during summer months.

The last category in Table 5.3, human-made reservoirs, is a point of contention among lake and river scientists. Which category do these reservoirs fit into—lakes or rivers? Human-made reservoirs tend to have a more longitudinal flow direction, like a river. However, large surface areas and deeper depths make these systems more similar to lakes. In addition, the increased depths of these lake systems set up conditions that lead to thermal stratification (discussed later in this chapter), which is definitely specific to lakes. Modeling of human-made reservoirs requires special considerations, as we will discuss later in this chapter.

TABLE 5.1.   Major Natural Lakes of the World (*The World Almanac and Book of Facts*, 2004)

Continent	Lake Name	Area (square miles)	Length (miles)	Maximum Depth (feet)
Africa	Albert	2,075	100	168
	Chad	839	175	24
	Nyasa (Malawi)	11,150	360	2,280
	Tanganyika	12,700	420	4,823
	Turkana (Rudolf)	2,473	154	240
	Victoria	26,828	250	270
Asia	Aral Sea[a]	13,000	260	220
	Baikal	12,162	395	5,315
	Balkhash	7,115	376	85
	Issyk Kul[a]	2,355	115	2,303
	Tonle Sap	2,500	—	45
	Urmia	1,815	90	49
Asia-Europe	Caspian Sea[a]	143,244	760	3,363
Australia	Eyre[a]	3,600	90	4
	Gairdner[a]	1,840	90	—
	Torrens[a]	2,230	130	—
Europe	Ladoga	6,835	124	738
	Onega	3,710	145	328
	Vanern	2,156	91	328
North America	Athabasca	3,064	208	407
	Erie	9,910	241	210
	Great Bear	12,096	192	1,463
	Great Slave	11,031	298	2,015
	Huron	23,000	206	750
	Manitoba	1,799	140	12
	Michigan	22,300	307	923
	Nettilling	2,140	67	—
	Nicaragua	3,100	102	230
	Nipigon	1,872	72	540
	Ontario	7,340	193	802
	Superior	31,700	350	1,330
	Reindeer	2,568	143	720
	Winnipeg	9,417	266	60
	Winnipegosis	2,075	141	38
South America	Maracaibo	5,217	133	115
	Titicaca	3,200	122	922

[a] Saltwater lake.

TABLE 5.2.   Water Volumes in The Great Lakes of North America (*The World Almanac and Book of Facts*, 2004)

Lake	Volume (km³)
Erie	484
Huron	3,543
Michigan	4,918
Ontario	1,638
Superior	12,234

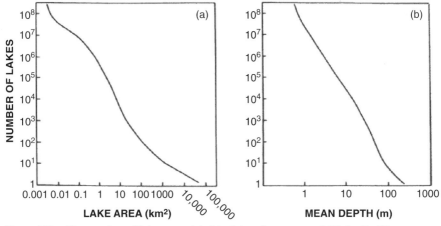

Figure 5.2.   The number of lakes correlated to (*a*) surface area and (*b*) depth. [From Wetzel (1990). Reprinted with permission.]

5.3   INPUT SOURCES

Input sources of pollutants to lakes are similar to those discussed in Chapter 1. We will be concerned with point and non-point sources and use pulse and step inputs to develop our fate and transport models. Common point inputs include industrial sites and feedlot sources, as well as effluent from domestic sewage plants. Non-point sources can include runoff from farming practices and agricultural settings. Underground non-point sources can include leachate from domestic and hazardous waste landfills or storage tanks. The physical characteristics of lakes can add unique input sources. Most lakes that are used for recreational activities experience pollution by outboard motors, which are notorious for releasing petroleum-related compounds. Although these compounds are not highly soluble, some of these compounds dissolve into the water and are spread throughout a lake. This is a non-point, but continuous, source of pollutants.

TABLE 5.3. Forces that Form Lakes

Type of Formation	Description
Tectonic basins	These lakes are depressions formed by the movement of deeper portions of the Earth's crust.
Fault lake	A process where depressions occur between the bases of a single fault displacement or in void formed by the dropping of one block (in the middle) or by the uplifting of two blocks on either side of a middle block. An example is Lake Tanganyika in Africa.
Graben lake	This similar to faulting, but a long trough is downfaulted. An excellent example is Lake Baikal in Asia.
Uplifting of marine sea floors	This isolates a portion of the sea but uplifting a portion of the sea floor and creating an area of trapped salt water. This uplifting usually results in a mountain range between the enclosed water body and the oceanic area. Examples include the Caspian Sea and the Sea of Aral.
Upwarping	This is similar to uplifting but only minor uplifting of the Earth's crust occurs (no mountain ranges are formed). This gentle uplifting processs can create a lake basin in the middle of uplifted areas. Examples include the Lake Okeechobee in Florida (United States) and Lake Victoria in Africa. Upwarping contributed to some of the formation of the Great Lakes in North America with the glacial ice sheets melted and released pressure on the Earth's crust.
Volcanic activity	These lakes occur in the craters or calderas of old volcanoes. An example is Crater Lake in Oregon (United States). Also, lava flows can dam rivers or streams and form lakes.
Landslides	Lakes can be formed by landslides that dam a stream. However, these lakes tend to be short-lived since the stream will eventually break through the earthern dam and form a new stream channel.
Glacial activity	Glaciers are one of the most important lake forming processes. Lake sizes formed by glacier activity range from very small kettle lakes to some of the largest lakes in the world, such as the Great Lakes in North America.
Glacial ice-scour lakes	These lakes are formed by ice moving over relatively flat rock surfaces that are jointed and contain fractures. In mountains where amphitheater-like formations (cirques) are present, glacier action results in cirque lakes (a lake at the end of the cirque). When the glacier action is along coastal regions, fjord lakes are formed in narrow, deep basins. These are common in areas of Norway and western Canada. Glacial action, in the form of retreating large ice sheets, is most known for the formation of large lakes such the Great Slave Lake and Great Bear Lake in Canada and the Great Lakes of the St. Lawrence drainage in North America.
Kettle lakes	Retreating glaciers can also deposit large pieces of ice in the glacial sediments that later melt to form small kettle lakes.
Morainal damming	Glaciers deposit large amount of sediment, or morain, along the sides and at the terminus. These morians can dam the valleys they are deposited in and dam adjacent valleys.

(Continued)

TABLE 5.3. Forces that Form Lakes (continued)

Type of Formation	Description
Lakes formed by river activity	The erosive power of some rivers is considerable and can create lakes along the course of the river. These types of lakes include plunge-pool lakes below waterfalls. Floodplain lakes form in low-lying areas adjacent to rivers, levee lakes, and oxbow lakes are formed when a river changes course and cuts off a meander.
Solution lakes	These lakes are formed in areas of the world containing extensive deposites of limestone ($CaCO_3$). Water slowly dissolves large pockets at or below the land surface and, over time, a lake is formed. Solution pockets (caves) formed below the surface can result in a collapse of the roof and form an exposed lake. Karst formations commonly have solution lakes. Solution lakes are found in the Adriatic, the Balkan Peninsula, the Alps of Central Europe, and in Michigan, Indiana, Kentucky, Tennessee, and Florida in the United States.
Human-made reservoirs	Humans have created lakes through the damming of rivers for thousands of years (beavers have done this much longer). The size of human-made reservoirs range from small ponds to large reservoirs approaching the surface area of the Great Lakes of North America.

Another unique and important pollutant source for lakes, due to their large surface area and relatively long residence times, is aerial inputs. Aerial application of pesticides to adjacent farmland can result in significant inputs of pollutants. These applications are usually treated as non-point seasonal inputs. Long-range transport of atmospheric pollutants can also result in inputs to lakes. In areas of the world where certain pesticides were never used or have been banned, scientists still detect inputs of these compounds to lakes. Such an example can be found for the large lakes in the United States. DDT was banned in the early 1970s, but surface lake waters and atmospheric samples above these waters still show measurable concentrations of these pollutants. Studies have found that long-range atmospheric transport of pesticides from Central America can reach the Great Lakes and remote alpine lakes in North America. This type of input can be treated as a non-point continuous input.

One last source of pollutants to lakes can be from within the lake itself. As we discussed in Chapter 2, particles in the water attract many inorganic and organic pollutants, and most of these particles aggregate and settle in quiescent regions of a lake. However, if these sediments are resuspended into relatively unpolluted lake water, desorption of the pollutants to the water can occur. Resuspension events of importance in most lakes include bioturbation (the mixing of lake sediments with water from biota in the lake), violent storm events, and dredging of harbors and shipping channels. Bioturbation is a constant process and would be treated as a continuous non-point input, while the latter two events would be treated as pulse non-point inputs.

The low degree of mixing in some lake systems combined with a relatively low input of atmospheric oxygen sets up another unique condition in lakes: eutrophication. In general, the limiting nutrient in freshwater aquatic systems is phosphate, but large inputs of any nutrient may result in the uncontrolled growth of algae in lake systems. The input of carbon substrates, from sources such as farm runoff and domestic sewage, often results in uncontrolled growth. During the daylight hours, large algal blooms are not a direct threat to a lake, but during respiration at night algae consume dissolved oxygen (DO) in the lake water. This creates a large oxygen demand on the system, usually one that cannot be met by the limited DO in lake water (maximum of ~12 mg DO/L) or by relatively slow diffusion of oxygen from the atmosphere and low mixing in lake systems. This sets up a condition known as *eutrophication*, and, as we will see in the next section, this can cause serious problems during months when a lake is thermally and chemically stratified.

5.4   STRATIFICATION OF LAKE SYSTEMS

Most lakes undergo some form of thermal stratification during the year. Stratification is a process in which differential heating or cooling occurs and two "climates" are set up in the lake. If a lake undergoes only one stratification event during the yearly cycle, it is referred to as a *monomictic* lake. The most common stratification scenario is illustrated in Figure 5.3*b*. Figure 5.3*a* shows the unstratified lake during

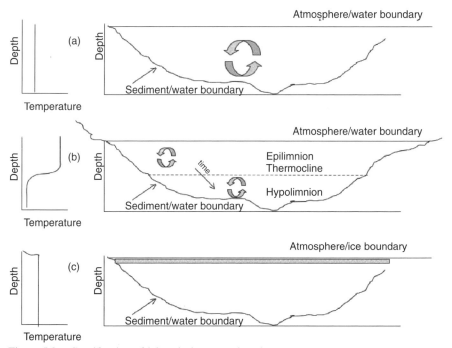

Figure 5.3.   Stratification of lakes during annual cycles.

the fall, winter, and spring, when the temperature is nearly uniform regardless of depth, due to a high degree of mixing (assuming that the lake surface does not freeze in winter). However, if we take a snapshot in time during the middle of summer, we would commonly find the temperature profile illustrated in Figure 5.3*b*. Here, a warm body of water, heated by the sun, develops at the surface, and a cooler, denser body of water resides in the bottom of the lake and is cooled by the Earth. The lake is divided into three zones by temperature, and thus density, differences that prevent mixing: the eplimnion (surface water), the thermocline (the area where a rapid change in temperature occurs), and the hypolimnion (the bottom of the lake). A representative temperature profile is shown to the left of each lake diagram.

Some lakes also freeze in winter, which sets up a different temperature profile (refer to Figure 5.3*c*). Here, the less dense ice floats on the lake surface, and although the water beneath the ice is not actively mixed, it is usually assumed to be consistent with respect to temperature and chemical concentrations with depth. If a lake freezes in winter and stratifies in summers, it has two thermal stratification events and is termed *dimictic*.

Let's return to the concept of summer stratification, since it can be one of the most important cycles in a lake. How does the stratification occur? During early summer, the sun heats the surface of the lake and a thin warm body of water forms. A strong wind from a weather frontal system can disrupt this initial stratification, but if the weather remains stable for a few days or weeks, the sun continues to heat the water and the depth of the warm water (epilimnion) increases and becomes a stable entity of the lake (refer to Figure 5.4*a*). As the summer heating continues, the epilimnion continues to increase in depth in the lake as illustrated in Figure 5.4*b*. Then, during autumn, the epilimnion, which contains the majority of the water in most lakes, cools to temperatures similar to that of the hypolimnion, and the wind associated with a passing weather front is sufficient to cause the two water bodies to mix. This mixing is referred to as the *fall overturn*.

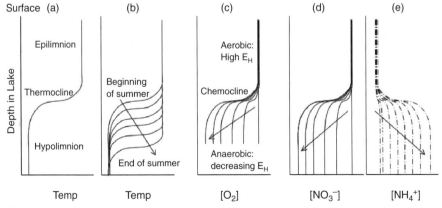

Figure 5.4.   Illustration of increase in depth of the thermocline and chemocline as summer heating progresses.

During summer stratification, if sufficient biodegradable organic matter is present in the hypolimnion, microorganisms will consume all of the dissolved oxygen in these waters. This oxygen cannot be replenished, due to the hypolimnion's isolation from the atmosphere. Thus, a chemocline will be present, associated with the thermocline. In most cases, the system will turn anoxic (devoid of oxygen) due to microbial activity, and a set of interesting biochemical reactions will occur that directly affects the chemistry of the lake water. A chemocline is illustrated in Figure 5.4c. The nitrate–ammonia transition shown in Figures 5.4d and 5.4e is an example of the chemistry that can occur in the chemocline. We discussed the sequence of reactions that occur under these conditions in Figure 2.12, and you should review this figure. Note that the terminal electron acceptor (nutrient) must be present in order for the E_H to be buffered (poised) at each level in Figure 2.12. Nitrate, sulfate, and carbon dioxide are common constitutants of lake water, so these redox reactions will almost certainly be present, and the $\Delta E^\circ_{(water)}$ associated with each half-reaction is shown in Table 5.4. Recall from our discussions in Chapter 2 concerning Figure 2.12 that a carbon substrate must simultaneously be oxidized when the terminal electron acceptor is reduced. Figure 5.5 shows the final equilibrium profiles of each redox couple after the system has become highly reducing (a low E_H value).

TABLE 5.4. $\Delta E^\circ_{(water)}$ **Values for Several Half-Reactions**

Balanced Half-Reaction	$\Delta E^\circ_{(water)}$ (V)
$O_2 + 4H^+ + 4e^- \Rightarrow 2H_2O$	+0.81
$NO_3^- + 10H^+ + 8e^- \Rightarrow NH_4^+ + 3H_2O$	+0.36
$SO_4^{2-} + 9H^+ + 8e^- \Rightarrow HS^- + 4H_2O$	−0.22
$CO_2 + 8H^+ + 8e^- \Rightarrow CH_4 + 2H_2O$	−0.25

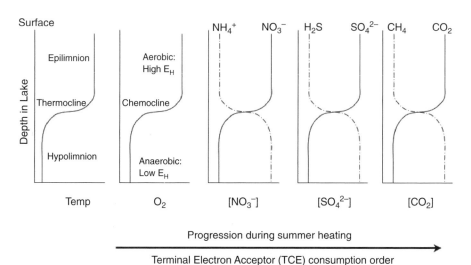

Figure 5.5. Terminal electron acceptor profiles.

So, what does this have to do with fate and transport modeling in lakes? First, if we are concerned with the release of a pollutant, either a pulse or step input, in a stratified lake, we are no longer modeling the total volume of the lake, since the hypolimnion and epilimnion do not mix during summer. If we did use the total volume of the lake, we would grossly underestimate the pollutant concentration in the lake water, since the pollutant emissions are now present in a smaller volume. We can account for this in our models by knowing the epilimnion–hypolimnion volumes. Second, the fate of pollutants trapped in the reduced (anoxic) water of the hypolimnion requires separate consideration. In the surface waters, pollutants can be oxidized by microorganisms and photochemical reactions, but in the hypolimnion many other types of transformation reactions occur for inorganic and organic pollutants under the anoxic conditions. Pollutants may be present in the hypolimnion from a previous release to the water column, or they may be released slowly from contaminated lake sediments. For example, substituted nitroaromatics (organic wastes from chemical and munitions industries) can be reduced to their respective anilines by biotic and abiotic pathways under anoxic conditions. Inorganic mercury, already an important toxin, can be biologically methylated to methylmercury, a much more toxic compound, in the hypolimnion. And remember, in the fall, the water from the hypolimnion will be mixed with surface water, re-exposing biota to these toxins. Fall overturn is usually a renewing experience for the lake, since many nutrients are returned to the waters from the sediments; however, depending on the pollutants and transformation reactions that can occur during the summer, it may also be a deadly time.

5.5   IMPORTANT FACTORS IN THE MODELING OF LAKES: CONCEPTUAL MODEL DEVELOPMENT

5.5.1   Definitions of Terms:

In order to describe a lake system mathematically, we must first make a list of variables (mathematical symbols) for several terms:

V is the volume of the lake (m^3).

Q_i is the inlet flow from the main inlet to the lake (m^3/yr).

Q_e is the outlet, or effluent, flow rate from the lake (m^3/yr) (We usually assume that Q_i is equal to Q_e and we represent both by simply Q).

C_i is the average pollutant concentration in the inlet to the lake (kg/m^3) (this value is zero in many cases).

C_e is the average pollutant concentration in the lake (kg/m^3) and the concentration in the effluent from the lake.

k is the first-order rate constant for removal of pollutant from the lake ($year^{-1}$).

W is the total mass flux of pollutant in the lake, which is equal to the sum ($Q_iC_i + Q_eC_e$).

Note the units used in each of the terms and note that they are compatible with each other. This is important in using the simulator package Fate®. These terms are used to develop the mass balance of pollutant in the lake and develop equations for the individual components of the mass balance (inflow, outflow, sources, and sinks of pollutant).

5.5.2 Detention Times and Effective Mixing Volumes

In this section we will develop a more conceptual, rather than mathematical, derivation of the chemical and physical processes important in the governing equation for the fate and transport of pollutants in lake systems. For a slightly more mathematical approach, refer to the background section of the lake module in Fate®.

In lake systems, it is useful to know or estimate how long water will stay in the system, since this provides an estimate of the minimum time the pollutant will stay in the system. This parameter is called the detention or retention time, and this brings us to the weakest assumption of commonly used lake models. In order to keep the mathematics relatively simple, we must assume that the lake is completely mixed with respect to pollutant concentrations. In some cases (small to medium-sized lakes), this is a valid assumption, but for others (large lakes) it is a weak assumption at best. When we assume that the lake is completely mixed, we can estimate the hydraulic detention time (t_0) by

$$t_0 = V/Q \tag{5.1}$$

which is expressed in years, in accordance with the units specified in Section 5.5.1.

Mixing in lakes, an applied form of entropy, is one of the most difficult parameters to estimate. The predominate mixing force is wind blowing across the surface of the lake and is commonly referred to as wind-driven advection (mixing due to the movement of water). The exact extent of mixing can be determined by costly and long-term monitoring projects. One extreme approach would be to release a known mass of dye at the inlet of the lake. Usually a fluorescent dye is used to enable detection of extremely small concentrations. A few European studies have used the radioactive tracer tritium. Of course, this will not work for very large lakes, since the large volume in these systems will dilute the dye to nondetectable concentrations and large lakes can have detention times of decades. After the dye has been placed into the lake, the effluent stream of the lake is monitored with respect to the dye concentration. If the slow increase and subsequent decrease in dye concentration is analyzed, the effective mixing volume can be calculated. Thus, the effective mixing volume is the volume of water actually mixing with the pollutant as opposed to the entire volume of the lake. Of course, if the lake is large the monitoring dye program could take months to years, or even decades, to complete. Note that this process is further complicated when stratification of the lake occurs. So, the dye technique for determining mixing is only of use in small ponds and lakes. Usually historical data or "experience" is used to estimate effective mixing volumes for larger lakes.

When the effective mixing volume is determined or estimated—for example, 78% of the total volume—this value can easily be used in place of V in Eq. (5.1) to calculate a more accurate estimate of the detention time of water and pollutant in the system. This approach can also be used for stratified lakes, where the depth of the hypolimnion can be measured and the volume of the water body receiving the pollutant can be calculated.

5.5.3   Chemical Reactions

In Chapter 2, a variety of potential degradation schemes were presented, including photochemical, biological, abiotic (chemical), and nuclear reactions. All of these are possible transformation reactions in lake systems. Whatever the type or types of reaction(s), all of these are usually represented by first-order kinetics, and we can add the individual rate constants (k values) together to obtain one overall first-order rate constant. This component of the fate and transport model is of the form

$$C_t = C_0 e^{(-kt)} \tag{5.2}$$

where C_t is the pollutant concentration at time t, C_0 is the initial pollutant concentration, e is the exponential function, k is the first-order rate constant, and t is time.

5.5.4   Sedimentation

In addition to washout of pollutants in the effluent from lakes, along with biological and chemical degradation, pollutants can be removed from a lake system by sorption to particles followed by subsequent settling to the lake bottom. This can be a significant removal mechanism for some pollutants, especially those that do not readily degrade through microbial or chemical means. In order to appreciate how pollutants can thus be removed from the water column of lakes, we will first look at the size of particles that can be present in aqueous systems. Table 5.5 shows the particle settling velocity as a function of particle size. As the particle size decreases, the surface area of the particle increases, and sorption processes become more important since more pollutant can sorb to the surface. In addition, smaller particles can contain more organic matter on the surface and be even more sorptive reactive. Hence, clay-sized particles can be very important in determining the fate of sorbed pollutants, and as you can see from the data in Table 5.5, they have the smallest settling velocities. This results in the particles and sorbed pollutants settling in the deepest and most quiescent (calmest) regions of the lake. This particle-settling veloc-

TABLE 5.5.   Particle Settling Velocity as a Function of Particle Size (Lapple, 1961)

Classification	Particle Diameter Range (μm)	Settling Velocity in Water (cm/sec)
Clay	<2	10^{-8} to 2×10^{-4}
Silt	2.0–20	2×10^{-4} to 2×10^{-2}
Fine sand	20–200	2×10^{-2} to 2
Coarse sand	200–2000 (0.2–2 mm)	2–20
Gravel	>2000 (2 mm)	>20

ity has been validated by monitoring results from lakes that find the highest concentration of polluted sediments in the deepest regions of lakes (referred to as pollutant focusing). In contrast, regions of higher energy flow and thorough mixing in lakes contain larger particles, which generally do not contain high levels of organic matter and therefore do not contain high levels of pollutants.

Settling velocities (ω) given in Table 5.5 were obtained by a relatively simple calculation, defined as Stokes' law:

$$\omega = \frac{(2/9)g(\rho_s/\rho_f - 1)r^2}{\eta} \tag{5.3}$$

where g is the acceleration due to gravity (length/time^2), ρ_s is the density of the spherical particle (mass/length^3), ρ_f is the density of the fluid (mass/length^3), r is the spherical particle radius (length), and η is kinematic viscosity of the fluid (length^2/time). The kinematic viscosity is the ratio of the dyanamic viscosity of a the fluid to the density of the fluid. Note that Eq. (5.3) assumes a spherical particle, but average particle radius can be used.

While Eq. (5.3) describes the settling of a particle, it is of little use, since pollutant concentration is not present in this equation. You should recall from Chapters 2 and 3 that sorption behavior of a pollutant is described by the distribution coefficient (K_d) for metals and the partition coefficient (K_p) for hydrophobic pollutants. Thus, we need an expression that incorporates particle removal and pollutant concentration:

$$r_A = -\frac{K_d \omega S}{H(K_d S + 1)} C \tag{5.4}$$

where r_A is the rate of decrease in pollutant A concentration per unit volume of water (mass/length^3-time), K_d is the distribution coefficient (or K_p is the partition coefficient), ω is the particle settling velocity (length/time), S is the suspended solids concentration (mass/length^3), H is the water depth (length), and C is the pollutant concentration in the water (mass/length^3).

Thus, we can account for pollutant removal by sedimentation in calm water. Waters with rapid currents that mix the water and suspended material will result in slower settling rates. Also note the units of the settling rate constant, concentration per time. These are the units of a zero-order, rather than first-order, rate expression, whereas our fate and transport models will use first-order expressions. Thus, unfortunately, the rate of pollutant removal cannot be directly substituted into our fate and transport models; still, we can estimate the removal of pollutant through sedimentation. Also note that this is assumed to be a steady-state process, since the concentration of pollutant and the suspended solids concentration in the water are assumed to be relatively constant. This is usually a reasonable assumption, except during storm events.

It is also important to note the rate of sediment accumulation in the bottom of a lake, since the settling sediment can bury previously contaminated sediment. Baker (1994) reports a range from 50 to 600 g of sediment per square meter per year for a variety of lake systems. This translates into an accumulation rate from millimeters to centimeters of sediment per year, which can result in significant deposits of sediment. While sediment accumulation and burial of contaminated sediment is important, it is also important to look at sediment resuspension rates. Sediment can be naturally resuspended through bioturbation (the mixing of sediment by bottom feeding fish and organisms living in the sediment) and by storm events. Wetzel (2001, p. 635) reports resuspension rates from 0.5 to 21 g/m^2 · day, which are very significant.

Considering these sedimentation rates, it is understandable that contaminated sediments can be buried and therefore removed from the aquatic system. In a sense, burial of sediments is a form of natural remediation. An example of this is shown in Figure 5.6 for chromium. This sediment profile is from Upper Mystic Lake, which is the water basin for the Aberjona Watershed, north of Boston, Massachusetts. As you can see from the profile, chrome used in the local tanning industry started in ~1900 and declined after 1925. The spike in chromium in the sediment deposited around 1959 is from rendering operations that utilized chrome-tanned hides in the

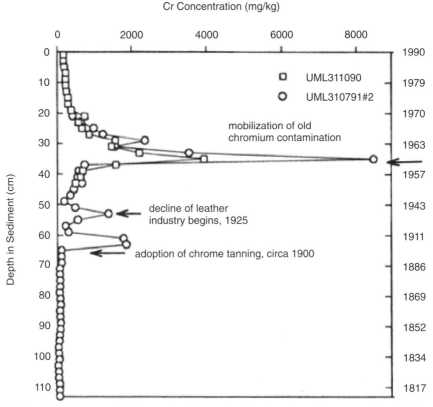

Figure 5.6. An example of natural sedimentation burying chromium-contaminated sediment. [From Spliefhoff and Hemond (1996). Reprinted with permission from the American Chemical Society.]

process of making glue. Note that today the chromium pollution has been isolated and buried by the natural sedimentation of unpolluted, or less polluted, material in the lake. If the sediment cap remains intact then the chromium has been effectively (and inexpensively) removed from the lake system.

5.6 TWO BASIC MATHEMATICAL MODELS FOR LAKES

As discussed in Chapter 3, the derivation of the fate and transport equations used in this textbook requires knowledge of linear algebra or differential equations. Since this textbook is designed for students who have only taken college chemistry and algebra, we will skip the derivation and simply state the governing fate and transport equation. A more mathematical derivation is given in the background section of Fate for the lake module.

By substituting the mathematical terms for mass input, volumetric inflow and outflow, mixing, and chemical reactions into the basic mass balance equation

Change in mass = Inflow mass − Outflow mass + Sources − Sinks

and with algebraic rearrangement, we can arrive at the first-order linear differential equation

$$V\frac{dC}{dt} + CV\left(\frac{1}{t_0} + k\right) = W \qquad (5.5)$$

where dC/dt ($\Delta C/\Delta t$) is the differential part of the equation and represents a change in concentration with change in time, V is lake volume, C is pollutant concentration, t_0 is hydraulic detention time, k is the first-order degradation rate constant, and W is the rate of input of pollutant into the system. It should also be noted that in deriving the model we assume that there are no pollutant sources in the lake or that these internal sources are insignificant (such as from contaminated sediments). Equation (5.5) must now be integrated using the Laplace transformation techniques discussed in Chapter 4, subject to the step or pulse boundary condition.

5.6.1 Continuous (Step) Model

For the continuous (step) model, $W(t)$ is not zero, but represents a constant input of pollutant per time. Under this boundary condition, upon integration we obtain

$$C(t) = \frac{W}{\beta V}\left(1 - e^{-\beta t}\right) + C_0 e^{-\beta t} \qquad (5.6)$$

where $C(t)$ represents the pollutant concentration as a function of time, $\beta = (1/(t_0) + k)$, and C_0 is the background concentration of pollutant in the lake. If the background concentration in the lake is negligible, the equation reduces to

$$C(t) = \frac{W}{\beta V}\left(1 - e^{\beta t}\right) \qquad (5.7)$$

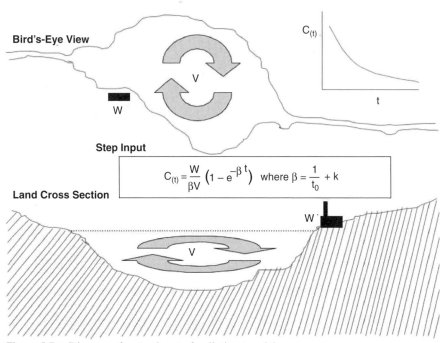

Figure 5.7. Diagram of a step input of pollution to a lake.

These two equations can be used to estimate the concentration of pollutant in a lake that receives a constant input of pollutant. Note that two opposing forces govern the effluent pollutant concentration. The addition of more and more pollutant mass by the source is opposed by washout of the pollutant in the effluent stream and by first-order degradation. Eventually the system will reach an equilibrium pollutant concentration if all parameters remain constant. Thus, as time approaches infinity, the pollutant concentration in the lake approaches

$$C = \frac{W}{\beta V} \tag{5.8}$$

An example of this pollution scenario is shown in Figure 5.7, and a concentration profile is shown in Figure 5.8. Note the shape of the plot. Compared to the pulse model, the highest concentration of pollutant for the step model is much later in time. In fact, since the pollutant is emitted slowly into the lake, the concentration slowly builds up in the lake and reaches a maximum.

Example Problem. A lake in a rural community has an average surface area of $5000\,m^2$ and a mean depth of $50\,m$. A stream exits the lake with an average annual flow rate of $45,000\,m^3/yr$. Aerial application of an insecticide in the area introduces

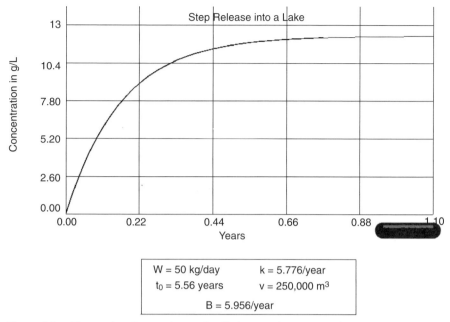

Figure 5.8.   The results of a step input of pollutant to a lake (results from Fate).

the compound into the lake. The average annual loading of this pollutant to the lake from the atmosphere and from agricultural runoff is estimated at 50 kg/day. Assuming a first-order removal of the insecticide from the lake (half-life = 43.8 days) and that the initial background concentration of insecticide in the lake is negligible, answer the following questions:

What is the detention time of water in the lake?

What is the equilibrium concentration of insecticide in the lake?

What is the concentration after 0.010 years?

Solution.   The volume of the lake is equal to the average surface area multiplied by the mean depth, and detention time is equal to this volume divided by the average annual flow rate:

$$\text{Volume} = (5000 \text{ m}^2)(50 \text{ m}) = 250,000 \text{ m}^3$$

$$\text{Detention time} = \frac{250,000 \text{ m}^3}{45,000 \text{ m}^3/\text{yr}} = 5.56 \text{ yr}$$

In order to calculate the equilibrium insecticide concentration, we must first convert the first-order half-life to a rate constant, k, expressed in units of reciprocal years. The half-life of 43.8 days is equal to a half-life of 0.12 years.

$$\ln \frac{C}{C_0} = - kt$$

$$\ln(0.5) = - k(0.12 \text{ yr})$$

$$k = 5.78/\text{yr}$$

Now,

$$C = \frac{W}{\beta V}$$

$$\beta = \frac{1}{t_0} + k = \frac{1}{5.56} + 5.78 = 5.96$$

$$C = \frac{W}{\beta V} = \frac{(50 \text{ kg/day})(365 \text{ days/yr})}{5.96(250,000 \text{ m}^3)}$$

$$= (1.26 \times 10^{-2} \text{ kg/m}^3)(1000 \text{ g/kg})(10^3 \text{ mg/g})(1 \text{ m}^3/1000 \text{ L})$$

$$= 12.5 \text{ mg/L}$$

Determine whether the first-order decay is an important removal process by constructing one plot using the first-order decay and another plot without the first-order decay.

Results of this exercise are shown in Figure 5.9. It is evident from these two plots that the decay rate is important in reducing the concentration of pollutant.

Finally, calculate the concentration after 0.010 years.

$$C = \frac{W}{\beta V}\left(1 - e^{-\beta t}\right)$$

$$\beta = \frac{1}{t_0} + k = \frac{1}{5.56} + 5.78 = 5.96$$

$$C = \frac{W}{\beta V}\left(1 - e^{-\beta t}\right)$$

$$C = \frac{(18250 \text{ kg/yr})(1000 \text{ g/kg})(10^3 \text{ mg/g})}{(5.96 \text{ yr}^{-1})(250,000 \text{ m}^3)(1000 \text{ L/m}^3)}\left(1 - e^{-5.96*0.010}\right)$$

$$C = 12.25\left(1 - e^{-5.96*0.010}\right)$$

$$C = 12.25 - 12.25(0.94)$$

$$C = 0.71 \text{ mg/L}$$

5.6.2   Instantaneous (Pulse) Pollutant Input Model

For the instantaneous (pulse) pollutant input model, we set $W(t)$ (the change in total mass of pollutant in the lake) equal to zero. As discussed previously, a pulse release could be any short-term, immediate release of pollutant to the system. Under these conditions, upon integration of Eq. (5.5) we obtain

$$C(t) = C_0 e^{-\left(\frac{Q_c}{V} + k\right)t}, \quad \text{or} \quad C(t) = C_0 e^{-\left(\frac{1}{t_0} + k\right)t} \tag{5.9}$$

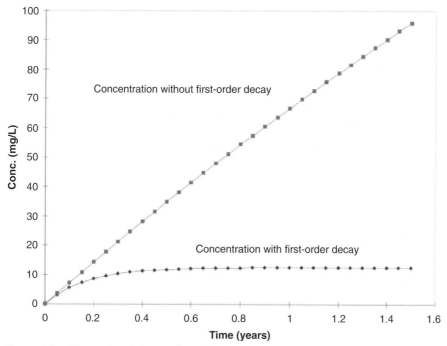

Figure 5.9. The results of the step input from Figure 5.7, showing the effect of degradation rates (results from Fate).

The parameters in this equation were defined earlier. An example of such a pollution scenario is shown in Figure 5.10, and a pollutant concentration profile is shown in Figure 5.11.

Note the shape of the plot in Figure 5.11. The highest concentration of the pollutant occurs immediately after its addition to the lake, and the pollutant is slowly removed from the lake by a generic first-order degradation reactions and by outflow through the effluent of the lake. The student should note the exponential shape of the plot and be able to relate this to Eq. (5.6). A decrease in the first-order degradation rate would prolong the time required for removal of pollutant from the lake. Using Eq. (5.6), note how increasing or decreasing of the other model parameters would affect $C(t)$ and the pollutant removal time.

Example Problem. We will use the same problem statement used in the step input example (given above), but for this pulse input example we will monitor the fate of the insecticide in the system if the input is ceased after 1 year. Thus, we can treat the input as a pulse. In this case, we will develop a formula to express the concentration of insecticide as a function of time. Then, we will be able to calculate how long it will take for the insecticide concentration to reach 0.100 mg/L, the detection limit for this compound.

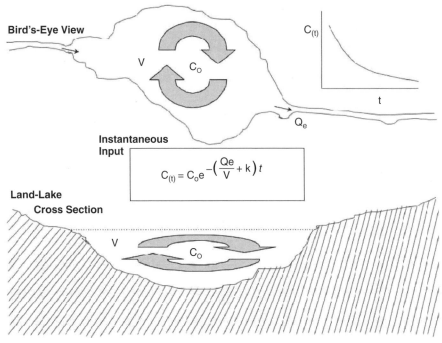

Figure 5.10.   Diagram of a pulse input of pollution to a lake.

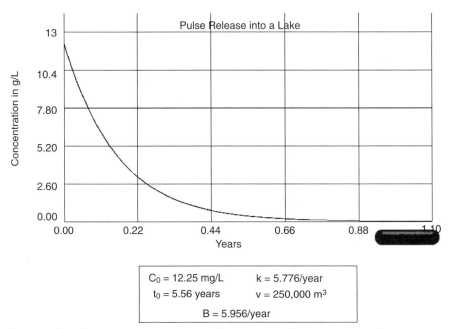

Figure 5.11.   The results of a pulse input of pollutant to a lake (results from Fate).

Solution.   The calculations of residence time and the first-order rate constant in the lake are identical to those shown for the step model (see above), since the dynamics of the lake system have not changed.

Now we must calculate the concentration after 1 year, since this will be the initial concentration for our expression showing the long-term removal of insecticide from the system.

The concentration after 1 year is

$$C = \frac{W}{\beta V}(1 - e^{-\beta t}), \quad \text{where } \beta = \frac{1}{t_0} + k$$

$$\beta = \frac{1}{t_0} + k = \frac{1}{5.56} + 5.78 = 5.96$$

$$C = \frac{W}{\beta V}(1 - e^{-\beta t})$$

$$C = \frac{(18,250 \text{ kg/yr})(1000 \text{ g/kg})(10^3 \text{ mg/g})}{(5.96)(250,000 \text{ m}^3)(1000 \text{ L/m}^3)}(1 - e^{-5.96*1})$$

$$C = 12.25(1 - e^{-5.96*1})$$

$$C = 12.25 - 12.25(2.58 \times 10^{-3})$$

$$C = 12.2 \text{ mg/L}$$

Now, for this problem calculate the insecticide concentration as a function of time. This is governed by Eq. (5.9):

$$C(t) = C_0 e^{-\left(\frac{Q_e}{V} + k\right)t}, \quad \text{or} \quad C(t) = C_0 e^{-\left(\frac{1}{t_0} + k\right)t}$$

where $C_0 = 12.2$ mg/L, detention time $t_0 = 5.56$, and $k = 5.78$ years. A plot of this equation is shown in Figure 5.7.

Next, calculate the time required to reach an insecticide concentration of 0.100 mg/L:

$$C(t) = C_0 e^{-\left(\frac{Q_e}{V} + k\right)t}, \quad \text{or} \quad C(t) = C_0 e^{-\left(\frac{1}{t_0} + k\right)t}$$

$$\frac{C(t)}{C_0} = e^{-\left(\frac{1}{t_0} + k\right)t}$$

$$\frac{0.100 \text{ mg/L}}{12.2 \text{ mg/L}} = e^{-\left(\frac{1}{5.56} + 5.78\right)t}$$

$$\ln \frac{0.100 \text{ mg/L}}{12.2 \text{ mg/L}} = -\left(\frac{1}{5.56} + 5.78\right)t$$

$$4.80 = 5.96t$$

$$t = 0.81 \text{ yr}$$

Thus, it is seen that the lake will recover relatively rapidly after the input of the insecticide is ceased.

5.7   SENSITIVITY ANALYSIS

The best modelers question their modeling results. One common form of determining the accuracy of a model is to perform a sensitivity analysis. In such an analysis, the modeler runs a series of model simulations to determine how much effect an error in a certain parameter will have on the output of the model. To illustrate this, we will perform a sensitivity analysis on first-order degradation constant using the basic (background) step scenario. In the background example problem in Fate®, we used a degradation rate constant of $5.78\,\mathrm{yr}^{-1}$. But what if this is in error? How sensitive is the effluent pollutant concentration to the rate constant? We ran a series of calculations using rate constants of 0.578, 2.58, 5.78, 28.9, and 57.8 and a time of 1.0 years. The results of this analysis are shown in Figures 5.9 and 5.12. We can now directly see that the results are strongly affected by the magnitude of the rate constant and that the relationship is exponential as expected. Similar sensitivity analyses can be performed for all of the parameters in each of the basic models.

The reader should be aware of the dangers and limitations of using "canned" programs such as Fate®. Always review the input data (boxes) in these programs and be sure you understand the units and how the calculations between the boxes are connected. A good mantra is "always question the results and check everything twice."

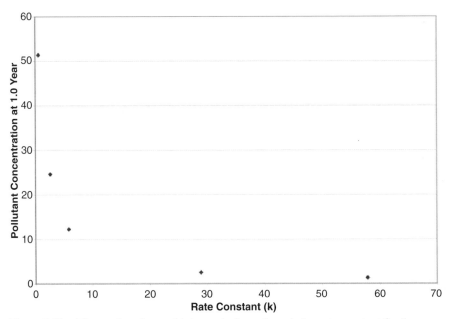

Figure 5.12.   The results of a sensitivity analysis on degradation rate constant for the background scenario in Fate®.

5.8 LIMITATIONS OF OUR MODELS

Incomplete Mixing. We have already discussed the limitations of our models with respect to incomplete mixing, but this can be a limitation of any modeling effort where the true mixing volume of the lake is not known. This unknown quantity will even plague modeling efforts of professionals who use numerical methods of analysis to calculate pollutant concentrations. Thus, the most accurate estimates of effective mixing volumes are needed.

No Internal Sources. If pollutant-sorbed sediment is present in the bottom of the lake, then resuspension of the sediments can act as a source of pollution to lake systems. Normally input from storm events and bioturbation is small or insignificant compared to the overall mass of pollutant in the system. However, it can be large during shipping or harbor dredging operations. This type of resuspension could be treated as a pulse input since dredging operations are of small time scales compared to the usual time scales of pollutant transport in lake systems.

Human-Made Reservoirs. These types of reservoirs pose a huge modeling problem, and our simple modeling approach will not work for complicated, multibasin systems. For example, consider the diagram of Lake Hartwell (Figure 5.13), a human-made reservoir along the Savanna River chain between Georgia and South Carolina (United States). Different mixing basins are separated in Figure 5.13 by the black rectangles which are located at constrictions in the lake. In order to effectively model this reservoir, we would need to develop a series of models, one for each mixing basin. The input for each reservoir would be the output from the previous reservoir. This could easily be handled using the numerical methods approaches described in Chapter 4.

5.9 REMEDIATION

No text on pollutant fate and transport would be complete without considering remediation of the contaminated system. Thus, each fate and transport chapter in this book (Chapters 5–9) will close with a section on remediation. Each environmental system discussed in this book presents a unique set of conditions, and thus problems, with respect to remediation. Issues specific to lakes include the hydraulic retention time, t_0, introduced in Section 5.5.2, which is determined by the total volume of the lake and the effluent flow rate. A lake with a high retention time will not naturally or rapidly purge itself of pollutants in the water column, and thus the pollution will be present for extended times. The large volumes of water that most lakes contain makes infeasible any direct *ex situ* (outside of the lake) pump-and-treat process on the lakeshore. While soluble pollutants remain in the water column, other pollutant will concentrate in the sediments. In general, a longer residence time in the lake enables pollutant sorption to particles in the water column, such that the ultimate resting place of the pollution is in the lake sediments. This is especially true

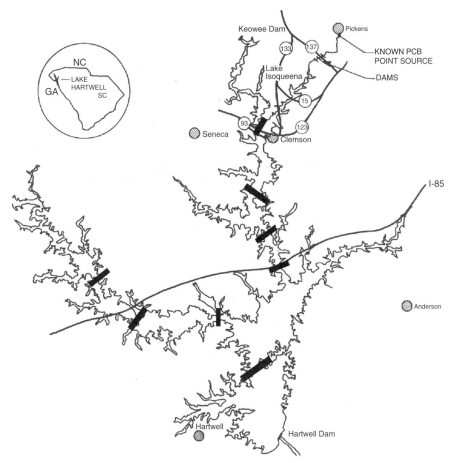

Figure 5.13. A diagram of the shoreline and basins contained in Lake Hartwell (South Carolina). Mixing basins are delineated by solid rectangles.

for metals and hydrophobic pollutants. Concentration of the pollutants in the sediments has good and bad consequences, as we will see in the following paragraphs.

So, what can we do to remediate a polluted lake? First and foremost is to remove or eliminate the source of pollution to the lake. This is relatively easy for point sources such as the effluent from an industry, but becomes more complicated for effluent from a sewage treatment plant, since treated sewage has to be disposed of somewhere. The problem becomes even greater when we are talking about nonpoint sources that are not easily identified or eliminated (such as agriculture runoff) or are uncontrollable (such as leachate entering a lake from an abandoned landfill). Nonetheless, identification, isolation, and eliminate of the source of the pollution should be the first step in any remediation plan.

One of the most common remediation efforts undertaken for lakes is to treat eutrophication, the uncontrolled or excessive microbial and algal growth that results from the presence of nutrients from agricultural runoff and sewage treatment facilities. Even the slightest amount of phosphorus can induce eutrophication; thus, elimination of the source is important. In general, the remediation approach for eutrophied lakes is to eliminate the nutrients and let nature take its course (cleansing by removal through burial of nutrients in the sediments and by removing nutrients by outflow from the lake). This approach is usually the only viable option given the large scale of the problem (volume of water and hydraulic detention time in the lake).

Remediation of polluted sediments has also been the focus of many lake remediation efforts. We will discuss remediation strategies from the most to least complicated (and costly). One obvious, but expensive, way to remediate contaminated lake sediments is to simply remove the sediment by a technique referred to as dredging. There are many types of dredging, but all involve physically removing the sediment from the lake, placing it on a barge, and transporting it to a treatment or containment facility. But before this "sure fix" is adopted, one must consider the scale of the problem. The cost of dredging uncontaminated sediment is approximately $5 per cubic yard, but this cost increases to $15 to $20 per cubic yard for contaminated sediment (National Research Council, 1977). And what do you do with the contaminated sediment once you have removed it since it is not a hazardous waste? Treatment costs (thermal degradation, biochemical treatment, extraction) can range from $100 to $1000 per cubic yard (National Research Council, 1997). You start to see the scale of the problem created by this "sure fix" solution. Thus, dredging the sediments of a lake is usually not a viable option.

Another proposed option is to bioremediate the sediment by adding nutrients and pollutant-degrading microorganisms to the contaminated areas. While this sounds good in theory, it also presents several problems. For example, engineered microbes that effectively degrade pollutants have been raised in the lab, but it is unclear whether these same microbes can effectively compete in the real world of a lake sediment. Furthermore, addition of nutrients to a lake can be a problem in itself, since these nutrients will most certainly migrate to the water column and could promote eutrophication of the lake. In addition, metals do not degrade via microbial action. To date, no human-engineered bioremediation action has been conducted on a lake, although nature is always working through these very same pathways to degrade organic pollutants in lakes.

The next less costly, but also much more viable, approach is to "cap" the contaminated sediment by placing a layer of pollutant-free substrate on the area of interest in the lake. This can be achieved by strategically placing anything from synthetic substrate to clay to sand in the lake. This speeds the process of natural burial and can effectively prevent the contaminated sediments from resuspending and contacting the water column. Of course, considerations that must be made in using this technique include sediment depth from the water's surface, the surface area needing capping, and hydraulic flow rates in the lake that my uncover the contaminated sediments.

The least expensive, and in many cases the most practical, remediation is natural recovery resulting from a combination of biodegradation and burial of the contaminated sediment. Burial occurs by natural sedimentation of suspended lake particles and material that washes into the lake during storm events. As noted in Section 5.5.4, some lakes have natural sedimentation rates of centimeters per square meter per year that can relatively quickly bury a contaminated area of the lake. This remediation approach may not be appropriate for every lake, but does prevent the release of pollutants into the water column during dredging activities. These releases can be significant, as up to 50% of the sorbed mass of pollutant can be released in a period of hours to days (Dunnivant et al., 2005). Sorption and desorption processes were discussed in Sections 3.3 and 3.4.

All of the remediation approaches discussed above must also include some form of long-term monitoring project to document the extent of pollutant occurrence in the lake, future releases of pollutants from the dredged or in-place sediments, and the overall effectiveness of the remediation effort. The extent of this monitoring can range from a few months to decades.

SUGGESTED PAPERS FOR CLASS DISCUSSION

Eisenreich, S. J., B. B. Looney, and J. D. Thornton. Airborne organic contaminants in the Great Lakes ecosystem. *Environ. Sci. Technol.* **15**(1), 30–38 (1981).

Holloway, T., A. Feore, and M. G. Hastings. Intercontinental transport of air pollution: Will emerging science lead to a new hemispheric treaty? *Environ. Sci. Technol.* **37**(20), 4535–4542 (2003).

Engstrom, D. R., E. B. Swain, T. A. Henning, M. E. Brigham, and P. L. Brezonik. Atmospheric mercury deposition to lakes and watersheds: A quantitative reconstruction from multiple sediment cores, In: *Environmental Chemistry of Lakes and Reservoirs*, Advances in Chemistry Series 237, Baker, L. A. ed., American Chemical Society, Washington, DC, Chapter 2, pp. 33–66, 1994.

Concepts:

1. List the major lake formation processes.

2. Give three examples of point sources of pollutants to lakes.

3. Give three examples of non-point sources of pollutants to lakes.

4. Draw a basic diagram showing a stratified lake in summer. Label each portion of the lake.

5. List the order of terminal electron acceptors from increasing to decreasing E_H values.

6. Draw a basic concentration versus time diagram for a pulse input to a lake.

7. Draw a basic concentration versus time diagram for a step input to a lake.

8. Define each term in Eq. (5.4.)

9. Write Eq. (5.6) and define each term.

Exercises

1. Sitting along the shore of a lake is a leaking storage tank of gasoline. The tank is releasing xylenes (found in gasoline) into the lake at a rate of 5 cubic inches per day. Since xylene is very volatile and moderately biodegradable, assume that it has a half-life of 18 hours when leaked into water. Create a plot by hand (but using Fate® to check your calculations) of concentration of the pollutant

in the lake over a period of time. Determine how long it will take the lake to reach an equilibrium concentration of the pollutant.

Volume: $313{,}348{,}796\,m^3$

Lake detention time: 10 years

Lake outlet flow rates: $44{,}650{,}012\,m^3$/year

2. A lake is surrounded by agricultural fields. The pesticide malathion, used on these crops, has a chemical degradation half-life of 6.5 days. As a result of its location, the lake receives high doses of malathion after aerial spraying of pesticides (by crop-dusting airplanes). If the initial concentration of malathion in the lake is 13 mg/L, find the amount of malathion in the lake one year after the contamination has ended, using the following parameters (check your answers with Fate®)

Retention time: 5.75 years

Flow Rate: $3.65 \times 10^3\,m^3$/yr

Lake Volume: $2.08 \times 10^4\,m^3$

Perform a sensitivity analysis using half-lifes of twice and half as long.

3. Lake Ontario receives an average of 140 kg DDT every year, or 0.384 kg/day, from the atmosphere. The pollutant enters as a step input and is derived from a chemical factories in South America. The volume of Lake Ontario is $1638\,km^3$, and the outflow into the St. Lawrence River, the effluent of the lake, is $7990\,m^3$/yr. The half-life of DDT is 31.3 years. Calculate the concentration of DDT in the lake after six months. Calculate and plot the time versus concentration graph for a time interval from 0.00 to 1.5 years. Check your answers using Fate®.

4. A chrome plating plant on a lake has been operating for years without trouble, but one year fishermen notice their catch becoming scarce. Wildlife biologists doing a survey of regional trout populations find that there are significant amounts of chromium ion (Cr^{3+}) in the fish bones, and conclude that the holding tank for the plant must be leaking. If this is true, what concentration of chromium ion species would you expect to find in the lake water if the tank started leaking 10 months ago? Use the data from the chart below to do the necessary calculations and use Fate® to check your results.

Lake volume $= 6.70 \times 10^6\,m^3$

Outlet flow $= 8.9 \times 10^6\,m^3$/yr

Cr^{3+} input per day $= 0.150$ kg/day

Note: Chromium does not have a rate constant for loss as it does not degrade. Thus, you will need to use a very high half-life value to graph this in Fate® (suggested value: 1,000,000).

5. Mining activity was common in western Montana, along the Rocky Mountains and nearby ranges during the twentieth century. For the majority of the

century, the major constituent used for the extraction of gold was cyanide. Assume that at one site a few barrels of HCN spilled into a nearby lake. The initial concentration of HCN in the lake was 17.5 mg/L. The outlet flow from the lake was 2.92×10^7 m³/year. The lake volume was 4.00×10^6 m³ of water. The half-life for HCN is 334 days. Calculate the concentration of HCN in the lake water after 1.00 and 5.00 years. Use Fate® to check your answers.

6. A frozen railroad track causes a train carrying radioactive cesium to wreck. Unfortunately, the track segment is located on a bridge over a lake, and cesium is released into the water, resulting in a uniform concentration of 6.00 µg/L in the lake water. The lake has a detention time of 5.56 years. Assuming complete mixing and a pulse release, calculate the concentration of cesium after 20.0 years. Cesium has a half life of 30.17 years. At what point in time does the cesium concentration become undetectable? (The best technologies can detect cesium at a concentration of 10^{-12} M.)

Spreadsheet Exercise

Create a spreadsheet that performs the same calculations as Fate® for both the step and pulse equation. Construct your spreadsheet so that it is interactive (so you can change numeric values for parameters and the plot automatically updates itself).

REFERENCES

Baker, L. A. *Environmental Chemistry of Lakes and Reservoirs*, Advances in Chemical Series 237, American Chemical Society, Washington, DC, p. 45, 1994.

Dunnivant, F. M., J. T. Coates, and A. W. Elzerman. Labile and non-labile desorption rate constants for 20 PCB congeners from lake sediment suspensions. *Chemosphere* **61**(3), 332–340 (2005).

Lapple, C. E. The little things in life. *Stanford Res. Inst. J.* (third quarter) **5**, 95–102 (1961).

National Research Council, *Contaminated Sediments in Ports and Waterways: Cleanup Strategies and Technologies*, National Academy Press, Washington, DC, pp. 10 and 140, 1997.

Ruttner, F. *Fundamentals of Limnology*, University of Toronto Press, Toronto, 1963.

Spliefhoff, H. S. and H. R. Hemond. History of toxic metal discharge to surface waters of the Aberjona watershed. *Environ. Sci. Technol.* **30**(1), 125 (1996).

Wetzel, R. G. Land–water interfaces: Metabolic and limnological regulators. *Verh. Int. Verein. Limnol.* **24**, 6–24 (1990).

Wetzel, R. G. *Limnology: River and Lake Ecosystems*, Academic Press, New York, 2001.

The World Almanac and Book of Facts, World Almanac Books, New York, 2004.

FATE AND TRANSPORT OF POLLUTANTS IN RIVERS AND STREAMS

CASE STUDY: THE RHINE RIVER

For our case study for Chapter 6, we return to the Rhine River pollution event (in western Europe) initially presented in Chapter 1. There are many historical river pollution events to choose from, but the Rhine River incident is classic in the sense that (a) it involved a major river passing through many highly populated areas and (b) the spread of pollution through a system was very well documented. Recall that on November 1, 1986, a storehouse owned by Sandoz Ltd. near Basel, Switzerland, caught fire and released pesticides, solvents, dyes, and various raw and intermediate chemicals (Capel et al., 1988). A map of the route of the Rhine River through Europe was presented in Figure 1.4. A rapid and valuable sampling effort was conducted by the researchers at the University of Zurich and the Swiss Federal Institute for Water Resources and Water Pollution Control to document the movement of pollutants through the river system. After analysis, the researchers used explanatory modeling to reverse fit the field data to a complicated river model to estimate the dispersion and movement of pollutants through the system. For modeling purposes, the release was treated as a pulse (short duration) release. Monitoring points for the various chemicals were set up at the stations labeled in Figure 1.4 (Capel et al., 1988). One of the most abundant pesticides released was disulfoton, a thiophosphoric acid ester insecticide. Figure 1.5 showed the movement and flushing of disulfoton through the Rhine (Capel et al., 1988). Note the bell or Gaussian shape of the concentration profile, which, as we will see in this chapter, is characteristic of a pulse release in a river. As the pulse of disulfoton moved downstream, it was diluted and degraded, as indicated by the broader peaks and lower concentrations shown in Figure 1.5. This was a case in which the model was fit to the data to better understand how pollutants move through the system and in order to better predict downstream concentrations of later accidental releases of pollutants.

In this chapter, we will learn the basic flow properties of river systems; source inputs of pollutants; how we model these systems for pulse and step pollutant inputs;

how dispersion, sedimentation, and degradation affect downstream pollutant concentrations; and how rivers can be remediated after pollutant releases.

6.1 INTRODUCTION

Rivers and streams make up approximately 0.014% of the terrestrial water on Earth. Human civilizations have always settled near bodies of fresh running surface water, not only for a source of water, but also because the water is constantly renewed and wastes can be instantly removed from the area (unless, of course, if one lives immediately downstream from another community!).

In our modern society, with chemical factories, railways, and highways in close proximity to natural waterways, unintentional releases of hazardous chemicals occur frequently. Once hazardous chemicals are in an aquatic system, they can have a number of detrimental effects that extend for considerable distances downstream from the pollutant source. In this chapter we will look at the concentration of a pollutant downstream of pulse (instantaneous) and step (continuous) releases. Examples of pulse releases can be as simple as small discrete releases such as pouring a liter of antifreeze off a bridge, or they can be more complex such as an accident that results in the release of acetone from a tanker-car. Step releases usually involve a steady input from an industrial process, drainage from non-point sources, or leachate from a landfill located near the stream. Once a pollutant is released to a system, the models we use assume that the pollutant and stream water are completely mixed (i.e., there is no cross-sectional concentration gradient in the stream channel). This is a reasonably good assumption for most systems. The models used here account for longitudinal dispersion (spreading in the direction of stream flow), advection (transport in the direction of stream flow at the flow velocity of the water), and a first-order removal term (biodegradation, chemical, and/or radioactive decay).

6.2 EXAMPLES OF RIVERS AND VOLUMETRIC FLOWS OF WATER

Specific physical features pertinent to modeling of a river system are depth, flow rate, and water velocity. Table 6.1 lists the major rivers of the world, and while these constitute most of the flow of fresh water, they represent a very small number of the total streams in the world. Given their size, many of these streams also require special fate and transport models, given that the input of pollutant is not instantly and evenly spread across the width of the stream channel, and cannot be approximated as such. To put the size of any given river in perspective, Wetzel and Likens (2000) used the drainage areas of streams to estimate the approximate number of streams of a given length and average water discharge. These data are summarized in Table 6.2. The first column refers to the stream order, which expresses how many streams come together to make up the stream of interest. The more streams join to form the final stream, the higher the order number. Note how few streams achieve the length, drainage area, or discharge rate of the major rivers listed in Table 6.1.

TABLE 6.1.   Drainage Area and Annual Flow of the Major Rivers of the World (Szestay, 1982)

Rivers by Continent	Drainage Area ($10^3\,km^2$)	Mean Annual Flow (m^3/sec)
Africa (all rivers)	30,300	136,000
Congo	4,015	40,000
Niger	1,114	6,100
Nile	2,980	2,800
Orange	640	350
Senegal	338	700
Zambezi	1,295	7,000
Asia (all rivers)	45,000	435,000
Bramahputra	935	20,000
Ganges	1,060	19,000
Indus	927	5,600
Irrawaddy	430	13,600
Mekong	803	11,000
Ob-Irtysh	2,430	12,000
Tigris-Euphrates	541	1,500
Yangtze	1,943	22,000
Huang Ho (Yellow River)	673	3,300
Europe (all rivers)	9,800	100,000
Danube	817	6,200
Po	70	1,400
Rhine	145	2,200
Rhone	96	1,700
Vistula	197	1,100
North America (all rivers)	20,700	191,000
Colorado	629	580
Mississippi	3,222	17,300
Rio Grande	352	120
Yukon	932	9,100
South America (all rivers)	17,800	336,000
Amazon	5,578	212,000
Magdalena	241	7,500
Orinoco	2,305	14,900
Parana	2,305	14,900
San Francisco	673	2,800
Tocantins	907	10,000

6.3   INPUT SOURCES

Input sources are very similar to those presented in Chapter 1 and discussed for lake systems in Chapter 4. To summarize, we will consider both point and non-point sources. Common point inputs include industrial and feedlot sources, as well as effluent from domestic sewage plants. Non-point sources can include runoff from farming

TABLE 6.2. The Approximate Number of Streams, Average Length, and Average Discharge Rate as Function of Drainage Area (Wetzel and Likens, 2000)

Order	Number of Streams	Average Length (km)	Average Drainage Area (km^2)	Average Discharge Rate (m^3/sec)
1	200,000	0.02	0.00018	0.000005
2	65,000	0.03	0.00091	0.000025
3	20,000	0.06	0.00414	0.00012
4	5,500	0.16	0.0129	0.00036
5	1,500	0.40	0.0906	0.0025
6	400	1.0	0.388	0.011
7	150	2.4	2.20	0.062
8	40	5.6	9.06	0.250

practices and agricultural settings. Underground non-point sources can include leachate from domestic and hazardous waste landfills or storage tanks. Many rivers are used for recreational activities, and outboard motors are notorious for releasing petroleum-related compounds. These compounds dissolve into the water and are spread throughout the reach of a river. This is a non-point but continuous source of pollutants.

Another source of pollutants to rivers can be from within the river itself. As we discussed in Chapters 2 and 3, particles in the water attract many inorganic and organic pollutants, and most of these pollutant-laden particles aggregate and settle in quiescent regions of a river. However, if these sediments are resuspended into river water with dilute pollutant concentration, desorption of the pollutants into the water will occur. Resuspension events of importance in most rivers include bioturbation (the mixing of river sediments with water by biota in the river), violent storm events, and dredging of shipping channels. Bioturbation is a constant process and would be treated as a constant non-point input, while the latter two events would be treated as pulse, non-point inputs.

6.4 IMPORTANT FACTORS IN THE MODELING OF STREAMS: CONCEPTUALIZATION OF TERMS

6.4.1 Definition of Terms

In order to describe a stream mathematically, we must first make a list of variables (mathematical symbols) for the important terms:

SA is the cross-sectional area of the stream (width, w, multiplied by average depth, d) (m^2).

Q_i is the flow rate at the beginning of the section of the stream to be modeled (m^3/yr).

Q_e is the flow rate at the end of the section of the stream to be modeled (m^3/yr) (we usually assume that Q_i is equal to Q_e and we represent both by simply Q).

$C(x)$ is the pollutant concentration profile as a function of distance (at a fixed time) downstream from the pollutant addition (kg/m^3 or similar units).

$C(t)$ is the pollutant concentration in the stream at a fixed distance as a function of time (kg/m^3 or similar units).

k is the first-order removal rate of pollutant from the river/stream (year^{-1} or in the Fate® model seconds^{-1}).

t is time (years or in the Fate® model, seconds).

M_0 is the total mass of pollutant in the river (kg or Ci).

W is the rate of continuous discharge of the waste (mass/time in kg/sec or Ci/sec).

Note the units used in each of the terms and note that they must be compatible with each other. Compatibility of units is essential in our calculation and is also important in using the simulator package Fate®. The terms given above will be used to develop the mass balance of pollutant in the stream and develop equations for the individual components of the mass balance (inflow, outflow, sources, and sinks of pollutant).

6.4.2 The Stream Channel

An illustration of a stream channel in Figure 6.1 shows an instantaneous and pulse input, the volumetric water flow rate (Q), the velocity (v), a first-order degradation (k), and the longitudinal dispersion term (E, although some texts use D to represent this term), which is the subject of the next section. Physical characteristics of the stream channel determine the magnitude of each of these terms. Of course, large streams usually have a large flow rate. Streams in steep terrain have higher velocities and more mixing. Streams in gentle sloping areas can have very tortuous flow paths and meander for large distances before reaching the receiving lake or ocean. As we will see in the next section, the physical features of the stream channel are responsible for the degree of mixing in a system.

6.4.3 Mixing and Dispersion in Rivers

All of the transport equations used in flowing media (water and air) are referred to as advective–dispersive. Advection refers to the bulk flow or movement of water (and pollutants). In the one-dimensional river model that we will develop, we are only concerned with advection and dispersion in the longitudinal (x) direction. Dispersion is a process primarily resulting from advection, which always results in a dilution of the pollutant by unpolluted water. A chemist can think of dispersion as a form of entropy, since the pollutant always decreases in concentration as the volume of solvent increases (thus, this dilution constitutes an increase of entropy in two ways). This also influences the cost of environmental remediation, since, if the

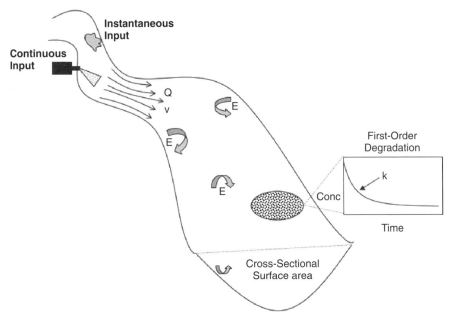

Figure 6.1.   Illustration of a stream channel with a pollution release.

concentration remains above acceptable limits (which are very low for some pollutants), remediation becomes much more difficult as the contaminated volume increases.

There are two fundamental forms of dispersion, hydrodynamic and diffusion-based. Turbulent mixing, a macroscale process, results from water velocity gradients along the flow path (represented by the curved arrows in Figure 6.1). Thus, dispersion is greater when higher flow rates are present. Diffusion-based dispersion is a microscale process, and in rivers it occurs at the water–solid interface of suspended particles and at the sediment–water interface. It is based on the movement of molecules in a concentration gradient (Fickian diffusion), which states that compounds move from high to low concentration, and although this process is very important in laboratory studies, it is of little concern in rivers, since the magnitude of the diffusion process is 100- to 1000-fold less than for turbulent dispersion. Thus, when the equation of fate and transport in streams is derived, diffusion-based dispersion is ignored.

Dispersion is characterized by the longitudinal dispersion (eddy) coefficient, E (m^2/s) in streams, and it can be estimated by the method of Fischer (Fischer et al., 1979):

$$E = 0.011 \frac{v^2 w^2}{du}$$

$$u = \sqrt{gds} \tag{6.1}$$

where v is the average water velocity (m/sec), w is the average stream width (m), d is the average stream depth (m), g is 9.81 m/sec^2 (acceleration due to gravity), and s is the slope or gradient of the stream bed (unitless). Note that increases in water velocity (influenced by slope of the stream bed) and stream width will increase longitudinal dispersion. Thus, reasonably accurate estimates of longitudinal dispersion can be made for a stream by using a topographic map of the area of interest (for slope) and with a few simple measurements of the stream channel.

Figures 6.2a–c show the mixing of two river waters in Switzerland, one coming from Lake Geneva and another draining a glacial valley. As these two streams start to mix in Figure 6.2a, they are separated by a concrete barrier just below the water surface. A close-up view of the mixing eddies is shown in Figure 6.2b. The continued mixing is still evident downstream from the observation point (reverse angle), shown in Figure 6.2c. As illustrated in these photographs, mixing currents in streams can be an important cause of pollutant dilution. Note that for this river, instantaneous mixing does not occur.

Values of E can be determined experimentally by adding a known mass of tracer to the stream and measuring the tracer concentration at various points as a function of time. As mentioned in Chapter 4, a fluorescent tracer is normally used, since it can be detected at extremely low concentrations. An equation defining dispersion [slightly different from Eq. (6.1)] is then fit to the tracer concentration-versus-time data set to calculate values of E for each section of the stream. Unlike in lake studies, this method is relatively fast and cost-effective for streams, but is rarely used today except when highly accurate estimates of E are needed. Most studies simply use Eq. (6.1).

The effect of E on pollutant concentrations in streams is shown in Figure 6.3. The length of the stream containing the plume of pollution, as a function of time, is illustrated by the dark rectangular bands. Note that as the pollutant is transported downstream, more and more dispersion occurs, and the band becomes longer and longer (and lighter in shading) as the pollutant concentration is diluted.

6.4.4 Removal Mechanisms

As discussed in Chapters 2 and 3, a variety of potential degradation schemes can be present, including photochemical, biological, abiotic (chemical), and nuclear reactions. All of these are possible transformation reactions in stream systems. Whatever the type or types of reaction(s), all of these are usually represented by first-order reactions, and we can add the individual rate constants (k values) together to obtain one overall first-order rate constant. This component of the fate and transport model is of the form

$$C_t = C_0 e(-kt) \tag{6.2}$$

where C_t is the pollutant concentration at time t, C_0 is the initial pollutant concentration, e is the exponential function, k is the first-order rate constant (reciprocal time units), and t is time (same time units as k).

Actual rate constants are usually determined in a laboratory setting via experimentation. These experiments start with a known initial pollutant concentration, and

(a)

(b)

Figure 6.2. (*a*) The junction (mixing) of two streams (one from a clear lake and another from a turbid stream). (*b*) Close-up of the mixing eddies. (*c*) Mixing of the two streams downstream.

(c)

Figure 6.2   *Continued*

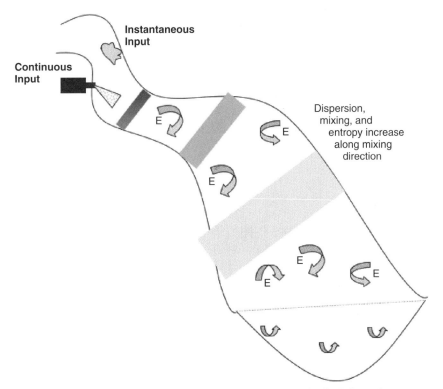

Figure 6.3.   A plane view of a stream channel, showing longitudinal dispersion.

pollutant concentration is monitored as a function of time. The rate constant of degradation is calculated by plotting the data ($\ln(C/C_0)$ versus time) and, if the plot is linear, measuring the slope of the line (k). If the $\ln(C/C_0)$ data plot results in a straight line, then the reaction is said to follow first-order kinetics and the rate constant, k, can be directly used in the fate and transport model. Processes with rates not describable by first-order kinetic expressions require that modelers derive equations that differ from the standard ones presented in this chapter.

Another important removal mechanism is sorption to particles that settle downstream in the river system. This may or may not be important for any given pollution event. Factors that determine the importance of sorption/settling are the aqueous solubility of the pollutant, which in turn is correlated to (a) the distribution/partition coefficient, (b) the presence of suspended matter in the river water, and (c) the rate of turbulence in the river system. As we discussed in Chapters 2 and 3, metal and hydrophobic pollutants tend to sorb to suspended particles, and the more organic matter present in the particle, the more the pollutant will sorb. Most streams have sufficient suspended matter present in the water for this sorption process to be important; however, streams by their basic nature tend to have high turbulence and therefore have less particle settling than lake systems. Storm events increase the ability of a stream to carry suspended particles, and thus they place tons of suspended matter in the stream in a short amount of time. As flood waters recede and flow in the stream decreases, particulate matter settles out and this process can take a significant mass of the pollutants with it to the bottom of the stream. But this presents an even more difficult process to understand and model, since sediments in river systems tend not to remain in one place but are redistributed during every storm event. In addition, as we saw in Chapters 2 and 3, resuspension of polluted sediments can result in re-release of pollutants to stream water not originally containing these pollutants. Modeling of these interrelated, very complicated processes is difficult.

We can break the process into individual steps in an effort to understand them. First, we can look at the sorption process that was presented in Section 3.3 where we discussed distribution and partition coefficients. If you do not recall the mathematical meaning of these coefficients, you should review this section. Next, we can relate these sorption reactions to the removal of the particles from the water by settling using the same concepts presented in Eq. (5.3) and (5.4), where we looked at the Stoke's settling velocity and rate of removal of pollutants. But Eq. (5.4) was developed for calm water such as that present in lakes. In streams, as the pollutant-laden particles settle, they are also transported downstream. So, we must balance these forces—movement down in the water column and movement down stream channel—in order to predict how far down the river system the particles will travel before they settle to the bottom of the stream and become incorporated into the sediment. We can estimate the distance downstream by using the same settling velocity calculated from Eq. (5.3) and summarized in Table 5.5, given in centimeters per second or meters per second, and imagining the particle settling in the water column as it is transported downstream. If we divide the average depth of the stream by the settling rate, we obtain a time unit. This is the average time that a particle of specified size will spend in the water column. Thus, if we multiply the stream velocity (in units of meters per second) by the settling time (in seconds), we obtain the distance downstream the particle will travel before it reaches the bottom. We can then estimate where a given particle size will be deposited in a system. You should work through the units of these calculations to confirm our statements and convince yourself of the math behind this concept. Keep in mind that turbulence in the system from mixing eddies will result in the deposition of the sediment at slightly longer distances that those predicted by our calculations.

Although these calculations cannot be readily incorporated into the simple models that we present in this text, you can understand the concepts behind the removal of pollutant-laden particles from stream waters. These same mathematical approaches are used by modelers to incorporate the removal of pollutants by sedimentation in more complicated models based on numerical methods of analysis.

6.5 MATHEMATICAL DEVELOPMENT OF SIMPLE TRANSPORT MODELS

As discussed in Chapter 4, the derivation of the fate and transport equations used in this textbook requires that the student has taken linear algebra or differential equations. Since this textbook is designed for students who have only taken college chemistry and algebra, we will skip the derivation and simply state the governing fate and transport equation. A more mathematical derivation is given in the background section of Fate® for the river and stream module. As we also discussed in Chapter 5 for lake systems, mathematical equations for each term are substituted into the mass balance equation (shown below), and it is integrated using Laplace transform techniques to yield a general solution for the instantaneous and pulse boundary conditions.

Change in mass = Inflow mass − Outflow mass + Sources − Sinks

6.5.1 Solution of the Differential Equation for the Instantaneous Input (Pulse)

The solution of the differential equation for the instantaneous input is

$$C(t) = \frac{M_0}{wd\sqrt{4\pi Et}} e - \left(\frac{x - vt^2}{4Et} - kt \right) \tag{6.3}$$

where $C(t)$ is the pollutant concentration (in mg/L, or µCi/L for radioactive compounds) at time t, M_0 is the mass of pollutant released (in mg or µCi), d is the average stream depth, w is the average width of the stream, E is the longitudinal dispersion coefficient (m²/sec), t is time (sec), x is the distance downstream from input (m), v is the average water velocity (m/sec), and k is the first-order decay or degradation rate constant (1/sec). Note that e represents the number "e" (the base of the natural logarithm). When there is no (or negligible) degradation of the pollutant, and k is set to approximately zero by entering a very long half-life into Fate®. An example of a concentration–time plot for a certain distance downstream, from the background problem from Fate®, is shown in Figure 6.4.

Example Problem. One curie of cesium-134 (^{134}Cs) is accidentally released into a small stream. The stream channel has an average width of 40 m and an average depth of 2 m. The average water discharge (Q) in the stream is 40 m^3/sec, and the stream channel drops 1 m in elevation over a distance of 10 km. Assuming that the ^{134}Cs is evenly distributed across the stream channel, estimate the distribution of ^{134}Cs as a function of distance downstream (using a maximum distance of 30 km) at 1, 3, 6, and 12 hours. Also estimate the ^{134}Cs activity (concentration) at a distance of 10 km at 6 hours after the release. ^{134}Cs has a half-life of 2.07 years.

Solution

1. Calculate the average stream velocity in meters per second.

$$\text{Cross-sectional area of stream channel} = \text{Width} \times \text{depth} = (40 \text{ m})(2 \text{ m}) = 80 \text{ m}^2$$
$$\text{Average velocity} = \text{Discharge/Cross-sectional area} = (40 \text{ m}^3/\text{sec})/(80 \text{ m}^2)$$
$$= 0.50 \text{ m/sec}$$

2. Calculate the rate constant, k, for ^{134}Cs. For a first-order reaction,

$$\ln \frac{C}{C_0} = -kt$$

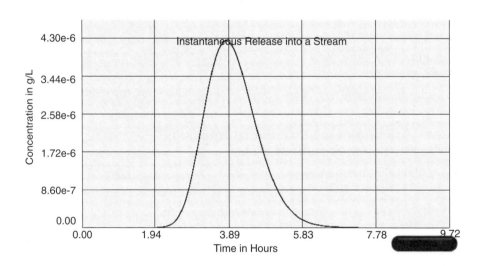

Figure 6.4. Example output from Fate® for an instantaneous pollutant input to a river.

where C is the concentration (or activity of ^{134}C) at time t, C_0 is the initial concentration (or activity) of ^{134}C, k is the decay rate constant, and t is time.

At the half-life ($t_{1/2}$), one-half of the original concentration remains. Substitution of this into the equation above yields

$$\ln\left(\frac{\frac{1}{2}C_0}{C_0}\right)=-kt_{1/2}$$

or

$$-\frac{\ln 0.5}{t_{1/2}}=-\frac{\ln 0.5}{2.05 \text{ yr}}=k=0.338 \text{ yr}^{-1}$$

$$(0.338 \text{ yr}^{-1})\left(\frac{\text{yr}}{365 \text{ days}}\right)\left(\frac{\text{day}}{24 \text{ hr}}\right)\left(\frac{\text{hr}}{60 \text{ min}}\right)\left(\frac{\text{min}}{60 \text{ sec}}\right)=1.07\times10^{-8} \text{ sec}^{-1}$$

Thus, the decay rate constant for ^{134}Cs is $1.07 \times 10^{-8} \text{ sec}^{-1}$.

3. Calculate the longitudinal dispersion coefficient, E (also referred to as the coefficient of eddy diffusion).

$$\text{slope}=\frac{1 \text{ m}}{10,000 \text{ m}}=10^{-4}$$

$$u=\sqrt{gds}=\sqrt{(9.81 \text{ m/sec}^2)(2 \text{ m})(10^{-4})}=0.044 \text{ m/sec}$$

$$E=0.011\frac{v^2w^2}{ud}=0.011\frac{(0.50 \text{ m/sec})^2(40 \text{ m})^2}{(0.044 \text{ m/sec})(2 \text{ m})}=50 \text{ m}^2/\text{sec}$$

4. Arrange data in the proper units:

$$M_0 = 1 \text{ curie} = 1 \times 10^6 \,\mu\text{Ci} \text{ (in the program this is entered as 1 Ci)}$$
$$w = 40 \text{ m}$$
$$d = 2 \text{ m}$$
$$E = 50 \text{ m}^2/\text{sec}$$
$$t = \text{variable, in seconds (sec)}$$
$$x = \text{variable, in meters (m)}$$
$$v = 0.50 \text{ m/sec}$$
$$k = 1.07 \times 10^{-8} \text{ sec}^{-1}$$

5. The plot showing the concentration profile at 12 hr is shown in Figure 6.5. Using the velocity and time, we see the peak concentration should be at 21.6 km, which is observed in Figure 6.5. The width of the Gaussian curve depends on the magnitude of E.

6. Calculate $C(t)$ at 10 km downstream, after 6 hr ($x = 10,000$ m and $t = 6$ hr $= 21,600$ sec)

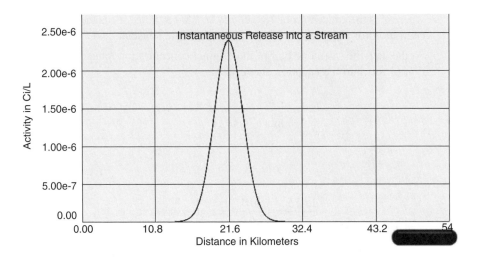

Figure 6.5.　Concentration versus distance from point source for the instantaneous example problem.

$$C(x,t) = \frac{M_0}{wd\sqrt{4\pi Et}}\exp\left(-\frac{(x-vt)^2}{4Et} - kt\right)$$

$$= \frac{1 \times 10^6\ \mu\text{Ci}}{(40\ \text{m})(2\ \text{m})\sqrt{4\pi(50\ \text{m}^2/\text{sec})(21,600\ \text{sec})}}$$

$$\exp\left(-\frac{(10,000\ \text{m}) - (0.50\ \text{m}/\text{sec})(21,600\ \text{sec})^2}{4(50\ \text{m}^2/\text{sec})(21,600\ \text{sec})}\right.$$

$$\left. -(1.07 \times 10^{-8}\ \text{sec}^{-1})(21,600\ \text{sec})\right)$$

$$= 3.39 \times 10^{-6}\ e^{-0.148} = 3.39 \times 0.862$$

$$= 2.92 \times 10^{-6}\ \text{Ci}/\text{m}^3$$

$$= 2.92 \times 10^{-9}\ \text{Ci}/\text{L} = 2.92\ \text{nCi}/\text{L}$$

6.5.2　Solution of the Differential Equation for the Step Input

The solution of the differential equation for the step input is

$$C(x,t) = \frac{W}{Q\sqrt{1 + \dfrac{4kE}{v^2}}}\, e^{\left(\frac{vx}{2E}\left(1 \pm \sqrt{1 + \frac{4kE}{v^2}}\right)\right)} \tag{6.4}$$

where $C(x, t)$ = the pollutant concentration (in mg/L or μCi/L for radioactive compounds) at distance x and time t (note that t is defined by the water velocity when x is fixed), W is the rate of continuous discharge of the waste (in kg/sec or Ci/sec), Q is the stream flow rate in m^3/sec, E is the longitudinal dispersion coefficient (m^2/sec), x is the distance downstream from input (m), v is the average water velocity (m/sec), and k is the first-order decay or degradation rate constant (1/sec). The positive root of the equation refers to the upstream direction ($-x$), and the negative root (what we use in Fate®) refers to the downstream direction ($+x$). An example of a concentration–distance plot for the background problem from Fate® is shown in Figure 6.6.

Again, when there is no (or negligible) degradation of the pollutant relative to the transport time, we set k to zero by entering a very long half-life. The longitudinal dispersion coefficient, E, is characteristic of the stream, or more specifically, the section of the stream that is being modeled and describes the degree of mixing in this section. Under these conditions, the governing equation reduces to

$$C(x) = \frac{W}{Q\sqrt{1 + \dfrac{4kE}{v^2}}} e^{\left[\left(\dfrac{vx}{E}\right)\right]} \tag{6.5}$$

Because we are considering a step input, the concentration of pollutant should vary with distance but not with time, and thus time can be dropped from the expression. Note that while time is not directly stated in Eq. (6.5), however, it is present in the combination of the stream velocity, v, and the distance downstream, x.

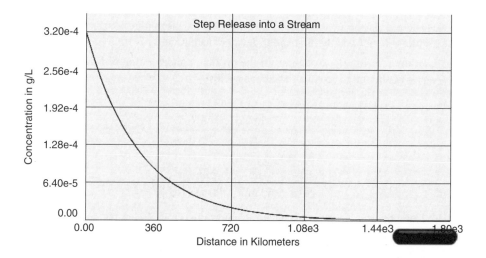

d = 2.3 m		
w = 20 m	v = 0.85 m/s	E = 11.24 m²/s
M₀ = 0.0125 kg	Q = 39.1 m³/s	k = 3.21e-6/minute

Figure 6.6.   Example output from Fate® for a step pollutant input to a river.

Example Problem. An abandoned landfill leaches water into an adjacent stream at a rate of 1500 L/min. The concentration of 2-chlorophenol in the water is 500 mg/L. The stream is 20 m wide and 2.3 m deep and has a water velocity of 0.85 m/sec. The regional slope of the stream channel is 1 ft per 1500 ft distance, and the first-order half-life of 2-chlorophenol is 2.5 days.

Construct the concentration profile of 2-chlorophenol in the stream. What is the concentration 25 km downstream from the point source?

Solution

1. Calculate the mass input to the stream in kilograms per second.

$$\text{mass input} = \left(\frac{1500\,\text{L}}{\text{min}}\right)\left(\frac{\text{min}}{60\,\text{sec}}\right)\left(\frac{500\,\text{mg}}{\text{L}}\right)\left(\frac{\text{g}}{1000\,\text{mg}}\right)\left(\frac{\text{kg}}{1000\,\text{g}}\right)$$

$$\text{mass input} = 0.0125\,\text{kg/sec}$$

2. Calculate the flow rate of the stream in cubic meters per second.

$$\text{Flow rate} = \text{Width} \times \text{depth} \times \text{water velocity} = (20\,\text{m})\,(2.3\,\text{m})\,(0.85\,\text{m/sec})$$
$$= 39.1\,\text{m}^3/\text{sec}.$$

3. Calculate the rate constant, k, for the first-order decay of 2-chlorophenol. For a first-order reaction,

$$\ln\frac{C}{C_0} = -kt$$

where C is the concentration at time t, C_0 is the initial concentration of 2-chlorophenol k is the first-order decay constant, and t is time in seconds. At the half-life ($t_{1/2}$), one-half of the original concentration remains. Substitution of this into the equation above yields

$$\ln\left(\frac{\frac{1}{2}C_0}{C_0}\right) = -kt_{1/2} \quad\text{or}\quad -\frac{\ln 0.5}{t_{1/2}} = -\frac{\ln 0.5}{2.5\,\text{days}} = k = 0.277\,\text{day}^{-1}$$

$$(0.277\,\text{day}^{-1})\left(\frac{\text{day}}{24\,\text{hr}}\right)\left(\frac{\text{hr}}{60\,\text{min}}\right)\left(\frac{\text{min}}{60\,\text{sec}}\right) = 3.21\times10^{-6}\,\text{sec}^{-1}$$

Thus, the decay rate constant for 2-chlorophenol is $3.21 \times 10^{-6}\,\text{sec}^{-1}$.

4. Calculate the longitudinal dispersion coefficient, E (the coefficient of eddy diffusion).

$$\text{slope} = \frac{1\,\text{m}}{1500\,\text{m}} = 6.67\times10^{-4}$$

$$u = \sqrt{gds} = \sqrt{(9.81\,\text{m/sec}^2)(2.3\,\text{m})(6.67\times10^{-4})} = 0.12\,\text{m/sec}$$

$$E = 0.011\frac{v^2 w^2}{du} = 0.011\frac{(0.85\,\text{m/sec})^2\,(20\,\text{m})^2}{(2.3\,\text{m})(0.123\,\text{m/sec})} = 11\,\text{m}^2/\text{sec}$$

5. Arrange data into proper units:

$$Q = 39.1\,\text{m}^3/\text{sec}$$
$$W = 0.0125\,\text{kg}/\text{sec}$$
$$W = 20\,\text{m}$$
$$d = 2.3\,\text{m}$$
$$v = 0.85\,\text{m}/\text{sec}$$
$$E = 11.24\,\text{m}^2/\text{sec}$$
$$x = \text{variable in meters (m)}$$
$$k = 3.21 \times 10^{-6}\,\text{sec}^{-1}$$

6. A plot of concentration versus distance from point source at a given time is shown in Figure 6.7.

7. Calculate $C(x)$ at 25 km.

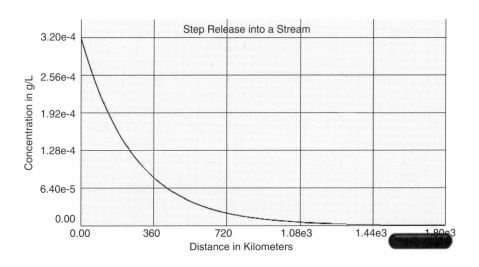

Figure 6.7. Concentration versus distance from point source for the step example problem.

$$C = \frac{W}{Q\sqrt{1 + \frac{4kE}{v^2}}} \exp\left[\frac{vx}{2E}\left(1 - \sqrt{1 + \frac{4kE}{v^2}}\right)\right]$$

$$C = \frac{0.0125 \text{ kg/s}}{39.1 \text{ m}^3/\text{sec}\sqrt{1 + \frac{4(3.21\times10^{-6} \text{ sec}^{-1})(11.24 \text{ m}^2/\text{sec})}{(0.85)^2}}}$$

$$\times \exp\left[\frac{(0.85 \text{ m/sec})(25,000 \text{ m})}{2(11.24 \text{ m}^2/\text{sec})}\left(1 - \sqrt{1 + \frac{4(3.21\times10^{-6} \text{ sec}^{-1})(11.24 \text{ m}^2/\text{sec})}{(0.85 \text{ m/sec})^2}}\right)\right]$$

$$C = 3.20\times10^{-4} \exp(-0.0944)$$

$$C = 2.91\times10^{-4} \text{ kg/m}^3$$

$$C = 0.291 \text{ mg/L (ppm)}$$

6.6   SENSITIVITY ANALYSIS

As in the other fate and transport chapters, we will perform a sensitivity analysis. As an example, we will do this for the water velocity using the basic (background) step scenario in Fate®. In the background example of Fate®, we used a water velocity of 0.85 m/sec. But what if this is in error? How sensitive is the effluent pollutant concentration to the water velocity? The results of a series of calculations using water velocities of 0.25, 0.50, 0.85, 1.00, 1.25, and 1.50 m/sec, observing the pollutant concentration at 500 m downstream, are shown in Figure 6.8. We can now directly see that the results are strongly affected by the magnitude of the water velocity and that the relationship is exponential, as expected from the presence of v in the exponential term of the transport equation. The higher water velocity causes pollutants to be transported through the system before they have time to be degraded by the first-order process, which results in higher water concentrations at each distance downstream. Similar sensitivity analyses can be performed for all of the parameters in each of the basic models.

As stated in every fate and transport chapter, the reader should be aware of the dangers and limitations of using "canned" programs such as Fate®. Always review the input data (boxes) in these programs and be sure you understand the units and how the calculations between the boxes are connected. Once again, remember our mantra for modeling: "Always question the results, and check everything twice."

6.7   LIMITATIONS OF OUR MODELS

One-Dimensional versus Two-dimensional Models and Inputs of Pollutant Plumes in Wide Streams. Neither step nor pulse pollutant inputs are evenly spread across the stream channel, as assumed by the basic models. For most

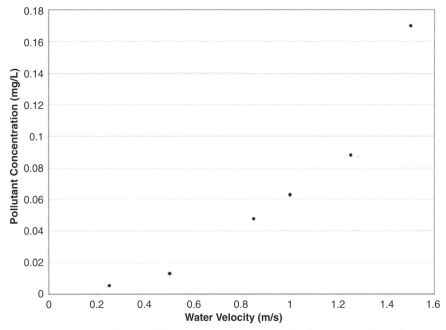

Figure 6.8.   Results of a sensitivity analysis of water velocity for a step pollutant input.

streams, this is not a problem, since the pollutant and stream water are well mixed within a few kilometers downstream of the point of input and most concentration profiles cover tens to hundreds of kilometers. However, for very large streams, the model would be modified to account for mixing (dispersion) in the y direction. This would be accomplished with a two-dimensional approach and could still be modeled using techniques similar to the ones presented in this chapter.

Volatilization of Pollutants. When dealing with organic pollutants, volatilization can be an important removal mechanism, since stream waters have considerable contact with the atmosphere. This removal is treated as a Henry's law partitioning process and treated as a first-order removal mechanism. Thus it is not really a limitation of the model, since it can be included in the kinetic "k" term.

6.8   REMEDIATION OF POLLUTED STREAMS SYSTEMS

As we saw in our case study for this chapter, rivers are highly dynamic systems and pollutants move rapidly through the system. In many cases, the movement is so rapid that we do not have time to respond and clean up the system before the pollution has moved on. If fact, pollution in a river system is a moving target, and by the time we have adequate technologies in place to respond to a pollution spill, the pollution has moved to another section of the stream. Often, the best we can do is predict the migration of pollution with the water column, warn residents to stay out of the water, and stop the intake of water into their drinking water facilities when the pollution is present in their section of the stream. Meanwhile, the ultimate destination or fate of the pollution is the lake or estuary system receiving the water from the stream. This presents a common problem for estuary systems: They receive all of the point and non-point pollution in a watershed. When you consider a large stream, the accumulated pollution can be huge. For example, take the Mississippi River in the United States. It has origins starting far to the north of its Louisiana delta, and it passes through many industrialized and agricultural areas as it flows south. Along the way it accumulates industrial pollutants and nutrients from sewage effluents and agricultural runoff. It is therefore not surprising that the mouth of the Mississippi outside of New Orleans has been effectively labeled a dead zone due to eutrophication and lack of dissolved oxygen (referred to as hypoxia).

There are a few exceptions to the rule of not attempting to treat pollution events in stream systems. For example, leaking barges or oil tankers can be surrounded by booms (floating containment curtains that trap the releases so they can be pumped into containment vessels).

Some of the remediation approaches given in Chapter 5 for sediments in lake systems are directly applicable to streams. However, streams are not quiescent like lakes, since the flow in most streams is significant and very dynamic. This prevents the use of remediation technologies such as capping, since even if we can place a protective layer of sediment on a contaminated region of a river, it can easily be removed during the next storm or flood event by turbulent flow. But as rivers make their long journey to their inevitable end, a lake or estuary, they tend to grow in size and width, become more shallow, and decrease in water velocity such that the settling of suspended particles is greatly promoted. This is good and bad. The good is that pollutants are removed from the water column and the biota are no longer exposed to their toxic effects. As more and more sediment is deposited, the pollution is eventually buried and removed from the ecosystem. This process is a natural remediation action in all lakes and estuaries. The downside of this process results from human use of these very same aquatic systems. Shipping channels slowly fill with sediment deposits and must be subsequently dredged to keep the channels open. You should immediately note the problem with this approach, given our previous discussions of pollutant sorption and sedimentation. The streambeds in these quiescent areas are the resting grounds of many toxin-laden particles, and dredging activities always mix some of the removed sediment material with the water column. This creates an ideal environment for the desorption of pollutants and re-release of these pollutants to the water column. Thus, our maintenance of shipping channels and harbors has an inherent environmental consequence associated with it. While our economy depends on these shipping channels, we re-expose ecosystems to buried pollutants when we dredge these channels.

It is important to note the scale of the dredging problem. The National Research Council estimates that approximately 14–28 million cubic yards of contaminated sediments must be managed annually (which is actually only 5–10% of the total sediments dredged in the United States) (NRC, p. 1, 1997). This raises a difficult problem: What do we do with this huge volume of sediment that is to be treated as hazardous waste? Recall the costs of dredging and treating these sediments from Chapter 5: $15 to $20 per cubic yard to remove and transport contaminated sediments and $100 to $1000 per cubic yard to treat the contaminated sediment. You do the math: The maintenance of shipping channels is a very expensive operation.

SUGGESTED PAPERS FOR CLASS DISCUSSION

Capel, P. D., W. Giger, R. Reichert, and O. Warner. Accidental input of pesticides into the Rhine River. *Environ. Sci. Technol.* **22**(9), 992–997 (1988).

Concepts

1. Name three point sources of pollution to streams.

2. Name three non-point sources of pollution to streams.

3. Name three step inputs of pollutants to streams.

4. Name three pulse inputs of pollutant to streams.

5. Discuss how mixing influences dispersion in streams.

Exercises

1. Look up the flow rate of a stream in your area using data from your country's geological survey. In the United States, data can be found from at www.usgs.gov. Using data in Table 6.2, determine the number of streams of this size.

2. Manually calculate the eddy dispersion coefficient for this stream using the background data for the stream model in Fate®. How would you expect an increase in water velocity, stream width, stream depth, or channel slope to affect E?

3. Using the background conditions in the step and pulse models in Fate®, perform individual sensitivity analyses on Q, k, E, and v.

4. A metal plating plant located on a river has an accident in which 102 kg of copper plating solution is released. Assume copper(II) ion is the pollutant in question. The river drops an average of 2.5 m over a distance of 1 km. Given the conditions below, create two plots that show the curve of concentration versus time and distance. Determine where the maximum concentration of copper will be at a time of 20 hours from the time of the spill. Check your answer with Fate®.

   Flow rate: 3400 ft³/sec

   Depth: 1.97 m

   Width: 20 m

   Velocity: 1.1 m/sec

   Half-life: 3000 yr

5. A train transporting a 1% solution of sodium cyanide was traveling along a river in the countryside when an accident occurred. The railroad tanker car released 26,500 kg of sodium cyanide (265 kg of cyanide) into the river. The half-life of hydrogen cyanide is 0.9 years. Create a plot of the concentration of the cyanide versus the distance from the point of release using the following parameters. Use a distance from the point source between 0.0 and 4.0 km.

   Depth of river: 2.0 m

   Width of river: 10 m

   Water velocity: 0.122 m/sec

   Stream coefficient of eddy diffusion: 3.104 m^2/sec

   Mass of cyanide: 265 kg

   It is important to have a sensitivity analysis for this scenario. What will happen if the degradation rate or half-life is twice as long and twice as short as the value given above? Does the concentration change with varying degradation rate? Use Fate® to check calculations and plots.

6. Dissolved cadmium is leaked continuously from a mining operation into a river. Determine the concentration of Cd^{2+} in the river 5000 km away from the source. Include a graph of the concentration of cadmium as it changes over the distance from 0.0 to 5000 km. The rate of mass input of cadmium is 0.002 kg/sec. The depth of the river is 2 m and the width is 6 m. The water velocity is 5 m/sec. The channel slope is 0.001 and gravity constant is 9.81 m/sec^2. Then, determine the concentration during drought conditions (assume depth, width, velocity, and flow rate are all half of their original values).

7. An explosion at a nuclear power plant damages the structure and threatens to cause a meltdown. Workers and an emergency crew shut down the reactor and keep radioactive waste from contaminating the steam released from the cooling towers, but are unable to prevent it from discharging into the nearby river. The waste contains primarily strontium-90. From measurements near the plant, workers estimate that about 17,000 Ci of radioactive material dissolved in the river. Eight kilometers downstream from the plant is an elementary school right on the water's edge. What will the peak radiation (in Ci/L) at the school be? If the spill occurs at 1 P.M., and school ends at 3:30 P.M., should the students be let out early? Use the data from the chart below to do the necessary calculations and use Fate® to check your results.

River depth:	2.3 m
River width:	35 m
Radioactivity:	17,000 Ci
Water velocity:	0.7 m/sec
Channel slope:	0.0003
Half-life:	28.8 yr
Graph distance:	8 km
Graph time:	3.15 hr

8. A pesticide factory located alongside a river has discovered a steady leak in one of its holding tanks. It has slowly been releasing malathion, a pesticide with a half-life of 5 days, at a steady rate of 0.117 kg/sec. At the factory, the river is 22.86 m wide and 1.7 m deep, has a velocity of 2.05 m/sec, and has a channel slope of 0.000878. What is the concentration at the site of the factory and at 1000 km downstream? Assume that the river dimensions do not significantly change as you go downstream. First calculate this scenario manually, and then use Fate® to check your answers.

9. You are an environmental scientist who has been hired to assess the damage caused by an accident in which an eyeglass lens manufacturer, located on a river, released acetone directly into the water. Fortunately, only one 5.0-L bottle was released. The river has an average depth of 10 m, a width of 45 m, and longitudinal dispersion coefficient of 49.7 m²/sec. The water velocity is estimated to be 3.0 m/sec. Acetone has a density of 0.786 g/mL and a half-life in water of 20.0 hr. Determine the concentration of acetone in the river due to this pulse release at distances of 2, 4, and 7 km from the site at times of 5, 15, 20, and 30 hr. Use Fate® to check your answers.

Spreadsheet Exercise

Create a spreadsheet that performs the same calculations as Fate® for both the step and pulse equation. Construct your spreadsheet so that it is interactive (so you can change numeric values for parameters and the plot automatically updates itself).

REFERENCES

Capel, P. D., W. Giger, R. Reichert, and O. Warner. Accidental input of pesticides into the Rhine River. *Environ. Sci. Technol.* **22**(9), 992–997 (1988).

Fischer, H. B., E. J. List, R. C. Y. Koh, I. Imberger, and N. H. Brooks. *Mixing in Inland and Coastal Waters*, Academic Press, New York, 1979.

National Research Council, *Contaminated Sediments in Ports and Waterways: Cleanup Strategies and Technologies*, National Academy Press, Washington, DC, 1997.

Szestay, K. River basin development and water management. *Water Qual. Bull.* **7**, 155–162 (1982).

The World Almanac and Book of Facts, World Almanac Books, New York, 2004.

Wetzel, R. G. *Limnology: River and Lake Ecosystems*, Academic Press, New York, 2001.

Wetzel, R. G. and G. E. Likens. *Limnological Analysis*, 3rd edition, Springer-Verlag, New York, 2000.

CHAPTER 7

DISSOLVED OXYGEN SAG CURVES IN STREAMS: THE STREETER–PHELPS EQUATION

CASE STUDY: ANY STREAM, ANYWHERE IN THE WORLD

The concepts and equations presented in this chapter are ecologically crucial and universally applicable. Virtually every stream in the world has inputs of sewage, or other organic wastes, whether from humans or from domestic and wild animals. The modeling concepts we develop in this chapter are thus applicable to every stream in the world. As we will see in this chapter, the waste presents two problems: (a) the spread of disease caused by microbes and viruses in the waste and (b) decomposition leading to consumption of valuable dissolved oxygen in the stream. While the spread of pathogens in the waste is an important problem, in this chapter we will mostly concern ourselves with modeling the consumption of dissolved oxygen. The disposal of sewage has plagued humans since our first permanent settlements. In fact, the accumulation of waste associated with permanent settlements has been suggested as one force promoting a nomadic lifestyle. Unfortunately, the way we overcame this problem through most of history was to build our cities next to streams or rivers and dump our waste directly into the flowing water. This solved our problem but was not very pleasant for those living downstream.

For example, consider the dissolved oxygen levels as a function of time (from the early 1800s until 1990) in three highly populated areas: New York City, London, and Western Europe (Figure 7.0). As you can see, once the population reached a certain level, the dissolved oxygen level in the streams receiving waste significantly dropped. After living with the problem for decades, we finally installed proper sewage treatment plants, and in the 1970s and 1980s the streams started to recover.

Chemically, the problem is as follows. When biodegradable waste is added to an aqueous system, two competing kinetics processes are set up: one chemical, by which the microbes in the waste and aquatic system consume dissolved oxygen (DO), and the other physical, in which the atmosphere replaces the dissolved oxygen. Thus, we are mostly concerned with the relative rates of DO consumption and re-aeration.

A Basic Introduction to Pollutant Fate and Transport, By Dunnivant and Anders
Copyright © 2006 by John Wiley & Sons, Inc.

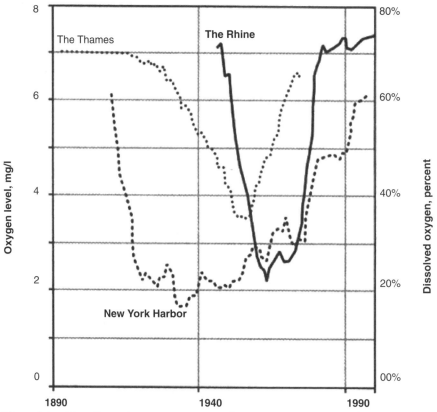

Figure 7.0. The effects of sewage outflow on dissolved oxygen levels in three major rivers. [Reprinted with permission from Chelsea Green Publishing Company; Meadows et al. (2004).]

In this chapter we will learn about biochemical oxygen demand, dissolved oxygen, sewage treatment plants, the kinetics of microbial decomposition, the kinetics of re-aeration, and how to remediate a stream or lake system after it has undergone eutrophication.

7.1 INTRODUCTION

This chapter is concerned with modeling the effects of the release of oxidizable organic matter to a flowing body of water. The most common form of organic waste is raw or untreated domestic sewage, but organic waste containing animal waste can have the same environmental effects. As we discussed in earlier chapters, the term *dissolved oxygen* (DO) refers to the chemical measurement of how much oxygen is dissolved in a water sample, usually expressed in mg/L. The biochemical oxygen demand (BOD) is an estimate of how much total DO is required to oxidize the organic matter in a water sample. Thus, the BOD of a water or wastewater is actually calculated as the change in DO from initial DO at saturation to the amount remaining after 5 days, and BOD is expressed in mg O_2/L. A plot of experimentally consumed oxygen (BOD) versus time is shown in Figure 7.1 for a domestic sewage sample. Before we discuss the modeling aspects of BOD in streams, represented by the Streeter–Phelps equation, it is important to gain an appreciation for the extent of the global sewage problem and the environmental issues surrounding wastewater and pathogens contained in the wastewater.

Our standard of living in the United States and other developed areas of the world is a direct result of having adequate water and wastewater treatment. As early as 1700 B.C., people began to obtain the luxury of running water and then to deal with the disposal of associated wastes. Though there is evidence of plumbing and sewage systems at many historical sites, including the *cloaca maxiumn* (great sewer) of the ancient Roman Empire, use of sewer and plumbing systems did not become widespread until modern times (Wastewater and Public Health, 2000). Along with providing drinking water and disposing of sewage come the challenge of preventing the rapid spread of disease within populations that utilize a common water source and treatment facility.

Examples of microorganisms and viruses associated with waterborne diseases are (1) bacteria responsible for typhoid fever, cholera, and shigellosis, (2) viruses causing hepatitis and viral gastroenteritis, and (3) protozoa that are the agents of the waterborne diseases cryptosporidiosis and giardiasis. These microbes can be killed or removed, and their associated diseases prevented, in domestic water supplies by a combination of sand filtration with chlorination or ozonation, placed at the end of

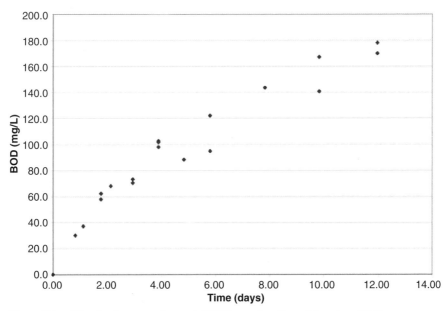

Figure 7.1.   Dissolved oxygen demand (BOD) as a function of time in a BOD_L experiment.

modern sewage treatment processes. While these diseases pose risks to human populations, the release of untreated sewage to waterways can also result in the immediate death of aquatic systems. Surface aquatic systems are aerobic, and the lifeforms contained in these systems are dependent on the constant presence of dissolved oxygen. Most streams are at or near saturation with respect to DO, with concentrations between 8–12 mg/L depending on the temperature and altitude of the stream. When readily oxidizable organic matter, such as domestic sewage, enters the stream, native microorganisms not only rapidly consume DO in the process of oxidizing this organic matter, but consume oxygen faster than it can be replenished through re-aeration from the atmosphere. Table 7.1 lists the saturated dissolved oxygen

TABLE 7.1. Solubility of Oxygen for Water in Contact with the Atmosphere (at 1 atm Containing 20.9% Oxygen)[a]

Temperature (°C)	Chloride Concentration (mg/L)				
	0	5000	10,000	15,000	20,000
0	14.6	13.8	13.0	12.1	11.3
1	14.2	13.4	12.6	11.8	11.0
2	13.8	13.1	12.3	11.5	10.8
3	13.5	12.7	12.0	11.2	10.5
4	13.1	12.4	11.7	11.0	10.3
5	12.8	12.1	11.4	10.7	10.0
6	12.5	11.8	11.1	10.5	9.8
7	12.2	11.5	10.9	10.2	9.6
8	11.9	11.2	10.6	10.0	9.4
9	11.6	11.0	10.4	9.8	9.2
10	11.3	10.7	10.1	9.6	9.0
11	11.1	10.5	9.9	9.4	8.8
12	10.8	10.3	9.7	9.2	8.6
13	10.6	10.1	9.5	9.0	8.5
14	10.4	9.9	9.3	8.8	8.3
15	10.2	9.7	9.1	8.6	8.1
16	10.0	9.5	9.0	8.5	8.0
17	9.7	9.3	8.8	8.3	7.8
18	9.5	9.1	8.6	8.2	7.7
19	9.4	8.9	8.5	8.0	7.6
20	9.2	8.7	8.3	7.9	7.4
21	9.0	8.6	8.1	7.7	7.3
22	8.8	8.4	8.0	7.6	7.1
23	8.7	8.3	7.9	7.4	7.0
24	8.5	8.1	7.7	7.3	6.9
25	8.4	8.0	7.6	7.2	6.7
26	8.2	7.8	7.4	7.0	6.6
27	8.1	7.7	7.3	6.9	6.5
28	7.9	7.5	7.1	6.8	6.4
29	7.8	7.4	7.0	6.6	6.3
30	7.6	7.3	6.9	6.5	6.1

[a] After Wipple, G. C. and M. C. Wipple. Solubility of oxygen in sea water. *J. Am. Chem. Soc.* **3**, 362 (1911).

concentration as a function of water temperature and salt content at 1.0 atmosphere of pressure. As you can see, increasing temperature and salt content decrease the amount of DO in a water.

A typical plot of the dissolved oxygen concentration as a function of distance from a point source, for a step release of sewage, is shown in Figure 7.2. Note the shape of the curve. Above the entry point of the sewage, the water is near saturation with respect to DO. Where the sewage enters the stream, the DO concentration plummets to near zero and often does drop to zero for a considerable length downstream of the input. As the organic matter is oxidized and as the stream re-aerates, the DO level slowly rises, eventually achieving natural background concentrations.

Figure 7.3 shows the dramatic effect of treating the sewage (for the same model input data used in Figure 7.2) before discharge into the stream. In this model simulation, the stream waters are not significantly affected by the small amount of BOD in the treated wastewater. In this chapter, we will learn to use the Streeter–Phelps equation, the mathematical model used to generate Figures 7.2 and

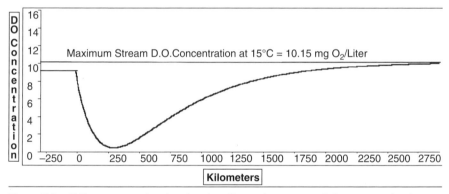

Figure 7.2.   DO sag curve plot showing the effects of sewage on DO concentration in a stream (output from Fate®).

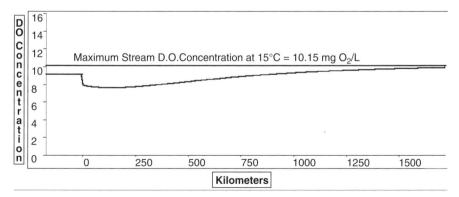

Figure 7.3.   DO sag curve plot of the same stream shown in Figure 7.2 but after sewage treatment has been installed (output from Fate®).

7.3. This model predicts the effects of sewage input to streams, evaluates the effectiveness of sewage treatment systems on stream quality, and can be used to determine which model parameters are the most important.

7.2 BASIC INPUT SOURCES (WASTEWATER FLOW RATES AND BOD LEVELS)

The Streeter–Phelps equation is designed to model the oxidation of any constant (step) input of oxidizable waste flowing to streams. Thus, the input source can be a variety of point and non-point sources, including the waste from domestic sewage plants, food-processing facilities, and inputs from livestock feedlots and agricultural settings. The only wastewater parameters that must be known are the flow rate of wastewater, the temperature of the wastewater, the DO of the wastewater (usually zero), and the concentration of BOD in the wastewater. All of these are relatively easy parameters to measure, with the possible exception of BOD. The BOD of a wastewater can be measured in most chemistry laboratories by taking a sample of the wastewater and diluting it so that it does not contain more than approximately 8.0 mg/L of BOD. (The total amount of organic matter in the bottle will not require more than 8 mg DO/L.) The appropriate dilution is determined based on experience with the wastewater or by trial and error in conducting the BOD test. Several 300-mL portions of the diluted wastewater are incubated at a constant temperature (usually 20°C) for 5–20 days. At predetermined times, the dissolved oxygen of samples is taken and plotted to determine the maximum DO needed to oxidize all of the organic matter in the sample as a function of time. A plot of a typical data set was shown in Figure 7.1. The maximum BOD (oxygen consumed) of the diluted sample is the flat portion at the top of the plot. Correcting for the dilution factor used to measure the BOD yields the total BOD of the (undiluted) wastewater. Since domestic sewage is the most common source of BOD to most streams, the components of a typical sewage treatment plant will be discussed briefly below.

The focus of modern sewage treatment is to remove turbidity, readily oxidizable organic matter, and pathogenic organisms. These three goals can easily be achieved at a minimal cost. First, we will give an overview of the treatment goals and how the wastewater treatment facility works, and then we will present a diagram of the system. Turbidity (suspended matter, microbes, etc.) is removed in settling clarifiers and in filtration systems. Dissolved organic matter is removed in biological contact units such as trickling filters and activated sludge lagoons. Most pathogens are naturally removed in the various treatment processes, but removal is ensured with the use of sand bed filtration, chlorination, and ozonation. One of the major design criteria for a wastewater treatment plant, and in fact a daily monitoring parameter for facility operators, is the experimentally determined BOD of the incoming and outgoing wastewater.

A diagram of a basic wastewater treatment plant is shown in Figure 7.4 and is explained in detail in the Flash program, *Water*, available at ftp://ftp.wiley.com/public/sci_tech_med/pollutant_fate/. We recommend that you download and watch this video after reading this chapter. Water enters the plant from the domestic sewer,

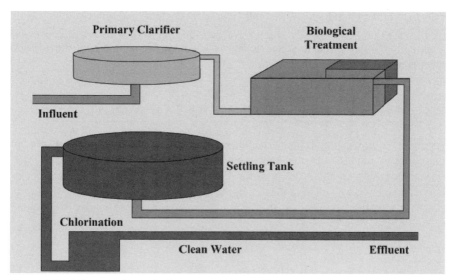

Figure 7.4. Diagram of a typical sewage treatment plant.

and first passes through a coarse bar screen to remove large materials that could damage pumps and clog pipes. Next, the water is passed through a finer filtration and grinder unit that removes paper and material the size of cigarette butts. Next, the water enters a primary settling tank (clarifier) where most of the remaining particles settle to the bottom of the tank and are pumped to the anaerobic digestor (discussed at the end of this section).

The relatively clear water leaving the primary clarifier is next processed in a biological contact unit. The most common biological units are the trickling filter and the activated sludge lagoon. The goal of both of these units is to convert dissolved organic matter to particulate matter by allowing microorganisms to absorb the dissolved organic matter and grow more microbial cells, such that they effectively become particulate matter. The trickling filter unit works by spraying the wastewater over rocks that contain mats of bacteria and algae that consume the dissolved BOD. The activated sludge lagoon is a highly mixed, dense slurry of microorganisms that also consume the dissolved BOD. Water leaving the biological contact unit enters settling tanks, where the microbial cells settle out. In the activated sludge lagoon, some of the cells are recycled to keep the microbial cell level high, and in the case of the tricking filter the settled particulate matter is transported to the anaerobic digester. Normally two or three settling tanks in series are used to produce very clear water. However, most governmental regulations require that the water be passed through a coal/sand filter to remove any additional particles and microorganisms. Next, water is chlorinated or contacted with ozone. Excess chlorine is removed prior to release to a stream. Water retention times in most treatment plants range from 10 to 20 hours.

Solids (particulate matter) that have been generated in the treatment process are digested in anaerobic digesters for 20–30 days, depending on the temperature of the digestion tank. The purpose of the digestion process is to reduce the microbial cells to relatively nondegradable solids (low BOD content) and reduce odors. Solids from digesters receiving only domestic sewage or food-processing waste are commonly composted and used as fertilizer or soil augmenters for farm fields and greenhouses.

7.3   MATHEMATICAL DEVELOPMENT OF MODEL

As discussed in Chapter 4, the derivation of the fate and transport equations used in this textbook requires that the student has taken linear algebra or differential equations. Since this textbook is designed for students who have only taken college chemistry and algebra, we will skip the derivation and simply state the governing fate and transport equation. A more mathematical derivation is given in the background section of Fate® for the Streeter–Phelps module.

The Streeter–Phelps equation can be represented by

$$D = \frac{k'\text{BOD}_\text{L}}{k_2' - k'}\left(e^{-k'(x/v)} - e^{-k_2'(x/v)}\right) + D_0 e^{-k_2'(x/v)} \tag{7.1}$$

where D is the dissolved oxygen concentration deficit (value below saturation) in $\text{mg O}_2/\text{L}$, k_2' is the re-aeration constant (in day^{-1}), BOD_L is the ultimate BOD (in mg/L), k' is the BOD rate constant for oxidation (day^{-1}), x is distance downstream from the point source (in miles or kilometers), v is average water velocity (in miles/day or kilometers/day but units must be compatible with distances, x), and D_0 is the initial oxygen deficit of the mixed stream and wastewater (in mg/L).

Before we discuss each parameter in Eq. (7.1), we will look at the overall solution of the equation. The equation is the net effect of two opposing reactions: the consumption of DO by microorganisms in oxidizing the BOD (represented by k') and re-aeration of the stream by dissolution of atmospheric oxygen (represented by k_2'). Both of these processes can be modeled by first-order reactions, hence the inclusion of exponential terms for the two k values. Of course, the model must incorporate the ultimate BOD (BOD_L) and the stream DO deficit [D_0 in Eq. (7.1)]. Note that the purpose of the equation is to calculate the amount of oxygen consumed by the waste (D). It is very important to note that D is *not* the remaining DO content of the stream water, but rather the amount of the original of DO that has been consumed. In order to calculate the resulting DO of the stream, we must subtract D from the original DO of the stream without BOD waste. We use the DO of the stream immediately prior to BOD entry for this value.

The term k_2' is the first-order rate constant associated with the re-aeration of the stream water. Exact measurement of this parameter is difficult since it is dependent on factors such as the stream depth, mixing in the stream, and the degree of water and air contact. For simplification purposes, a set of values has been tabulated by the Engineering Board of Review for the Sanitary District of Chicago (1925) and can be used based on a qualitative description of the stream. These values have been summarized by Metcalf and Eddy (1972) and are given in Table 7.2. Note that the actual k_2' values used in Eq. (7.1) are the log to the base e (natural log).

TABLE 7.2. Table of Re-Aeration Constants

Water Body	Ranges of k_2' at 20°C (base 10)	Ranges of k_2' at 20°C (base e, for calculations)
Small ponds and backwaters	0.05–0.10	0.12–0.23
Sluggish streams and large lakes	0.10–0.15	0.23–0.35
Large streams of low velocity	0.15–0.20	0.35–0.46
Large streams of normal velocity	0.20–0.30	0.46–0.69
Swift streams	0.30–0.50	0.69–1.15
Rapids and waterfalls	>0.50	>1.15

BOD_L is the ultimate BOD or maximum oxygen required to completely oxidize the waste sample. This value is determined or estimated through the BOD experiment discussed earlier. Normally BOD values are determined on a 5-day basis, which corresponds to the O_2 consumed during the first 5 days of degradation (oxidation). However, since we may be concerned with a travel time in the stream exceeding 5 days, we need to know the ultimate BOD (BOD_L). This value can be determined experimentally or estimated from the BOD_5 value using the following equation:

$$BOD_L = \frac{BOD_5}{1 - e^{-k'(x/v)}} \tag{7.2}$$

The k' term in this equation and in Eq. (7.1) is the DO uptake constant, a measure of the rate of the chemical reaction by which microorganisms use dissolved oxygen to degrade organic matter. This is obtained from a 20-day BOD experiment. The data are transformed by what is known as the Thomas slope method, and the slope of the line is equal to the rate constant, k', in day^{-1}.

The average water velocity is represented by v. This value is easily measured and is usually given in the problem statement. As noted above, the initial oxygen deficit (D_0) is calculated by subtracting the initial DO content of the stream-waste mixture from the dissolved oxygen level in the stream immediately upstream from the waste input. The plotted value in Fate® is a result of subtracting the oxygen deficit (D) at each point, calculated using Eq. (7.1), from the stream DO concentration above the waste input ($x < 0$). The net result is the remaining DO concentration in the stream.

The dissolved oxygen curve for a BOD-contaminated river can be divided into several zones based on dissolved oxygen content and physical appearance of the stream. These are shown in Figure 7.5. The area of the stream above the entry of sewage is referred to as the Zone of Clean Water (Zone 1 in Figure 7.5) and has the physical appearance of any natural clean system (stable fish, macro-invertebrate, and plankton populations). At the point of sewage entry, microbial consumption of the waste and oxygen begins. This marks the beginning of the Zone of Degradation (Zone 2 in Figure 7.5) that continues down the stream until the oxygen level falls to 40% of the initial value (as compared to the Zone of Clean Water). The Zone of

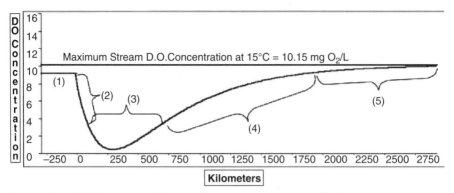

Figure 7.5. A DO sag curve labeled with its various zones of pollution and recovery.

Degradation is also characterized by the presence of more turbid water, an increase in CO_2 levels, and the presence of organically bound nitrogen. Physical characteristics of this zone include a high presence of green and blue-green algae (cyanobacteria), fungi, protozoa, tubiflex worms, and blood worms. Larger aquatic plants die off in this region. Zone 3 in Figure 7.5 begins when the oxygen level falls below 40% of the initial dissolved oxygen concentration and ends where it rises above 40% again. This is the Zone of Active Decomposition. Water in this zone tends to be gray or black and releases gases typical of anoxic enviornments (H_2S, CH_4, and NH_3). Bacteria and algae are usually the dominant life forms present. As the dissolved oxygen level rises above 40% of the initial value, the Zone of Recovery begins (Zone 4 in Figure 7.5). Carbon dioxide levels decrease, but nitrogen is still present as NH_3 and in organically bound forms. Biological characteristics include a decrease in the number of bacteria and an increase in the presence of protozoa, green and blue-green algae, and tubiflex and blood worms. Finally, the Zone of Cleaner Water (Zone 5 in Figure 7.5) is achieved where the chemical and biological characteristics of the stream are similar to those of Zone 1 (Zone of Clean Water).

With respect to these zones, one point is of special interest: the point at which the dissolved oxygen concentration (D) reaches its minimum value, which is referred to as the critical dissolved oxygen concentration (D_c). This point can be characterized by either (1) the time after which this minimum value occurs (the critical time, t_c) or (2) its distance downstream from the point source (the critical distance, x_c).

The time required to reach the critical distance can be calculated by

$$t_c = \frac{1}{k_2' - k'} \ln \frac{k_2'}{k'} \left[1 - \frac{D_0(k_2' - k')}{k' \text{BOD}_L} \right] \tag{7.3}$$

where D_0 is the initial oxygen deficit (O_2 saturation value minus initial stream-waste mixture value).

The critical distance is calculated by

$$x_c = v t_c \tag{7.4}$$

where the water velocity, v, can be given in miles/time or km/time.

The critical dissolved oxygen concentration (D_c), the minimum DO concentration in the stream, can be calculated by

$$D_c = \frac{k'}{k_2'} \text{BOD}_L e^{-k'(x_c/v)} \tag{7.5}$$

Example Problem. A city discharges 25 million gallons per day (mgd) of domestic sewage into a stream whose typical rate of flow is 250 cubic feet per second (cfs). The velocity of the stream is approximately 3 miles per hour. The temperature of the sewage is 21°C, while that of the stream is 15°C. The 20°C BOD_5 of the sewage is 180 mg/L, while that of the stream is 1.0 mg/L. The sewage contains no dissolved oxygen, but the stream is 90% saturated upstream of the discharge. At 20°C, k' is estimated to be 0.34 per day while k_2' is 0.65 per day.

1. Determine the critical oxygen deficit and its location.
2. Also estimate the 20°C BOD_5 of a sample taken at the critical point. Use temperature coefficients of 1.135 for k' and 1.024 for k_2'.
3. Plot the dissolved oxygen sag curve.
4. Determine the dissolved oxygen concentration at 1000 km from the point source.

Solution

1. Determine the dissolved oxygen in the stream before discharge.

   Saturation concentration at 15°C (from table on worksheet) = 10.2 mg/L

   Dissolved oxygen in stream = 0.90 (10.2 mg/L) = 9.2 mg/L

2. Determine the temperature, dissolved oxygen, and BOD of the mixture using the mass balance approach. Note that units should be compatible.

Flow rate of stream (conversion from cubic feet per second to liters per day):

$$\left(\frac{250 \text{ ft}^3}{\text{sec}}\right)\left(\frac{7.48 \text{ gal}}{\text{f}^3}\right)\left(\frac{60 \text{ sec}}{\text{min}}\right)\left(\frac{60 \text{ min}}{\text{hr}}\right)\left(\frac{24 \text{ hr}}{\text{day}}\right) = 161.6 \times 10^6 \text{ gal/day}$$

$$\left(\frac{161.6 \times 10^6 \text{ gal}}{\text{day}}\right)\left(\frac{3.79 \text{ L}}{\text{gal}}\right) = 612 \times 10^6 \text{ L/day} = 612 \text{ million liters/day}$$

Flow rate of sewage effluent (from gallons per day to liters per day):

$$\left(\frac{25 \times 10^6 \text{ gal}}{\text{day}}\right)\left(\frac{3.79 \text{ L}}{\text{gal}}\right) = 94.8 \times 10^6 \text{ L/day} = 94.8 \text{ million liters/day}$$

Note: Flow rates should be expressed in million gallons/day, million liters/day, or million cubic feet/day.

Temperature of mixture:

Net change in temperature $(\Delta T) =$ Stream input + Sewage input − Output effect

$0 =$ (stream flow rate)(stream temp.) + (sewage flow rate)(sewage temp.)

$\quad$ − (mixture flow rate)(mixture temp)

$0 = (612 \times 10^6 \text{ L/day})(15°C) + (94.8 \times 10^6 \text{ L/day})(20°C)$

$\quad - (612 \times 10^6 \text{ L/day} + 94.8 \times 10^6 \text{ L/day}) \, T_{\text{mixture}}$

upon rearrangement yields:

$$T_{\text{mixture}} = \frac{(612 \times 10^6 \text{ L/day})(15°C) + (94.8 \times 10^6 \text{ L/day})(20°C)}{612 \times 10^6 \text{ L/day} + 94.8 \times 10^6 \text{ L/day}} = 15.7°C$$

Dissolved oxygen of mixture:

Net change in DO = Stream input + Sewage input − Output

$0 =$ (stream flow rate)(stream D.O.) + (sewage flow rate)(sewage DO)

$\quad$ − (mixture flow rate)(mixture DO)

$0 = (612 \times 10^6 \text{ L/day})(9.2 \text{ mg/L}) + (94.8 \times 10^6 \text{ L/day})(0.0)$

$\quad - (612 \times 10^6 \text{ L/day} + 94.8 \times 10^6 \text{ L/day}) (\text{DO}_{\text{mixture}})$

upon rearrangement yields

$$\text{DO}_{\text{mixture}} = \frac{(612 \times 10^6 \text{ L/day})(9.2 \text{ mg/L}) + (94.8 \times 10^6 \text{ L/day})(0.0 \text{ mg/L})}{612 \times 10^6 \text{ L/day} + 94.8 \times 10^6 \text{ L/day}}$$

$$= 7.97 \text{ mg/L}$$

BOD$_5$ of mixture:

Net change in BOD$_5$ $(\Delta \text{BOD}_5) =$ Stream input + Sewage input − Output

$0 =$ (stream flow rate)(stream BOD$_5$) + (sewage flow rate)(sewage BOD$_5$)

$\quad$ − (mixture flow rate)(mixture BOD$_5$)

$0 = (612 \times 10^6 \text{ L/day})(1.0 \text{ mg/L}) + (94.8 \times 10^6 \text{ L/day})(180 \text{ mg/L})$

$\quad - (612 \times 10^6 \text{ L/day} + 94.8 \times 10^6 \text{ L/day}) \, \text{BOD}_{5\,\text{mixture}}$

upon rearrangement yields:

$$\text{BOD}_{5\,\text{mixture}} = \frac{(612 \times 10^6 \text{ L/day})(1.0 \text{ mg/L}) + (94.8 \times 10^6 \text{ L/day})(180 \text{ mg/L})}{612 \times 10^6 \text{ L/day} + 94.8 \times 10^6 \text{ L/day}}$$

$$= 25.0 \text{ mg/L}$$

BOD$_L$ of mixture (at 20°C):

$$\text{BOD}_L \frac{\text{BOD}_5}{1 - e^{-k'(x/v)}} = \frac{25.0 \text{ mg/L}}{1 - e^{-(0.34/\text{day})(5 \text{ days})}} = 30.6 \text{ mg/L}$$

3. Correct the rate constants to 15.7°C: Rate constants are not linearly related to changes in temperature; therefore, we must correct them using an exponential

relationship. Typically, these can be corrected using the two constants and equations given below. Note that 20°C is used as the reference point since this is where the original data for the k' values were collected.

$$k' = 0.34(1.135)^{15.7-20} = 0.197 \, \text{day}^{-1}$$
$$k'_2 = 0.65(1.024)^{15.7-20} = 0.587 \, \text{day}^{-1}$$

4. Determine the critical time (t_c) and critical distance (x_c): In the table note that the saturation value for O_2 at 15.7°C is 10.1 mg/L; however, the stream is at 90% of the saturation value (9.2 mg/L). Thus, the initial oxygen deficit is

$$D_0 = (\text{initial stream } O_2 \text{ value} - O_2 \text{ of the mixture})$$
$$= (9.2 - 7.97) = 1.23 \, \text{mg } O_2/L$$

$$t_c = \frac{1}{k'_2 - k'} \ln \frac{k'_2}{k'} \left[1 - \frac{D_0(k'_2 - k')}{k'\text{BOD}_L} \right]$$

$$t_c = \frac{1}{0.587/\text{day} - 0.197/\text{day}} \ln \frac{0.587/\text{day}}{0.197/\text{day}}$$
$$\left[1 - \frac{2.13 \, \text{mg}/L(0.587/\text{day} - 0.197/\text{day})}{0.197/\text{day}(30.6 \, \text{mg}/L)} \right]$$

$$t_c = 2.42 \, \text{days}$$

$$x_c = vt_c$$
$$= \left(\frac{3 \, \text{miles}}{\text{hr}} \right) \left(\frac{24 \, \text{hr}}{\text{day}} \right)(2.42 \, \text{days}) = 174.2 \, \text{miles}$$

or

$$= \left(\frac{3 \, \text{miles}}{\text{hr}} \right) \left(\frac{1.61 \, \text{km}}{\text{mile}} \right) \left(\frac{24 \, \text{hr}}{\text{day}} \right)(2.42 \, \text{day}) = 280 \, \text{km}$$

5. Determine D_c: To calculate the critical oxygen deficit, D_c, we must first convert the water velocity into units of miles/day: 3 miles/hr = 72 miles/day.

$$D_c = \frac{k'}{k'_2} \text{BOD}_L e^{-k'(x_c/v)}$$
$$= \left(\frac{0.197/\text{day}}{0.587/\text{day}} \right)(30.6)e^{-(0.197/\text{day})(174.2 \, \text{miles})/(72 \, \text{mile}/\text{day})} = 6.37 \, \text{mg}/L$$

Thus, the DO will be depressed 6.37 mg/L from its saturation value. The minimum O_2 concentration of the stream will be the saturation value minus the D_c, or $9.2 - 6.37 = 2.83 \, \text{mg } O_2/L$.

6. Determine the BOD_5 of a sample taken at distance x_c.

$$BOD_5 = BOD_L e^{-k'(x/v)}$$
$$= (30.6 \text{ mg/L})e^{-0.197/\text{day}(174.2 \text{ miles})/(72 \text{ mile/day})} = 19.0 \text{ mg/L}$$
$$20°C \ BOD_5 = BOD_L[1 - e^{-(k')(5)}]$$
$$= 19.0 \text{ mg/L}[1 - e^{-(0.34/\text{day})(5 \text{ days})}] = 15.5 \text{ mg/L}$$

7. The oxygen sag curve in km from the point source is shown in Figure 7.6.

7.4 SENSITIVITY ANALYSIS

Each modeling chapter has discussed the need to question the model and input parameters. Model parameters for a sensitivity analysis of the Streeter–Phelps equation include the stream flow rate (v), the biological oxygen consumption rate (k'), and the stream re-aeration rate (k_2'). This will be the focus of a homework problem.

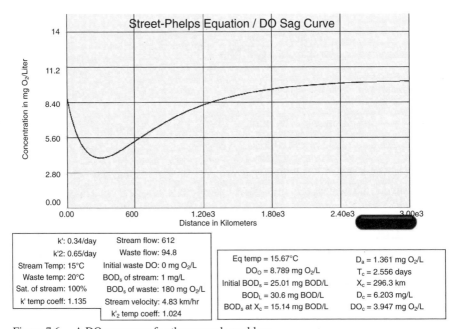

Figure 7.6. A DO sag curve for the example problem.

7.5   LIMITATIONS OF OUR MODEL

Average Re-aeration Rates for Streams. The basic Streeter–Phelps equation used in this chapter only allows for an average re-aeration constant to be used. This is appropriate for many large streams, but smaller streams alternate between riffle areas (small rapids) and calm pools. In order to account for these varying re-aeration areas, one would need to use a numerical methods approach. However, by using an average re-aeration constant, adequate estimates of the DO curve for most streams can be obtained.

Sedimentation of BOD Particles. The basic Streeter–Phelps equation does not allow for sedimentation of raw sewage to the bottom of streams, which could occur in large slow-moving waters. Thus, if a total BOD is used, but only a portion of the BOD is actually in the stream water, consuming DO, the oxygen deficit (D_0) would be overestimated. An extended version of the Street–Phelps equation is available to account for the sedimentation of sewage, but it requires measurement of additional (and less readily available) parameters (Metcalf and Eddy, 1991).

7.6   REMEDIATION

The first step in remediating a system that has been contaminated with biodegradable organic matter is to remove the source of the pollution. We have been successful in doing this in developed countries by installing relatively simple sewage treatment plants, even in highly populated areas where our waste generation is excessive. But in order to completely restore a system to pristine conditions, additional treatment is usually needed. The typical sewage treatment plant has sufficient technology to remove essentially all of the degradable organic matter and most of the nitrogen (ammonium, nitrite, and nitrate). However, these two nutrients are usually not the limiting nutrient in aquatic systems. A limiting nutrient is simply the microbial nutrient that is in limiting supply (limits the growth) in a given system. For freshwater aquatic systems this is usually phosphorus in the form of phosphate. Therefore, by adding phosphate, we increase the growth potential, as well as DO consumption, of microbes in the system. Most sewage treatment systems today do not remove phosphate, although Europe is making considerable headway at doing this. In the next 5–10 years, the U.S. EPA is expected to impose phosphate removal requirements on sewage treatment plants. Although it is amazing the EPA has no federal phosphate emission limit, a few states have created them. Phosphate removal is relatively simple and low cost, because iron phosphate can be precipitated in the settling basins of treatment plants by adding iron chlorine to the influent waters to the plants. Iron phosphate is highly insoluble and is incorporated in the waste sludge. This is actually an advantage when the sludge is used as a soil augmentation (fertilizer). Sewage treatment plants in locations prone to eutrophication often have taken such steps.

Once the source of the waste has been removed, we simply must wait for the system to recover. In most cases, because of the enormous volumes of water that would need to be treated for most contaminated aquatic system, we cannot directly treat the water in the lake or river. In theory, small lakes could be aerated using mechanical aerators or diffusers, but this can be expensive over extended time periods. Streams can recover very rapidly, since they naturally re-aerate and flush the nutrients out of the system. But this can create a even greater problem for the downstream receiving system, usually estuaries. Many estuaries are the spawning grounds for fishery industries, and the constant input of nutrients, especially limiting nutrients, can result in eutrophication and destruction of the ecosystem. Such a case is the dead zone at the mouth of the Mississippi River in the United States. As more advanced sewage treatment facilities are installed and as we better handle non-point sources of organic waste, streams and estuaries will continue to recover.

Concepts

1. Explain the difference between DO and BOD.
2. Draw a basic BOD plot (DO consumed versus time).
3. What is the difference between BOD_5, BOD_{20}, and BOD_L?
4. Name three sources of BOD to a stream.
5. There are two main reasons for installing sewage treatment plants, one human and one ecological. Name these.
6. Draw a basic sewage treatment plant, label each treatment unit, and explain what it does.
7. Draw the basic shape of the Streeter–Phelps (DO sag) curve and label the defined zones.
8. Describe each zone of the DO sag curve.
9. Perform a sensitivity analysis on v, k_2', and k' using the background example in Fate®.

Exercises

1. A cattle stockyard borders a small creek. Due to its close proximity to the creek, there is considerable seasonal runoff. With the given stream parameters below, along with the rate constants for the BOD of the runoff, manually create a Streeter–Phelps graph of the change in DO concentration versus distance ($BOD_5 = 2600$ mg/L). Next, create a plot of the same situation but with treated runoff ($BOD_5 = 65$ mg/L). Assume that the treatment plant is 98% efficient. Use Fate® to check your answers.

   k_1': 0.52 day^{-1}

   k_2': 0.71 day^{-1}

   Temperature of the stream: 12°C

   Temperature of the waste: 18°C

Flow rate in (stream): $500 \, m^3/min$

Flow rate out (waste): $83.6 \, m^3/min$

Stream velocity: $3.96 \, km/h$

2. A flood event forces a sewage treatment plant out of commission. Thus, all of the domestic sewage flows directly into the large stream running adjacent to town. Create a plot showing the concentration of DO in $mg \, O_2/L$ versus distance from the plant using the following parameters:

BOD rate constant (k'): 0.34/day

Reaeration rate constant (k_2'): 0.65/day

Stream temperature: 17°C

% Saturation: 100%

Waste temperature: 22°C

Stream flow: 1.2 million gallons per day

Waste flow: 0.19 million gallons per day

Initial waste DO: 0 mg/L

BOD_5 of stream: 1 mg/L

BOD_5 of waste: $180 \, mg \, O_2/L$

Stream velocity: 1.9 km/hr

k_2' temp coefficient: 1.024

k' temp coefficient: 1.135

Now, for your sensitivity analysis, plot the previous graph assuming the plant removes 98% of the initial BOD_5. Use Fate® to check your answers.

3. A vegetable processing plant is assessing the risk of a pond breach, for a waste holding pond that is adjacent to a river. They hire you to determine the concentration of dissolved oxygen (in mg/L) in the stream if a breach were to occur. They wish to know the effects on DO given two cases: (1) no BOD removal performed within the plant and (2) removal of 98% of the BOD. The effluent from the pond has a BOD of 50 mg/L and a flow of $175 \, m^3/sec$. The stream has a BOD of 1 mg/L, 100% DO saturation, a flow of $300 \, m^3/sec$, and a velocity of 3 m/sec. The initial concentration DO of waste is zero. The temperature of the stream is 15°C and the temperature of the waste is 20°C. The k' temperature coefficient is 1.135; the k_2' temperature coefficient is 1.024. Use $k' = 0.12$ /day and $k_2' = 0.25$/day in order to calculate the DO at 500 km downstream from the spill and 3000 km downstream. Use Fate® to plot the entire graph of distance (km) versus concentration O_2 (mg/L).

4. Dairy product processing plants produce waste that is extremely rich in organic matter. Consider a cheese-processing center in the Midwest that releases its waste into a large stream without treatment. The cheese waste is released at 21°C, has a BOD_5 of 12,000 mg/L, and a flow of 480,000 gallons per day. There is no DO in the waste released from the plant. The large stream that the waste is released into is at a temperature of 15°C and has a flow of 90 million

gallons per day, a velocity of 10 km/h, a BOD_5 of 1 mg/L, and a DO of 9.0 mg/L. At 21°C, k' is 0.4/day and k_2' is 0.75/days. (The temperature coefficients for k' and k_2' are 1.135 and 1.024, respectively.) Plot the DO curve in km from point source. Does the system become anaerobic? Where?

Now, consider that the cheese plant installs a new cleaning system that eliminates 95% of the BOD from its waste. Make a new plot showing the DO curve after treatment.

Use Fate® to assist you in making the plots and to check your work. The data from this problem are organized in the table below.

	Stream	Waste
Temperature	15°C	21°C
BOD_5	1.0 mg/L	12,000 mg/L
Flow	90 million gal/day	480,000 gal/day
DO	9.0 mg/L	0 mg/L
Velocity	10 km/hr	—

	Value	Temp. Coefficients
k'	0.4/day	1.135
k_2'	0.75/day	1.024

5. Until the late 1900s, domestic sewage treatment was basically nonexistent. Presently, wastewater treatment in industrialized countries may be at its height; almost nothing gets through the rigorous treatment, except when accidents happen. Yet accidents do occur: On one river with a flow of 216 ft³/sec and a velocity of 2.2 km/hr, the wastewater treatment facility fails. Normally, the treatment plant has a 98% removal rate of BOD, but on this day everything passes through without treatment. The plant treats, on average, 25 million gallons of water per day. The river has a re-aeration constant of 0.47 day⁻¹ and a BOD rate constant of 0.103 day⁻¹. The water coming into the plant has a temperature of 16°C and the treated water leaving the plant is 20°C. The temperature coefficients for the stream and the waste are 1.135 and 1.024, respectively. The BOD_5 of the stream is 1 mg/L and the BOD_5 of the treated waste is 4 mg/L. The stream has a DO saturation of 90%. Using the Streeter–Phelps equation, compare the DO of the stream with and without treatment at 100, 400, 600, 1000, 1500, and 2500 km downstream. Use Fate® to check your answers.

Spreadsheet Exercise

Create a spreadsheet that performs the same calculations as Fate®. Construct your spreadsheet so that it is interactive (so that you can change numeric values for parameters and the plot automatically updates itself).

REFERENCES

Craun, Gunther. *Waterborne Diseases in the United States*. CRC Press, Boca Raton, FL, 1986.

Meadows, D., J. Randers, and D. Meadows. *Limits to Growth: The 30-Year Update*, Chelsea Green Publishing Company, White River Junction, VT, 2004.

Metcalf and Eddy, Inc., *Wastewater Engineering*, 2nd Edition, McGraw-Hill New York, 1972.

Metcalf and Eddy, Inc., *Wastewater Engineering*, 3rd Edition, McGraw-Hill, New York, 1991.

Nauman, H. A comparison of the prevalence of infectious gastrointestinal disease and the quality of individual, surface source water system in Lincoln County, OR. Masters Thesis; Oregon State University, 1983.

Streeter, H. W. and E. B. Phelps. *A Study of the Pollution and Natural Purification of the Ohio River*. United States Public Health Service, U.S. Department of Health, Education, and Welfare, 1925.

Sawyer, C. N. and P. L. McCarty. *Chemistry for Environmental Engineering*, McGraw-Hill, New York, 1978.

Snoeyink, V. L. and D. Jenkins. *Water Chemistry*, John Wiley & Sons, New York, 1980.

Standard Methods for the Examination of Water and Wastewater, 20th edition, American Water Works Association, Washington, DC, 1998.

Tchobanoglous, G. and F. L. Burton (Metcalf and Eddy, Inc.). *Wastewater Engineering: Treatment, Disposal, and Reuse*. McGraw-Hill, New York, 1991.

Wastewater and Public Health. Online Internet Source. Accessed February 24, 2000, http://danpatch.ecn.purdue.edu/~epados/septics/disease.htm

CHAPTER 8

FATE AND TRANSPORT CONCEPTS FOR GROUNDWATER SYSTEMS

CASE STUDY: THE TEST AREA NORTH DEEP WELL INJECTION SITE AT THE IDAHO NATIONAL ENVIRONMENTAL AND ENGINEERING LABORATORY (INEEL)

There are an endless number of polluted soils and aquifers that could be used as case studies for this chapter, but we will use the one described in Chapter 1 located at the INEEL, a Department of Energy site in Idaho (United States). This U.S. government owned-and-operated hazardous waste site was placed on the U.S. EPA's National Priorities List (NPL) on November 21, 1989. The injection well (TSN-05) was drilled in 1953 to a depth of 93 m (305 ft) with a diameter of 30.5 cm (12 inches). The groundwater surface at the site is approximately 63 m (206 ft) below the land surface. During its years of operation, the well received approximately 133,000 L (35,000 gallons; 193,000 kg) of liquid and dissolved trichloroethylene (TCE), organic sludges, treated sanitary sewage, metal filing process waters, and low-level radioactive waste streams. Basically the well was the means of disposal for any liquid or semiliquid waste (domestic and hazardous) produced at the remote and isolated site. Although several pollutants of concern were disposed of and detected in the groundwater at TAN, our main concern here is the TCE groundwater plume.

Figure 1.6 showed iso-concentration circles (isopleths; lines of equal TCE concentration) for the area surrounding the disposal site that were obtained by fitting a step-model to field measurements of TCE concentrations in the groundwater (explanatory modeling). Assuming, as expected, that TCE is present in the aquifer as a dense nonaqueous phase liquid (DNAPL), it solubilizes slowly, and the input of TCE can therefore be considered a continuous source (step input over an extended period of time). You will note that the isopleth near TSN-05 injection well represents a concentration of 1000 parts per billion (ppb), while the lowest concentration shown in this figure is 3 ppb (the maximum allowed drinking water concentration). If no remedial action was to be taken in 1994 (the proposed year of remediation if any was to be attempted) and the DNAPL continued to release TCE to the ground-

A Basic Introduction to Pollutant Fate and Transport, By Dunnivant and Anders
Copyright © 2006 by John Wiley & Sons, Inc.

water, another step model predicted that the TCE plume would expand as shown in Figure 1.7 by the year 2044.

In this chapter we will learn the chemical and physical processes responsible for pollutant fate and transport in groundwater systems such as the one in Idaho. As we will see, special approaches have been developed for groundwater in order to sample and study the sites. Polluted groundwater systems present special challenges to the environmental chemist and engineer and have accounted for a significant portion of the Superfund budget.

8.1 INTRODUCTION

Groundwater is the most prominent reservoir of liquid freshwater, although glaciers constitute the most common reservoir of freshwater overall. Data given in Figure 2.1 show that globally there are approximately $9.5 \times 10^7 \, km^3$ of groundwater, which accounts for 98% of the liquid freshwater on Earth. Yet groundwater is somewhat more difficult to access, with the exception of flow from natural springs, than is water in lakes and rivers. The residence time of water in the ground is also very long, from 200 to 1000 years or longer. Thus, a contaminated aquifer can take much longer to recover by natural means than would relatively rapidly flowing surface waters. In addition, contaminated groundwater systems are much more difficult to access for treatment and remediation efforts than are other aquatic systems.

Water in the ground can be grouped into two categories, based on whether it is present in saturated or unsaturated media. In general, saturated areas are those where the soil or fractured rock does not contain air spaces and freely drains water, whereas unsaturated soil or fractured rock contains air spaces and retains most or all of its water. Water flows in both saturated and unsaturated media. Geologic media can also be classified into two basic categories, consolidated (solid or fractured rock) and unconsolidated (sediments). Figure 8.1a shows a cross section of the Earth's surficial crust. The source of all groundwater is from the land surface (the recharge area on the left-hand side of the figure), and it flows downward until it contacts an impermeable layer of soil/rock (the aquiclude in Figure 8.1a). Water can enter the soil rapidly in major recharge areas, such as the one shown in Figure 8.1a, or by slow percolation through the soil. Rivers and lakes also contribute water to the subsurface. Water-saturated layers of the subsurface may be separated by layers of permeable or impermeable soil or rock deposits. Saturated layers between two impermeable soil layers are referred to as confined aquifers. The uppermost unconfined layer of water is referred to as the surface aquifer and is the most likely portion of the subsurface to be contaminated, since it is located the closest to the land surface.

There are many laws that regulate the purity or quality of groundwater. In the United States, the major law is the Safe Drinking Water Act (SDWA), originally passed by Congress in 1974 and amended in 1986 and 1996. As noted in the title, it is primarily concerned with the use of water removed from the ground for drinking water purposes. There are other sections of the SWDA and other laws regulating the disposal of waste into the ground and the use of water for irrigation purposes.

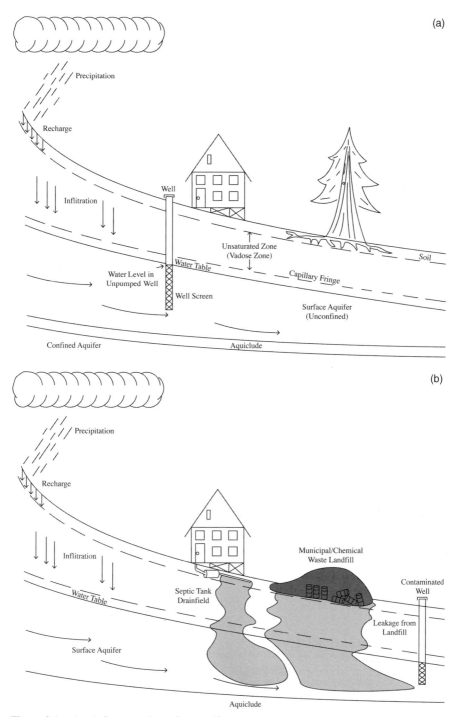

Figure 8.1. (*a–c*) Cross section of an aquifer.

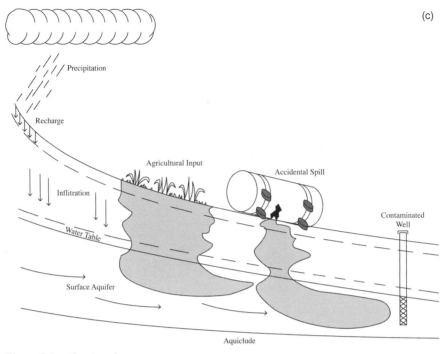

Figure 8.1.   *Continued*

Many areas of the world use groundwater as a source of drinking water, and while much of the water is pristine, there are also many areas that are contaminated. One major natural form of contamination is arsenic, which is leached into the water from mineral deposits. Of course there are many man-made pollutants present in groundwater in urban areas or near industrial complexes (discussed in the next section). The SDWA establishes several "barriers" against the pollution of water. These include regulations to protect the water supplies and for treatment of drinking water, ways to maintain distribution systems, and guidelines for providing information to the public. Under the SDWA, the U.S. Environmental Protection Agency sets limits on the amount of each pollutant or microorganism that can be present in drinking water. These are referred to as primary pollutants and have maximum contaminant limits (MCLs). For each pollutant, the concentration deemed safe for consumption is determined based on the risk assessment principles discussed in Chapter 10.

8.2   INPUT SOURCES

Sources of pollution to groundwater may include any liquid that can drain into the unsaturated or saturated zone (direct spills of liquid or pollutants washed into the subsurface by rain). Examples are shown in Figure 8.1b,c and include liquids from leaking surface or subsurface storage tanks, material storage (coal, road salt, and ore leachate), drainage and leachate from farming operations, contamination entering the aquifer from rivers and lakes, subsurface mining operations, pollutants disposed of in waste injection wells, saltwater intrusion, and leachate from domestic and hazardous waste landfills. Two groundwater pollutants that are in the news today and will be in the news for years to come are perchlorate (from solid rocket fuel) and methyl tertiary butyl ether (MTBE; an additive to gasoline). As noted in the introduction, groundwater systems are much more complicated than river or lake systems. In rivers and lakes we have one distinct advantage in pollution treatment or remediation efforts: the flushing of the system and removal of the polluted water by rapid flow through the system. In comparison, flow rates in groundwater are very slow, and access to the system for treatment of the water or soil is highly limited.

Many sources of pollution produce one or more biodegradable pollutants that can result in the consumption of available dissolved oxygen. Thus, the same set of microbial communities and degradation schemes discussed in the lake chapter for anaerobic waters can, and often do, occur in groundwater systems. This results in a changing oxidation state (E_H) of the water. Near sources such as landfills or hazardous waste injection wells, organic waste is oxidized and the dissolved oxygen is consumed. This results in an anaerobic zone in the aquifer, which will proceed through the usual order of terminal electron acceptors (O_2, NO_3^-, SO_4^{2-}, and finally CO_2). As unpolluted and oxygenated water mixes with the polluted water, the aquifer will establish a facultative zone (an area containing only a small amount of oxygen). As more and more clean water mixes with the polluted water, dissolved oxygen levels will slowly return to normal. However, this sequence of mixing and terminal electron acceptors can occur over a great distance in the subsurface if a large amount of contamination is present, and the water in this region will be unfit for consumption. Thus, the waste entering the system must be characterized or a history of the waste inputs must be known in order to predict the chemical reactions occurring in the aquifer. In addition, as we illustrated in Chapter 3, the pH and E_H of the system (as controlled by microbial degradation of pollutants) can greatly affect sorption (K_d or K_p) and degradation reactions (k). As we will see in this chapter, each of these parameters can be the governing factor determining the fate and transport of pollutants in groundwater systems.

8.3   MONITORING WELLS

Taking environmental samples to monitor the concentration of a pollutant in river, lake, and atmospheric systems is relatively easy, although there are strict procedural guidelines for sampling and analysis. In these systems, you can see and feel the sample matrix. However, the sampling of groundwater is very different. For simplification purposes, we will limit our discussions primarily to saturated groundwater systems, but remember that contamination also can occur between the ground surface and the flowing aquifer in the unsaturated zone.

First, one has to determine the depth of the water table (boundary of the saturated zone; Figure 8.1) and then determine which way the water is flowing. Although today there are nonintrusive techniques that do this, we usually resort to the very intrusive and potentially destructive method of installing a monitoring well, since this also enables us to obtain water samples to analyze for pollutants. A monitoring well consists of a metal or PVC casing (tube), installed from the land surface to a given depth in the aquifer. There is no good way to install such a well without significantly disrupting the nature of the aquifer. In environmental sampling, we always want to take a "representative and unbiased" sample, so that the water we obtain is characteristic of any water in the system. Obviously, if you violently drill a hole into a system, grinding and mixing dirt and rock fragments with the water, it is difficult to obtain a representative sample. But for now, this is the only technically feasible and economical method we have of obtaining groundwater samples.

We will next present a few of the well-installation techniques used today, since installing groundwater monitoring wells is often an expensive part of a characterization process and since our modeling results depend on obtaining samples to validate our modeling efforts. Also, as noted earlier, the sampling of groundwater requires special considerations. The types of drilling techniques we will cover include the cable tool, direct rotary drills, and auger and coring systems.

8.3.1 Cable Tool Percussion Method

The cable tool percussion method of drilling wells was probably used in the first attempts at installing a deep well. The technique goes back at least 4000 years and was developed by the Chinese, who successfully drilled wells to approximately 915 m (3000 feet). A photograph of a modern system is shown in Figure 8.2. The drilling process works by repeated lifting and dropping of the drill bit (shown in Figure 8.3). Water is usually used to suspend the broken rocks and media, but this is not always necessary. Periodically, when sufficient debris has accumulated at the bottom of the well being drilled, a drill bit is replaced with a bailer and the solids are removed. The cable tool technique can be used in almost any geologic media, but works especially well in coarse glacial sediments (till), boulder deposits, or rock strata that are highly disturbed, broken, fractured, or cavernous. Well depths typically run from 90 m (~300 ft) to 1520 m (5000 ft). Major advantages of the cable tool technique include the following (Johnson Filtration Systems, 1986):

- Drill rigs are relatively inexpensive.
- Drill rigs are simple in design and use.
- Machines have low energy requirements.
- Recovery of samples of geologic media is possible.
- Wells can be drilled in areas with little water supply.
- Wells are not contaminated by drilling equipment (muds, discussed later).
- Drill rigs can be operated in all temperature regimes.
- Wells can be drilled in media where loss of circulation is a problem (discussed in the next section, rotary drill rigs).
- Water yield from the formation can be determined at any point in the drilling operation.

Figure 8.2.   A cable tool drill rig (U.S. DOE-INEL photograph).

Disadvantages include relatively slow drilling rates and high casing costs resulting from heavier diameter wall casing needed.

8.3.2   Direct Rotary Drill Method

A much faster drilling technique is the direct rotary drill method. In this approach, the drill bit is directly spun (motor driven) to break the geologic media. The drill

Figure 8.3. A cable tool drill bit and bailer (U.S. DOE-INEL photograph).

shaft is hollow, and some form of fluid is forced down the inside of the shaft, removing the cuttings from the well by forcing them up the well between the drill shaft and the borehole. You may have seen this type of drilling technique in use, or in movies on television, since it is used in wells for water, gas, and petroleum exploration. A typical rotary drill rig is shown in Figure 8.4. Drilling fluids must be used to remove cutting, to prevent the drill bit from becoming lodged in the borehole.

Figure 8.4.   A direct rotary drill rig (U.S. DOE-INEL photograph).

Air, water, and synthetic "mud" (a clay mineral, bentonite) have been used for this purpose. Synthetic mud contains clay in suspension and, in some cases, surfactants. Air and clean water are preferable for installing monitoring wells for environmental sampling, since synthetic mud is transported into the geologic medium during the installation process. Once there, the mud cannot be completely removed upon well development (discussed later) and affects the movement of pollutants. Water and synthetic mud are necessary, however, in traditional direct rotary drilling in unconsolidated media, since they also serve to hold the formation open (prevent collapsing) around the drill bit and borehole. With this in mind, it follows that air can only be used in semiconsolidated or consolidated media, which will not be subject to collapse, as would unconsolidated material. One added feature of the rotary drill technique is that it can drill monitoring wells under permanent facilities by drilling at steep angles, as illustrated in Figure 8.5.

The major advantages of the direct rotary drill method include the following (Johnson Filtration Systems, 1986): relatively high drill rates in all types of media, minimal casing required during the actual drilling process, and rapid rig mobilization and demobilization.

Figure 8.5.   A direct rotary drill rig being used to drill a diagonal well (U.S. DOE-INEL photograph).

Major disadvantages of this technique include high cost of drilling rigs, high level of maintenance required by drilling rigs, special procedures required for collection of media samples, and possible plugging of the drill bit and borehole caused by loss of pressure of the drilling fluid in fractured formations.

8.3.3   Earth Augers

The most common type of drilling method for installation of shallow monitoring wells is the hollow stem auger. A small, hand-operated system is shown in Figure 8.6, but larger truck-mounted systems are common. These systems are only used in unconsolidated media. The hollow stem allows soil samples to be collected and analyzed for pollutant contamination. This technique is used both for soil sampling and for installation of monitoring wells. These systems allow relatively rapid installation, but are limited to shallow depths.

8.3.4   Well Casing, Grouting, and Sealing the Well Casing

In most cases, once a borehole is drilled it must be lined with casing to prevent collapse of the geologic media. Commonly available casing materials include

Figure 8.6.   A handheld auger drilling system. (Courtesy of Forestry Suppliers, Inc. http://www.forestry-suppliers.com).

aluminum, carbon steel, stainless steel, and PVC. For monitoring wells, the type of casing used depends on the type of contamination. For example, PVC casing would not be used for monitoring the subsurface around a leaking underground organic chemical tank, since the liquid chemical might dissolve the PVC casing. PVC casing would also sorb any hydrophobic pollutant from the water. Similarly, metal casing would unlikely be used in wells for monitoring metal contamination.

Grouting is a term used to describe the filling of the space between the well casing and the borehole (referred to as the annular space). This filling is needed to prevent water from entering the well from the surface and in order to isolate different regions of the well. Some monitoring operations, especially in the unsaturated zone of a formation, use lysimeters for obtaining water samples. A lysimeter is a porous cylinder (usually ceramic) that contains two tubes that lead back to the land surface. A vacuum can be pulled on the lysimeters, allowing the collection of water over long periods of time. When sufficient time has passed, one tube of the lysimeter is pressurized, which lifts the waters in the lysimeter to the surface through the second tube for sampling. Lysimeters are placed at areas in the unsaturated subsurface where water is expected to collect during infiltration from rain or flooding. Collection areas can be located at different depths in unconsolidated media or at fractures in consolidated media. In any event, water must be prevented from entering at the land surface or flowing between different lysimeters. To prevent the inflow of water from another region of the subsurface, a clay medium, usually bentonite (one form of grout), is placed above and below the lysimeter, and sand or silica is placed immediately around the lysimeter to allow water to freely pass to the collection point. The addition of grouting materials to a borehole is illustrated in Figure 8.7. Above the lysimeter, more bentonite is placed to seal the lower lysimeter from water flowing above. Another lysimeter is placed at the next collection point, and so on, until each fracture or sampling locale has a lysimeter and the well is full. Many wells/boreholes are sealed at the surface with cement, but given the high pH associated with the $Ca(OH)_2$, cement would not be used near any sampling location.

Wells at monitoring stations are drilled to the desired depth and then typically a porous section of pipe (referred to as a screen) is placed in the borehole to allow collection of aquifer water. Screen depths, where water is expected to be collected, can range from a few centimeters to several meters, depending on the section of the aquifer that you desire to sample. Sand or silica is placed between the screened casing and the borehole, and grout, usually bentonite, is placed above the screened area. Again, cement is avoided except at the land surface.

8.3.5 Well Development

As noted at the beginning of this section, the installation of monitoring wells is an intrusive and destructive process, and the area immediately adjacent to the borehole/well needs to be returned to its original condition, or as close to it as possible. Thus, well development refers to (1) the repair of damage to the subsurface (geological) formation and (2) the alteration of the aquifer to allow water to flow freely to the well for collection. First and foremost, it includes removal of any drilling fluid

Figure 8.7.   The addition of grouting materials to a borehole (U.S. DOE-INEL photograph).

(other than air) that was used in the drilling process. As noted earlier, the installation of most environmental monitoring wells does not allow the use of synthetic muds or surfactants. Yet it is necessary to remove any drill cutting (rock chips and disrupted material) that may be present in the formation and contaminate subsequent samples. In order to restore the aquifer to near-normal conditions, the well can be "overpumped" (pumped at a high water removal rate for hours or days). Another technique is to reverse the flow several times to dislodge particles in the formation. In general, states, federal agencies, or concerned clients have an established procedure for developing the well. It should be noted that you do not simply walk up to a monitoring well and take a sample. Usually, three to five well casing volumes of water must be removed before a water sample is taken for chemical analysis. The purpose of this is to ensure that you have a representative sample from the aquifer and not stagnant water from the well casing.

8.3.6 But How Good Is Our Well?

Groundwater monitoring wells have been used for decades, but as amazing as it may seem, the effectiveness of well grouting was only recently evaluated. Dunnivant et al. (1997) found that standard well installation procedures for consolidated media were highly effective in isolating different productive zones within a borehole. An example of this effectiveness is shown in Figure 8.8. The dark areas of the borehole represent the location of bentonite used to seal the borehole and prevent water from flowing from above or below. This lysimeter installation was later flooded with at least 1 m of water for 50 days, yet no water passed directly down the borehole. Water and tracer (^{75}Se) did migrate through the subsurface via fractures in the basalt media and arrived at some lysimeter locations but not at others. The first lysimeter placement at 6.4 m shows a typical tracer profile, with time, for a pulse input. The next two lysimeters in the borehole at 10.4 m and 18.9 m did not receive water or tracer over the entire experiment (50 days of water application). The next lysimeter at 37.8 m did receive water, apparently from the beginning of the flooding when no tracer was present. The next two lysimeters at 42.7 m and 52.7 m received water and tracer. Again, these and other experimental measurements confirm that water did not flow in the borehole between lysimeter placements, which validates the grouting technique used. These and other borehole studies during the experiment by Dunnivant et al. (1997) found that the standard grouting technique was effective in sealing the borehole in consolidated media. Another study has found standard grouting techniques to also be effective in unconsolidated media (Christman et al., 2002).

8.4 CHEMISTRY EXPERIMENTS USED TO SUPPORT MODELING EFFORTS

As we discussed in Chapters 2 and 3, chemistry can play important roles in fate and transport processes. In our previous discussions on river and lake systems, we mostly noted these roles in a qualitative manner, with the exception of the degradation rate constant, k. However, in groundwater systems, pollutants dissolved in the water (the mobile phase) are in constant contact with the soil/rock matrix (the stationary phase). Hence, all of the chemistry related to sorption phenomena plays a very important role. Recall that we summarize sorption reactions in terms of the distribution coefficient (K_d) for metals or partition coefficient (K_p) for hydrophobic pollutants. Also recall that for metal pollutants the pH, E_H, salinity, complexation, and content of

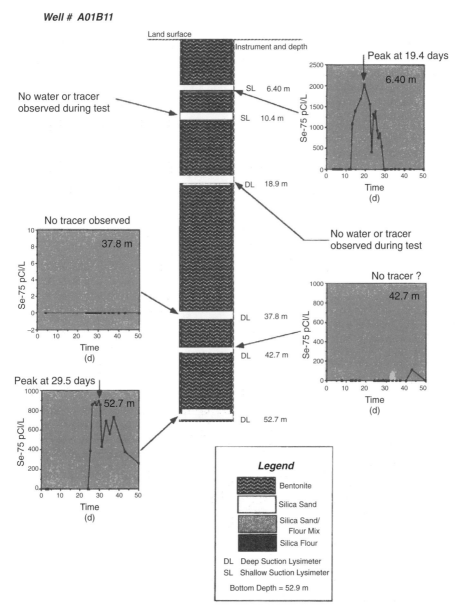

Figure 8.8.   Illustration of the effectiveness of grouting in lysimeter placements (Wood and Norrell, 1996).

natural organic matter (NOM) determine the degree of adsorption. For hydrophobic pollutants, the most important factors are the presence of natural organic matter and clay particles. However, we will see in a few paragraphs that the type of NOM, either dissolved or sorbed, is also important.

8.4.1 K_d and K_p Values

The sorption coefficient is perhaps the most important factor determining the fate and transport of a pollutant in groundwater systems. We will most often use a generic K, since we will be discussing both the adsorption of metals and the partitioning of hydrophobic pollutants. For metal pollutants, there is no one equation that can be used to estimate the distribution coefficient. The extent of adsorption for metal pollutants must be evaluated experimentally. For hydrophobic pollutants, predictive equations correlating adsorption with pollutant solubility have been developed independently by a number of researchers. Three of these are shown below

$$\log(K_{OC}) = 3.803 - 0.557 \log(S_1) \qquad \text{(Chiou et al., 1979)}$$
$$\log(K_{OC}) = 0.440 - 0.540 \log(S_2) \qquad \text{(Karickhoff et al., 1979)}$$
$$\log(K_{OC}) = 4.273 - 0.686 \log(S_3) \qquad \text{(Means et al., 1980)}$$

where K_{oc} is the organic carbon partition coefficient between the soil organic matter and the water, and

S_1 is the pollutant aqueous solubility in micromoles/L

S_2 is the pollutant aqueous solubility in mole fraction

S_3 is the pollutant aqueous solubility in mg/L

Recall that K_p can be calculated from K_{oc} by

$$K_p = K_{OC} \times \text{fraction of organic carbon}$$

Thus, if you know the aqueous solubility of the pollutant, either an experimental value or an estimated value from SPARC (http://ibmlc2.chem.uga.edu/sparc/; see Chapter 3), you can estimate the partition coefficient. But the user of these equations should note that each equation was developed for a particular set of compounds and experimental conditions, and minor or major errors may be present in your estimate of K. A much more accurate method is to measure K_d or K_p using the experimental setup discussed at the end of Chapter 3 or using the column study approach discussed later in this chapter.

8.4.2   Relationship Between K and the Groundwater Fate and Transport Equation

No matter how you obtain a value for K, it is not directly entered into the fate and transport equation. Groundwater fate and transport equations calculate movement of water and pollutants in terms of how much water passes through the system or unit volume of media. Thus, we need the K term expressed in terms of water volume. To begin understanding how this works, we will only discuss the movement of water through a unit volume of soil or fractured rock. Unlike in rivers and lakes, where a unit volume of the system was essentially all water, in groundwater systems, most of the unit volume is soil or rock. In groundwater flow and pollutant movement, we talk in terms of the number of pore volumes that need to be passed through a unit volume of media to move the pollutant. One pore volume is equal to the volume of water contained in the pores of one unit aquifer volume. In theory, it should take one pore volume for a tracer that does not react with the soil or rock media (referred to as a conservative tracer) to pass through the system. For a sorbing pollutant, the volume of water needed to push the pollutant completely through the system will be greater than one pore volume. We can mathematically define R, the retardation coefficient, as

$$R = \frac{V}{V_1}$$

where V is the volume of water needed to pass the tracer or pollutant completely through the system, and V_1 is the volume of water in the unit volume of media (the pore volume).

This approach for determining R works fine for laboratory settings, but is of little use in the real world, where we would have to collect and measure the volume of all water passing through a polluted aquifer. For these real-world applications, we can define R in a different manner using properties of the aquifer itself. This new approach requires a few more terms. First, we need the porosity (n) of the aquifer, which is similar to the pore volume but represented as a fraction of the total aquifer volume. Porosity for a saturated system is defined by the water capacity of the groundwater system and is calculated as the volume of water contained in a certain volume of aquifer divided by that aquifer volume. Typical values range from 0 for solid rock to ~0.4 for highly porous media. Another term we need is the bulk density (ρ_b) of the soil or fractured rock, since it is considerably different from that of water (for which we usually ignore density) and in order to cancel out the units of K and obtain a unitless value for R (L of water per kg of soil). Finally, we can express R, the retardation coefficient, in terms of K by

$$R = 1 + \frac{\rho_b K}{n}$$

Now, we have a way of relating pollutant movement to water movement by measuring two easily obtained physical parameters of the aquifer and K. Again, K can be determined by a direct experiment or by the column experiment, described next, which more accurately reflects the aquifer properties and conditions.

8.4.3 Column Studies for Evaluating Pollutant Transport in Subsurface Media

In designing a column study for a polluted aquifer, it is important to use soil/rock media characteristic of the site, and, as we will see, water collected from the site. If variations in geologic media are present, several column experiments may be necessary. As an example of a column study, we will present the work of Dunnivant et al. (1992a,b), who investigated the transport of cadmium and a polychlorinated biphenyl in a sandy aquifer material containing dissolved and sand-adsorbed natural organic matter. The goal of this investigation was to evaluate the effect of mobile (dissolved) NOM on pollutant transport through the soil columns. We will use pore volumes and R to describe the movement of conservative tracers and pollutants through the soil columns.

In soil column experiments, step or pulse inputs of pollutants can be used. In the study by Dunnivant et al. (1992a), a step (continuous) input was used. Before a soil column can be used to study pollutant movement, it must be characterized with a conservative tracer to make sure that dispersion is low. Excessive dispersion in the column, resulting from poor column packing or construction, will negate the interpretation of the sorption data. Bromide is commonly used as a conservative tracer, and the results from this experiment are used to estimate the pore volume of the column and dispersion. Dilute bromide solution is added as a step input, and small measured volumes of elution waters (water exiting the column) are collected and analyzed for bromide. The resulting data set, referred to as a breakthrough curve since it shows the breakthrough of chemical from the column, is tested using a computer program and an estimate of dispersion is obtained. If the column is acceptable, it is then used to test the transport of a pollutant through the column.

After an acceptable column has been constructed and tested with the bromide tracer, the column is saturated with NOM at a specified dissolved NOM level. This is achieved by pumping a NOM solution through a column of soil identical to the soil used in the pollutant experiment but not containing pollutant. This step is necessary in order to simplify the experimental design, since NOM and the pollutants sorb to the soil at different rates and we want to have only one variable in the experiment (the sorption and desorption of pollutants in the soil column). In general, NOM saturation experiments take 3–5 days (Dunnivant et al., 1992b).

Finally, water containing the cadmium, and in some cases NOM, is passed through the column. Discrete volumes of eluent water are again collected, the volumes measured, and the samples analyzed for Cd^{2+}. Figure 8.9 contains several breakthrough curves (BTCs) for Cd^{2+} as a function of dissolved NOM. First, note the BTC to the right. This is the BTC for Cd^{2+} dissolved in water not containing NOM and is our reference point. Note the axes of the plot. The y-axis is the fraction of Cd^{2+} collected, and the x-axis is expressed not in time but in pore volumes. The mid-point of this plot (0.50 on the y-axis) corresponds to R for Cd^{2+} (corresponding to an R value of 68, from which we calculate a K_d of 14.1). We can also calculate K_d since we can also measure the porosity and bulk density of the soil column. Many studies, including the one by Dunnivant et al. (1992a), also measure K_d using the procedures described in Chapter 3 and compare it to the column-calculated value. Obviously, agreement between the two techniques gives more credence to the approach and results. These concepts will become clearer if you complete the K_d and column study laboratory exercises in Chapter 13.

We noted in the introduction to this section that it is important to use soil and water from the site under investigation. The importance of this is illustrated in Figure 8.9. The site under investigation contained varying amounts of dissolved NOM; thus, it was important to determine the effect of the presence of NOM in the dissolved phase since NOM is known to bind to metal pollutants. In Figure 8.9, note that as the level of NOM is increased from 5.2 mg/L to 20.4 mg/L to 58.1 mg/L, the cadmium moves faster and faster through the column, as indicated by the fewer pore volumes of water needed to remove the Cd^{2+} from the column. The R values for

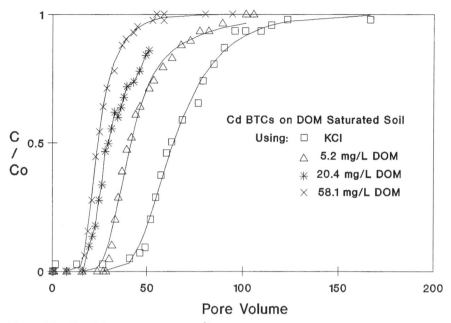

Figure 8.9.   Breakthrough curves for Cd^{2+} on a NOM-saturated soil column as a function of dissolved NOM concentration. [Reprinted with permission from Dunnivant et al. (1992a). Copyright 1992 American Chemical Society.]

these BTCs are 47, 36, and 26, respectively. Note the error that would have been introduced to the laboratory study if no NOM had been included in the water used in this study, while NOM was present in the aquifer. An R value of 68 would have been assumed, whereas in reality an R value as low as 26 could have been accurate for behavior of cadmium in the aquifer. Hence, the researchers would have significantly underestimated the mobility of cadmium in the system.

Recall again what the breakthrough curves represent. Initially, as water containing pollutant is passed through the column, pollutants sorb onto the soil and no detectable pollutant exits the column. Eventually, the sorption sites on the soil are saturated and pollutant starts to exit the column. In theory, the BTC should be a perfectly vertical line starting at a C/C_0 value of 0.0 and going immediately to 100.0. However, this does not occur in laboratory experiments or in the environment due to two factors: the slow kinetics of sorption to the soil and the dispersion that occurs in the aquifer media. Finally, all of the sorption sites are covered and the effluent pollutant from the column reaches the inlet concentration. But what happens if we remove the input of pollutant? This is illustrated in Figure 8.10, where the inlet water to the pollutant-saturated column is changed to NOM-containing water without pollutant. This point is indicated by the upward arrow in Figure 8.10. Note the predictable trend. NOM increases the mobility of cadmium during the adsorption process by shifting the BTC to the left as compared to the eluent water not containing NOM and increases mobility in the desorption process by shifting the BTC

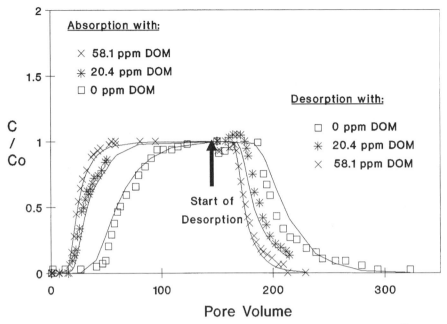

Figure 8.10. Breakthough curves for the adsorption and desorption process for Cd^{2+} as a function of NOM concentrations. [Reprinted with permission from Dunnivant et al. (1992a). Copyright 1992 American Chemical Society.]

to the left. Increasing the NOM results in a greater leftward shift and increased mobility. These types of experiments illustrate the need to characterize the water and soil at the site under investigation in order to closely match the chemistry of the systems.

The transport of the polychlorinated biphenyl 2,2′,4,4′,5,5′-tetrachloro-biphenyl (TCB) was also investigated in the soil columns. These results are shown in Figure 8.11. The transport of the pollutant (TCB) in water not containing NOM is indicated by the stars (the rightmost plot). Note the much larger pore volume scale, as compared to the Cd^{2+} experiment, which indicates a much higher R value and, in turn, a much higher K_p value for the TCB. This is common; hydrophobic pollutants tend to have higher K values as compared to metal pollutants, especially in organic-rich soils or sediments. Again, as mobile (dissolved) NOM is added to the elution water, the hydrophobic pollutant becomes more mobile. R values were 1018 (0.0 mg/L NOM; $K_p = 245$ L/kg), 625 (5.2 mg/L NOM), 456 (10.2 mg/L NOM), and 293 (20.4 mg/L NOM). Note that the length of experiments measuring pore volumes of 1000 can require weeks to months if they are to yield representative transport predictions.

One final point should be made concerning column studies and the need to use laboratory conditions that are representative of the field site. Some chemical parameters, such as pH and E_H, can significantly affect the fate of a pollutant. Recall the Swiss study mentioned in Chapter 2 documenting the reduction of substituted nitrobenzenes by reduced NOM. If reducing E_H conditions are present in the field,

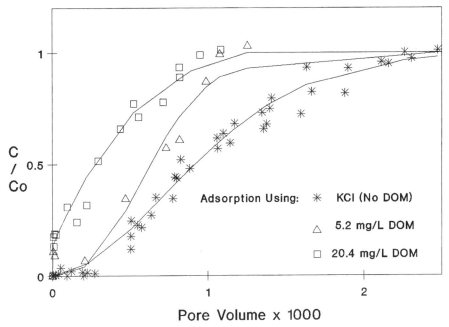

Figure 8.11.   Breakthrough curves for a PCB as a function of NOM on a soil column. [Reprinted with permission from Dunnivant et al. (1992a). Copyright 1992 American Chemical Society.]

you must incorporate these into your K determination or column study in order to obtain representative results. And in regard to transformation reaction rates, surface-catalyzed degradation rates in the environment tend to be orders of magnitude faster than those by dissolved catalysts. Hence, these types of reactions can be extremely important.

8.5   DIRECTION OF WATER FLOW (THE THREE-POINT PROBLEM)

Determining the direction of pollutant transport is easy in river and atmospheric systems but more complicated in groundwater systems. Yet the direction of flow must be known in order to calculate the gradient (magnitude of water flow), which is defined as the difference in water elevation (head) between two points along the flow line, divided by the longitudinal distance between the two points. Ideally you would only need two wells, one up-gradient and one down-gradient, but this would require that the wells' locations define a line parallel to water flow. Such a representation is shown in Figure 8.12*a*, in which the wells are located on the same water flow line. Rarely, if ever, will you come across such an ideal situation in the real world. Typically, you need at least three monitoring wells of known location (on an *x–y* grid) to yield measurements of depth to the water table, as shown in Figure

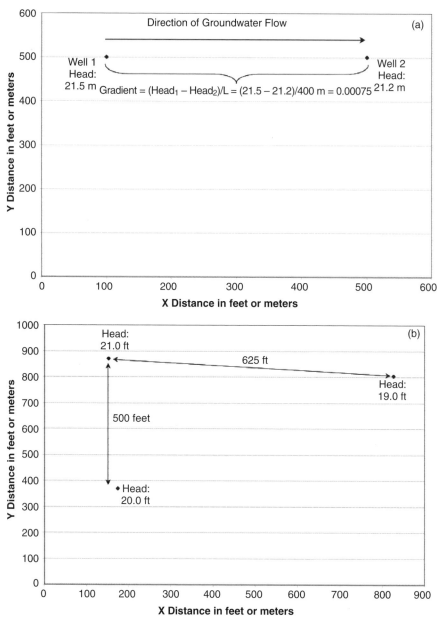

Figure 8.12.   (*a–c*) The three-point problem.

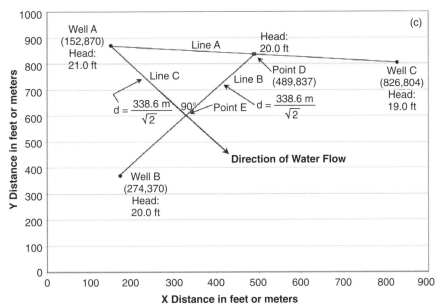

Figure 8.12   *Continued*

8.12*b*. The depth to the water table values are converted to depth above sea level or related to another reference point. With this information, you have sufficient data to solve a problem commonly referred to as the "three-point problem." There are two ways to solve this problem, graphically and with an exact mathematical calculation. We will illustrate the graphical method and refer you to Fate® for the mathematical method.

Figure 8.12b gives the general layout of three wells and heads in each well. Well A has a location of (152, 870) with a head of 21 m, Well B has a location of (174, 370) with a head of 20 m, and Well C has a location of (826, 804) with a head of 19 m. To determine the direction of groundwater flow, you arrange the wells on graph paper as shown in Figure 8.12c and draw the following lines between them. First, a line is drawn between the highest and the lowest wells (Line A in Figure 8.12c). Next, you must find the point (Point D) along Line A that has the same head as the intermediate-head well (Well B). This point is located by assuming that the water table between the two wells has constant gradient. For our example, therefore, this point is halfway between the two wells (20.0 m is halfway between 21.0 m and 19.0 m) and is at Point D (489, 837). Next, a line is drawn from the well with the highest head (Well A) to Line B (the line between Well B and Point D) such that these two lines intersect at a 90° angle (refer to Figure 8.12c). (In some cases you will draw the line from the lowest well. It will be clear when to do this.) This line gives the direction of water flow and, therefore, the direction of pollutant migration.

In order to find the hydraulic gradient, we must first find the distance between Well A and Point D. To accomplish this we use the Pythagorean theorem, where the distance *d* is calculated by

$$d^2 = (x_D - X_A) + (y_D - y_A)^2$$
$$= (489 - 152)^2 + (837 + 870)^2$$
$$= 113,569 + 1089$$
$$= 114,658$$
$$d = 338.6 \text{ m}$$

Using $a^2 + b^2 = c^2$, we can obtain the distance of the short legs of the triangle, where the distance from Well A to Point E (the intersection of Line B and C) or the distance from Point D to Point E is

$$d = \frac{338.6 \text{ m}}{\sqrt{2}}$$

Thus, the hydraulic gradient (Δhead/Δdistance) is

$$\text{gradient} = \frac{(21.0 - 20.0 \text{ m})}{\dfrac{338.6 \text{ m}}{\sqrt{2}}} = 0.00418$$

The direction of water flow can also be determined using the Three-Point Module in Fate®. This is found under the groundwater fate and transport section where an identical problem to the one worked above is illustrated. Fate uses a vector analysis approach to solve the three-point problem. The hydraulic gradient (dh/dL) is used in Darcy's law to determine the flow rate of water through porous media

$$Q = KA\left(\frac{dh}{dL}\right)$$

where Q = the flow rate (m³/day), K = the hydraulic conductivity or coefficient of permeability (m/day), A = the cross-sectional area (m²), and dh/dL = the hydraulic gradient. As a point of reference, groundwater velocities range from 0.01–42 m/day for sand and gravel to 150–200 m/day for gravel with cobbles.

8.6 PHYSICAL PARAMETERS IMPORTANT IN POLLUTANT FATE AND TRANSPORT

In Section 8.4, we mentioned the importance of bulk density and porosity in relating K_d and K_p to retention (R) in a aquifer or column. These physical measurements are also important in estimating water velocity. In rivers and lakes, logically, the water flows downhill, and it is also generally easy to tell how fast the water is flowing with a tool as simple as a tennis ball. But in aquifers, where we cannot see the slope of the water table, it is difficult and in many cases impossible to predict how fast the water is moving. This is further complicated by the fact that water is not the major space-occupying substance in the subsurface, where soil or rock is the predominate media. Thus, when measuring the volume of water, you must account for the media porosity. There are three major factors that contribute to the movement of pollutants in the subsurface environment: sorption phenomena; chemical, bio-logical, and nuclear degradation; and dilution of the pollutants by mixing and dis-persion. We have already discussed sorption phenomena and degradation rates. In this section we will concentrate on dispersion, a highly complicated and very unpre-dictable phenomenon that presents a major challenge for prediction of fate and trans-port in groundwater.

8.6.1 Sources of Dispersion in Geological Media

Dispersion occurs primarily because of small- and large-scale mixing phenomena. These are illustrated in Figure 8.13 for porous media. As polluted water flows through the porous media, it does not travel as a unified volume, but mixes with unpolluted water. The mixing is primarily random (Figure 8.13a), although disper-sion also is affected by the heterogeneity of the subsurface, with respect to miner-alogy, size of particles, or diameter or curvature of fractures. This is illustrated in Figure 8.13b for porous media containing "lenses" of different media (usually clay sedimentary deposits) dispersed through the aquifer. We account for these unpre-dictable mixing events with a "fudge factor" in our modeling efforts, defined as D_x, the longitudinal dispersion coefficient.

Diffusion results in dilution of the polluted water. More importantly from a modeling standpoint, it also increases the extent of the zone of polluted water. To

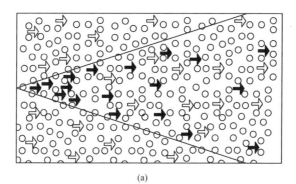

(a)

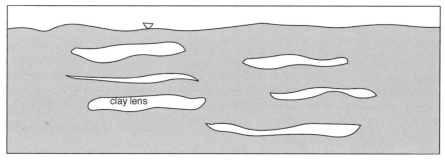

(b)

Figure 8.13.   Sources of dispersion in geologic media.

understand how dispersion and the dispersion coefficient are used, we will use the following case study.

8.6.2   A Case Study: The INEEL Water and Tracer Infiltration Experiment

The following case study was conducted at the Idaho National Engineering and Environmental Laboratory (INEEL) located near Idaho Falls, Idaho (United States). This facility is operated by subcontractors for the U.S. Department of Energy (DOE). Its primary historical function was to design and test nuclear reactors as well as to conduct secret projects of the U.S. Department of Defense. These operations generated hazardous and radioactive wastes that had to be disposed of on-site. In order to understand the hazardous waste situation that exists today, we must try to place ourselves in the shoes of the scientists who worked at the INEEL during the Cold War. The scenario presented next does not in any way excuse the actions taken by the DOE during these times, but does shed light on their thinking at that time.

In the midst of the Cold War, nearly everything at national laboratories was secret and everyone was under suspicion. The problem facing workers at the INEEL was that they needed a secure, easily accessible place to dispose of hazardous and radioactive waste that was located near their construction and testing facilities and could not be seen from the highways crossing the INEEL (for security reasons at the time). Given that the INEEL is located on the very flat Snake River Plain, the only potential hidden places are low spots. This did not seem to be a problem at the time, since the average precipitation rate was so low as to not create flooding events; however, as it turned out, the disposal site that was finally selected was a very bad choice.

Waste was buried in barrels and containers, as shown in Figure 8.14, in unlined trenches and covered with local soil (unlined cover). For many years this did not present a problem, but in 1962, 1969, and 1982, heavy late spring snowfalls in the nearby mountains were followed by warm winds and rapid snow melts that resulted in the flooding of a river (the Big Lost River) adjacent to the burial site. This resulted in the burial site being flooded three times for periods of days to weeks. The flooding event in 1969 is shown in Figure 8.15. The obvious concern regarding these flooding events was the release and movement of radioactive materials, both into the groundwater and out of the disposal site. The Snake River Plain Aquifer, a major pristine source of water to the region, is located below the site, but fortunately at a depth of ~550 ft. The goal of the following experiment was to determine whether contamination could reach the aquifer during such a flooding event and to estimate water velocities, chemical tracer velocities, and dispersion values at the site for future pollutant modeling efforts. Before we discuss the experiment, we should mention that the disposal site at the INEEL has now been made more resistant to flooding by an earthen berm placed around the site, as well as by more control and diversion of the Big Lost River.

A site for the experiment, approximately 1.5 km south of the disposal site and with similar geology, was chosen so as not to interfere with current operations at the burial site. A 183-m-diameter basin was constructed and instrumented with lysime-

Figure 8.14.   Disposal of radioactive waste at the INEEL (U.S. DOE-INEL photograph).

Figure 8.15.   Flooding of the waste site in 1969 (U.S. DOE-INEL photograph).

ters and wells for monitoring and collecting water and chemical tracers (refer to Figure 8.16). Tracers included a semipulse addition of ^{75}Se, ^{85}Sr, ^{160}Tb, which have a sufficiently short decay half-life such that no detectable radiation would be present at the site after approximately two years. The basin was flooded with tracer-free water for 6 days, followed by ~11 days of tracer input, followed by ~19 days of

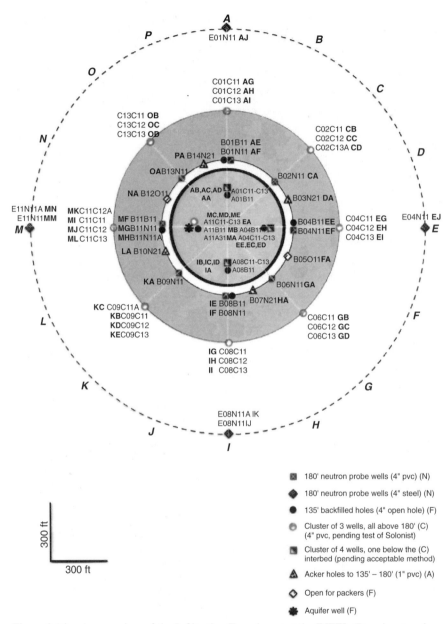

Figure 8.16. An overview of the Infiltration Experiment at the INEEL (Dunnivant and Newman, 1995).

tracer-free water to flush the tracers through the system. There were 101 monitoring locations for water and tracer.

An experiment of this size is rare, to say the least, but was necessary due to the scale of potential contamination of a major pristine water source. Many smaller experiments were conducted to answer fundamental research questions as well as to address the obvious problem, the transport of radioactive waste to the aquifer. These experiments are summarized in Dunnivant et al. (1998). The geology at the burial area is primarily composed of undulating, fractured basalt flows with some soil covering. Fortunately, although unknown to the scientists working at the burial site during the Cold War, there were several sedimentary interbeds (buried soils) located between the land surface and the Snake River Plain Aquifer. The scientists conducting the infiltration experiment (and the U.S. DOE) now hoped that these interbeds had impeded the flow of water and pollutants to the aquifer during the three mentioned flooding events. But first let's look at some results from the experiment, specifically the breakthrough curves for the conservative tracer ^{75}Se.

Of the 101 sampling sites, water was found and recovered from only 30, while the conservative tracer (^{75}Se) was detected in only 26 of these. In fact, during the two-year planning of the experiment, some scientists had expected that no water or tracer would be recovered from any monitoring site due to the complex, fractured nature of the geologic media underlying the site. Figures 8.17 and 8.18 contain two of the BTCs for ^{75}Se. The lines in these BTCs represent the model fit (explanatory modeling from Chapter 1) from which estimates of water velocity and dispersion were obtained, fulfilling one of the major goals of the experiment.

In an experiment conducted in the natural environment and at this scale, it is not uncommon to obtain unexpected results. It has often been proposed that unconnected and dead-end fractures exist in fractured media, but to our knowledge no one has extensively demonstrated their flow characteristics during a tracer study. Researchers found a variety of BTCs, suggesting the complexity of groundwater flow at the site. A summary of the results are shown in Figure 8.19.

As mentioned, one goal of the experiment was to obtain BTCs to estimate water velocity and dispersion in the subsurface at the INEEL. These estimates were

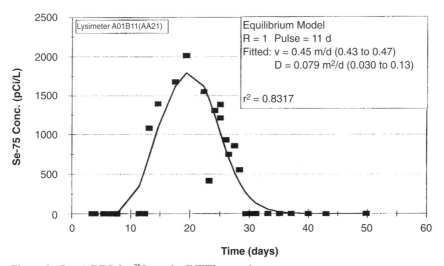

Figure 8.17.   A BTC for ^{75}Se at the INEEL experiment.

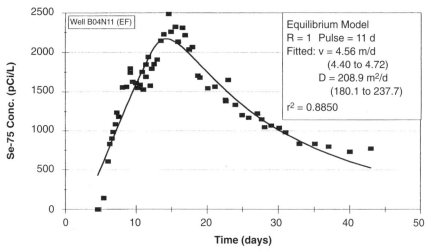

Figure 8.18.   A BTC for ^{75}Se at the INEEL experiment.

successful from BTCs such as the one shown in Figures 8.18 and 8.19a, where water apparently flowed directly from the basin to the fracture being monitored. Unexpectedly, water was also observed during the experiment in some fractures in which tracer never appeared at the monitoring location (Figure 8.19b). This may be due to the monitoring of a dead-end fracture that initially filled with tracer-free water and was never drained to receive tracer. Another unexpected finding was the lack of water and tracer in lysimeter locations immediately below the water basin. These fractures were apparently isolated within an otherwise solid basalt flow section. In addition, nonclassical BTCs (non-bell-shaped) were observed, in which the first water observed contained the highest concentration of tracer (Figure 8.19c). This may be explained by overflow of a dead-end fracture into a fracture where a monitoring station was located. The complexity of the BTCs strongly illustrates the need to characterize polluted sites individually, although costs of such extensive characterization may be prohibitive.

Thirteen estimates of dispersion were obtained from the experiment that can now be used to estimate past and future pollution transport at the site. In order to understand dispersion in the subsurface environment, we must distinguish between two terms, dispersion coefficient and dispersion. The goal of estimating a dispersion coefficient is to provide a method of estimating the dispersion (given in meters, m) for any distance along the flow path. In theory, the degree of dispersion or mixing and dilution of pollutants should increase as you move downstream. To obtain values of dispersion, we simply divide the dispersion coefficient by the water velocity. The units of dispersion are meters, so, for the example, the longitudinal dispersion represents the length of the spread of pollutants.

The explanatory modeling effort used to model the transport of water and conservative tracers in the INEEL experiment fit the general transport equation (Section 8.7, below) for groundwater to the breakthrough curves. This yielded 13 estimates

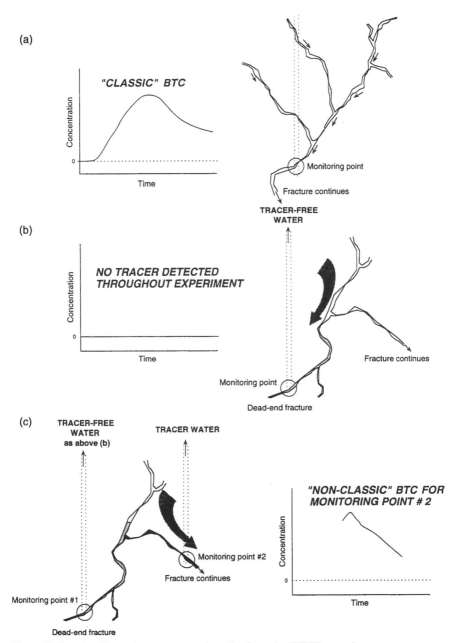

Figure 8.19.   An illustrative summary of results from the INEEL experiment.

of water velocity, with units of meters/day (m/day) and dispersion coefficients with units of meters squared per day (m²/day). However, experimental data do not always confirm the theory. Consider the data from the INEEL experiment plotted in Figure 8.20. In general, the lysimeter data follow a trend: increasing dispersion with increas-

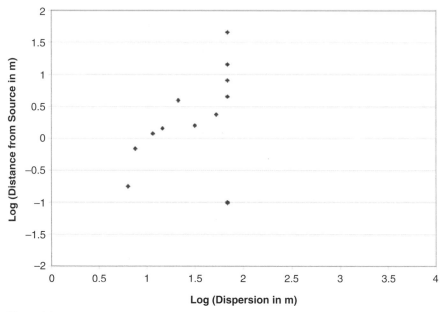

Figure 8.20. Summary of dispersion values for the INEEL experiment.

ing distance from the point source. But this trend falls apart when the interbed (a semi-impermeable layer) is reached at 68.7 m (a log value of 1.84 in Figure 8.21). The dispersion along the interbed ranges from 0.10 m to 45.8 m, over a factor of 15. This is a common observation in field experiments, where a large range of dispersion values are often obtained for similar distances from the point source. Most researchers average the values for future modeling efforts. The data from the INEEL experiment were averaged to yield one estimate for the lysimeters (2.16 m) and another for the interbed locations (16.2 m).

Regarding the main question of the INEEL experiment, "Did pollutants travel to the Snake River Plain Aquifer during the flooding events?", researchers of the experiment concluded that this was unlikely. Neither water nor tracer in the experiment appeared to penetrate the first continuous sedimentary interbed located beneath the basin. The Cold War burial site is also underlain by at least two semi-continuous, relatively thick, sedimentary interbeds. And it should be noted that the experiment flooded the basin for 50 days while the flooding events of the burial sites lasted a few weeks at the longest. Scientists rarely have an absolute answer, and results of the infiltration experiment are still being debated, but most scientists feel that pollutants have not reached the aquifer from the burial site. Again, the point of considering this experiment in such detail was to note the complexity of determining dispersion at a site.

8.6.3 Towards a Universal Estimate Technique for Dispersion

Hydrologists and modelers have long sought a way of estimating dispersion for a given aquifer based on data from other aquifers. The desire for such a generalized model is understandable, given the huge effort and expense of characterizing every polluted site, as done in the INEEL experiment. However, very limited success has been achieved in developing such a predictive method. Gelhar et al. (1992) was one of the first to attempt to use this approach, when they compiled data from 59 field studies of dispersion into one data plot. These data are shown in Figure 8.21, with the solid triangles representing data from porous media (sand, etc.), the hollow squares representing data from fractured rock, and the hollow circles representing data from the INEEL experiment (fractured media). Several points should be made concerning this plot as compared to other published versions of the Gelhar et al. (1992) data. First, we have intentionally chosen to use the same scale on the x- and y-axes. Some other versions of this plot have extended the x-axis to give the standard elongated rectangle along the x-axis, but this seriously biases the interpretation of the data. After plotting the data in the elongated rectangle, researchers often then draw a regression line through the data to "create" a trend of increasing dispersion with increasing distance from the point source. As the reader can clearly see from our plot in Figure 8.21, there is little to no evidence to support such a linear trend in either the porous or fractured media data sets. Although dispersion does seem to generally increase with increasing distance from the point source, we would not call this a predictable trend, and we definitely should not draw a regression line through it as a few hydrologists have done. For example, say you want to predict the dispersion at your site at a distance of 100 m (a log value of 2 on the x-axis in Figure 8.21). You could choose a dispersion from approximately 0.00178 m (a log value of -2.72) to approximately 316 (a log value of $+2.5$). The "acceptable" values from such a "correlation" vary over five orders of magnitude! A child with a dartboard could produce more accurate estimates of dispersion. But here we see the problem: There is a great need to develop such a "correlation" in order to save money and resources on site characterization, and even when we spend large amounts of money to characterize a site (i.e., the INEEL experiment), we still obtain a range of dispersion values (a factor of 15).

The lack of correlation in Figure 8.21 raises another issue. When fate and transport modeling is performed, its goal is to feed risk assessment calculations, and the most important result from the fate and transport modeling is the pollutant concentration in the water. So, which dispersion value would you select taking into consideration this ultimate goal? If you are from the party responsible for the cost of cleanup (remediation), you would be financially wise to use the high estimate of dispersion, since this will result in lower pollutant concentrations and less needed cleanup. However, if you are the person who might drink the water in the future, you would be wise to choose the lower dispersion estimate in the fate and transport calculations, since it will result in higher pollutant concentrations and thus increase the chance that the pollutant will be cleaned up. Who is correct? There is no single, correct answer to this question, but it is best to be conservative and err on the safe side (use low estimates of dispersion) in most cases.

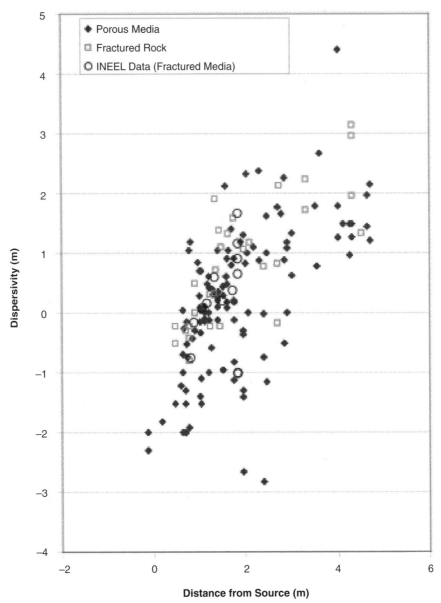

Figure 8.21. A summary of dispersion data from various field investigations. [Data from Gelhar et al. (1992).]

So, how does one obtain estimates of dispersion for a polluted site? There is not one absolute answer to this. Unfortunately, you cannot usually perform the extensive characterization experiments at the polluted site, due to cost and fear of increasing the environmental damage from the contamination by pumping pollutants off site. But when possible (when sufficient accuracy is needed to warrant the expense), dispersion and water velocity estimates can be obtained through experimentation at a site with geology similar to that of the polluted area (generally near the polluted site). As a worst-case scenario, one could use data from Figure 8.21 or similar plots, but fate and transport prediction must use the complete range of dispersion values present in the plot at each distance from the point source. One of the best, and perhaps most common, ways of estimating site-specific dispersion coefficients is to perform explanatory modeling on the contaminated site based on historical input source data and results from an extensive groundwater monitoring program. Then the estimated dispersion coefficients, water velocities, and K values can be used to perform future fate and transport modeling.

8.7 MATHEMATICAL MODELS

Instantaneous (Pulse) Pollutant Input.   If we assume the spill contaminates the entire thickness of the aquifer, as shown in Figure 8.22, an equation can be obtained using Laplace transformation to predict the pollutant concentration as a function of time or distance from the point source:

$$C(x,t) = \frac{M}{A\sqrt{4\pi \dfrac{D_x}{R} t}} e^{\dfrac{\left(x - \frac{v}{R}t\right)^2}{4\frac{D_x}{R}t} - kt}$$

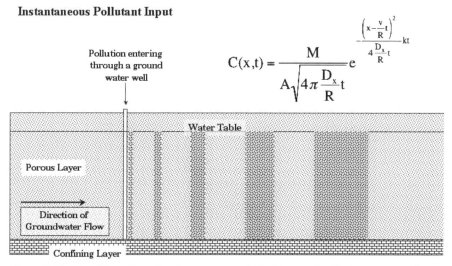

Figure 8.22.   An illustration of an instantaneous (pulse) input of pollution to an aquifer.

where x is distance from the source (m), t is time (days), M is the mass of contaminant added to the aquifer (mg), A is the cross-sectional void volume contaminated by the pollution (m^2), D_x is the dispersion coefficient (m^2/day), R is the retardation factor (unitless), v is velocity (m/day), and k is the first-order reaction rate (1/day). First, note the use of the dispersion coefficient (dispersivity) instead of the dispersion. Unfortunately, dispersion coefficients must be used, since we need a way of estimating dispersion as water moves away from the point source. We noted the dangers of this approach in the previous section, but all of the fate and transport equations used today rely on this approach, even if there little to no experimental basis for its use. This coefficient defined by

$$D_x = D \times v$$

where D_x is the called the dispersivity or dispersion coefficient, D is the dispersion, and v is the water velocity. Because of the many causes of dispersion, discussed previously, dispersivity is one of the most difficult parameters to measure accurately. Dispersivity values tend to increase with the scale (distance) over which they are measured because the degree of heterogeneity within the aquifer generally increases with scale.

Example Problem. A 100-m^3 tanker containing a 100-mg/m^3 solution of 2,4-dinitrotoluene crashes and empties its entire contents above an aquifer. The cross-sectional area of the spill is 50 m^2. The aquifer has a porosity of 30%, a bulk density of 1.6 g/cm^3, a velocity of 10 m/yr, and a dispersion coefficient (D_x) of 10 m^2/yr. The distribution coefficient of 2,4-dinitrotoluene for this aquifer material has been measured to be 2.5 mL/g. 2,4-dinitrotoluene biodegrades through a first-order reaction at a rate of 0.693 yr^{-1}. Calculate the concentration 10 m down-gradient from the lagoon 10 years after the input.

Solution

1. Calculate the retardation factor from the distribution coefficient.

$$R = 1 + \frac{\rho_b K_d}{n}$$

$$= 1 + \frac{\left(1.6 \, \frac{g}{cm^3}\right)\left(2.5 \, \frac{mL}{g}\right)}{0.3} = 14.33$$

2. Correct the cross-sectional area of the spill site to the void volume (rather than the total volume).

$$\text{Cross-sectional area} \times \text{Porosity} = 50 \, m^2 \times 0.3 = 15 \, m^2$$

3. Calculate the total mass of dinitrotoluene spilled.

$$\text{Volume spilled} \times \text{Concentration of solution}$$

$$= (100 \, m^3)\left(1000 \, \frac{mg}{m^3}\right) = 100,000 \, mg \text{ or } 100 \, g$$

4. If necessary, calculate the first order degradation rate from the half-life.

$$-\frac{\ln 0.5}{t_{1/2}} = k = \frac{-0.693}{0.09625(\text{yr})} = 0.693 \text{ yr}^{-1}$$

5. Arrange data into proper units

   Groundwater velocity, $v = 10\,\text{m/yr}$

   Retardation factor, $R = 14.33$

   Mass of contaminant, $M = 100{,}000\,\text{mg}$

   Dispersion coefficient, $D_x = 10\,\text{m}^2/\text{yr}$

   Reaction rate constant, $k = 0.693\,\text{yr}^{-1}$

   Cross-sectional area, $A = 15\,\text{m}^2$

6. Input data into the Fate® program or governing equation and obtain/draw a graph.

7. Calculate the concentration 10 m down-gradient from the lagoon 10 years after the input.

$$C(x,t) = \frac{M}{A \times \sqrt{4\pi \dfrac{D_x}{R} t}} \times e^{\dfrac{\left(\left(x - \frac{v}{R}\right) \times t\right)^2}{4\frac{D_x}{R} t} - kt}$$

$$= \frac{100{,}000\,\text{mg}}{15\,\text{m}^2 \sqrt{4\pi \left(\dfrac{10\frac{\text{m}^2}{\text{yr}}}{14.33}\right) \times 10\,\text{yr}}} \times e^{\dfrac{\left(10\,\text{m} - \frac{10\frac{\text{m}}{\text{yr}}}{14.33} \, 10\,\text{yr}\right)^2}{4 \times \frac{10\frac{\text{m}^2}{\text{yr}}}{14.33} \times 10\,\text{yr}} - \left(0.693\frac{1}{\text{yr}}\right)10\,\text{yr}}$$

$$= 1971\,\text{mg/m}^3 \; or \; 1.971\,\mu\text{g/L} \; 2,4\text{-dinitrotoluene.}$$

Step Pollutant Input. For the initial condition $C(x, 0) = 0$, where the concentration equals zero everywhere, and the boundary condition $C(0, t) = C_o$, where the concentration at the source remains constant at the value of C_o, the basic advective–dispersive groundwater equation may be solved using Laplace transformations to yield (see Figure 8.23)

Step Pollutant Input

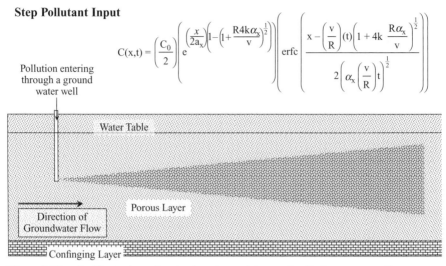

Pollution entering
through a ground
water well

Water Table

Porous Layer

Direction of
Groundwater Flow

Confinging Layer

Figure 8.23.   An illustration of a step (continuous) input of pollution to an aquifer.

$$C(x, t) = \left(\frac{C_0}{2}\right) e^{\left(\frac{x}{2D_x}\left(1-\left(1+\frac{R4kD_x}{v}\right)^{\frac{1}{2}}\right)\right)}$$

$$\left(erfc\left(\frac{x-\left(\frac{v}{R}\right)(t)\left(1+4k\frac{RD_x}{v}\right)^{\frac{1}{2}}}{2\left(D_x\left(\frac{v}{R}\right)t\right)^{\frac{1}{2}}}\right) + e^{\left(\frac{x}{D_x}\right)}erfc\left(\frac{x+\left(\frac{v}{R}\right)(t)\left(1+4k\frac{RD_x}{v}\right)^{\frac{1}{2}}}{2\left(D_x\left(\frac{v}{R}\right)t\right)^{\frac{1}{2}}}\right)\right)$$

where C_o is the initial concentration of the contaminant, x is distance from the source, D_x is longitudinal dispersivity (dispersion coefficient), k is the first-order reaction rate, v is velocity, t is time, and erfc is the complementary error function.

The final term in the above equation,

$$e^{\left(\frac{x}{D_x}\right)}erfc\left(\frac{x+\left(\frac{v}{R}\right)(t)\left(1+4k\frac{RD_x}{v}\right)^{\frac{1}{2}}}{2\left(D_x\left(\frac{v}{R}\right)t\right)^{\frac{1}{2}}}\right)$$

is generally considered insignificant and is ignored. This term is ignored in Fate©.

Again, note that the dispersion coefficient, not dispersion, is used in these equations.

The complementary error function (erfc in the equation given above) is the area between the midpoint of the normal curve and the value that you are taking the error function of. The basic purpose of the complementary error function is to calculate the area under the bell-shaped curve representing pollutant concentrations through time or with distance downstream following a pulse input of pollutant.

Example Problem.   A waste lagoon containing 100 mg/L benzene is contaminating the underlying aquifer. The aquifer has a porosity of 30%, a bulk density of 1.6 g/cm³, a velocity of 10 m/yr, and a dispersivity of 10 m²/day. The distribution coefficient of benzene for this aquifer material has been measured to be 5 mL/g. Benzene biodegrades through a first order reaction at a rate of $0.025 \, \text{yr}^{-1}$. Calculate the concentration 10 m down-gradient from the lagoon 10 years after the input.

Solution

1. Calculate the retardation factor from the distribution coefficient.

$$R = 1 + \frac{\rho_b K_d}{n}$$

$$= 1 + \frac{\left(1.6 \frac{\text{g}}{\text{cm}^3}\right)\left(5 \frac{\text{mL}}{\text{g}}\right)}{0.3} = 27.67$$

2. If necessary, calculate the dispersivity from the dispersion coefficient and the velocity.

   This step is not necessary here.

3. If necessary, calculate the first order degradation rate from the half-life.

$$-\frac{\ln 0.5}{t_{1/2}} = k = -\frac{-0.693}{27.72 (\text{yr})} = 0.025 \, \text{yr}^{-1}$$

4. Convert data into proper units:

$$C_0 = 100 \, \text{mg/L}$$
$$v = 10 \, \text{m/yr}$$
$$D_x = 10 \, \text{m}$$
$$R = 27.67$$
$$k = 0.025 \, \text{yr}^{-1}$$

5. Input data into the Fate® program or governing equation and obtain/draw a graph.

6. Calculate the concentration 10 m down-gradient from the lagoon 10 years after the input.

$$C(x, t) = \left(\frac{100 \frac{\mu g}{L}}{2} \right) e^{\left(\left(\frac{10 \text{ m}}{2(10 \text{ m})} \right) \left[1 - \left(1 + \frac{(27.27)(4)\left(0.025 \frac{1}{\text{yr}} \right)(10 \text{ m})}{10 \text{ m}} \right)^{\frac{1}{2}} \right] \right)}$$

$$\left(\text{erfc} \left(\frac{10 \text{ m} - \left(\frac{10 \frac{\text{m}}{\text{yr}}}{27.67} \right)(10 \text{ yr}) \left(1 + (4)\left(0.025 \frac{1}{\text{yr}} \right) \left(\frac{(27.67)(10 \text{ m})}{10 \frac{\text{m}}{\text{yr}}} \right) \right)^{\frac{1}{2}}}{2 \left((10 \text{ m}) \frac{10 \frac{\text{m}}{\text{yr}}}{27.67} (10 \text{ yr}) \right)^{\frac{1}{2}}} \right) \right)$$

$= 0.78 \, \mu g/L$ benzene

8.8 SENSITIVITY ANALYSIS

While chemical factors, such as sorption and degradation, can determine the down-gradient pollutant concentration, the most difficult parameter to estimate is hydro-dynamic dispersion. Sensitivity analysis should include a variety of values for distribution and partition coefficients and chemical, biological, and nuclear degradation rates. In terms of dispersion estimates, Figure 8.21 should be used for your initial estimate and should be combined with water velocity to obtain estimates of dispersion coefficients. Note that your sensitivity analysis should use large ranges of dispersion coefficients, in some cases from two to four orders of magnitude.

8.9 LIMITATIONS OF OUR MODELS

In this chapter, we presented the classic one-dimensional general transport equation in order to keep the mathematics simple and concentrate on the important processes. It should be noted that professional modelers use two- and three-dimensional models and also use numerical methods of analysis, which allow for variations in aquifer conditions such as water velocity, sorption phenomena, degradation rates, and, most importantly, dispersion. Obtaining an accurate estimate of dispersion or the dispersion coefficient is the major limiting factor in groundwater modeling.

8.10 REMEDIATION

Contaminated groundwater systems are one of the most difficult environmental media to remediate, due to their location below the land surface (out of direct observation), the large volumes of soil and water involved, and the slow water velocities inherent in these aquifers. Many technologies and approaches have been developed specifically for groundwater remediation with mixed and site-specific successes and failures. There are numerous books on groundwater remediation; given the expansive research in this area, we will only briefly summarize some of the common approaches to this problem. In general, we will begin with the least expensive and complicated and progress to the more expensive.

A summary of suggested and common remediation actions are given in Figure 8.24 (EPA, 1988), and this list has been expanded upon since its original publication. These actions are divided into natural attenuation, containment, and active restoration. The first two categories, while similar, are two distinct approaches adopted by EPA. Both of these can be undertaken in a variety of cases and for a variety of reasons. For example, even when a risk assessment shows no immediate risk, governing bodies (EPA, local government, or the public) sometimes decide that some control of the site is necessary. Natural attenuation or containment may also be employed when there is no acceptable remediation plan at present and the site needs to be removed from possible human contact, or when remediation of the site using one of the acceptable technologies may create a worse problem than waiting for the development of a whole new remediation approach. Notice that all remediation plans involve long-term monitoring of the site.

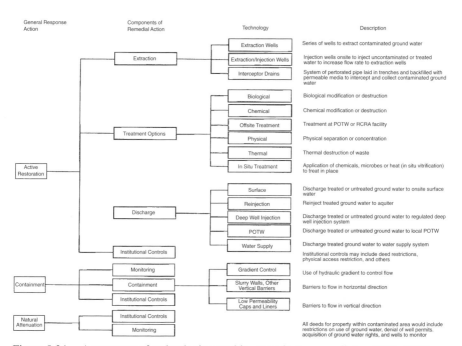

Figure 8.24.   A summary of technologies used in groundwater remediation (EPA, 1988, p. 5–10).

Today, natural attenuation may be expanded to a form of active restoration, by employing the addition of nutrients to aid in the microbial removal of pollutants. These projects still take years to decades to complete, but operating costs are relatively inexpensive compared to other approaches such as removal and treatment of the soil. Institutional controls, a common listing in Figure 8.24, are basically easements placed on the land deed specifying and limiting how the land can be used or developed in the future.

In some cases, containment may be the only viable and affordable option. Containment does nothing to treat the site, but limits the further spread of pollution and is intended to protect local citizens from exposure to the toxins while a more permanent solution can be designed. Containment uses ways to limit or stop water and therefore pollutant migration from the site boundaries. This can be accomplished by controlling the gradient (placing water pumping and extraction wells around the contaminated area), surrounding the underground site with walls of impermeable media (grout and clay), and capping the site to prevent water from entering from the ground surface and spreading the pollution to new areas.

There are a seemingly endless number of active treatment technologies for polluted groundwater and soil sites. These are divided into extraction of polluted groundwater (termed "pump and treat"), direct treatment options that may involve the removal of the contaminated soil and water, and institutional controls that limit use of the water from the site. The first approach, pump and treat, was originally thought of as the cure-all of polluted groundwater. The concept is simple: Polluted groundwater is pumped out relatively slowly, treated at the ground surface, and then the "clean or cleaner" groundwater is pumped back into the polluted aquifer in order to displace more pollution. The water extraction and injection wells are placed so as to contain the polluted groundwater plume. By using partition and distribution coefficients, it is relatively easy to estimate how many aquifer volumes of groundwater must be removed and treated for the pollution to be effectively removed. However, after decades of use, this technology has proved ineffective in many, though not all, cases. Sites that were once thought to be remediated have been later found to still contain unsafe levels of pollutants in their water. This is due to the slow desorption of pollutants from the soils. After the pump and treat system has been shut off, the pollution re-occurs by desorbing from the polluted soil. Thus the pump and treat approach has been augmented with the addition of *in situ* biological and chemical treatment processes to increase pollutant degradation and/or removal.

There are many *ex situ* treatment technologies that can be used on the polluted groundwater after it has been brought to the surface. These are too numerous to even mention here but include physical, biological, and chemical treatment. After treatment, the water may be re-injected or disposed of in other manners depending on the local restrictions.

Polluted aquifers are one of the most difficult and costly sites to remediate. In fact, most of the Superfund effort has focused on remediating groundwater. Yet, there is much to be accomplished in the future given the difficult nature of these systems.

SUGGESTED PAPERS FOR CLASS DISCUSSION

Dunnivant, F. M., P. M. Jardine, D. L. Taylor, and J. F. McCarthy. Cotransport of cadmium and hexa-chlorbiphenyl by dissolved organic carbon through columns containing aquifer material. *Environ. Sci. Technol.* **26**(2), 360–368 (1992).

Dunnivant, F. M., M. E. Newman, C. W. Bishop, D. Burgess, J. R. Giles, B. D. Higgs, J. M. Hubbell, E. Neher, G. T. Norrell, M. C. Pfiefer, I. Porro, R. C. Starr, and A. H. Wylie. Water and radioactive tracer flow in a heterogeneous field-scale system. *Groundwater* **36**(6), 949–958 (1998).

Concepts

1. Contrast the following pairs of terms:

   Saturated versus unsaturated media

   Porous versus fractured media

   Consolidated versus nonconsolidated media

2. Research the Safe Drinking Water Act (SDWA) on the Internet. Write a one-page summary of the act with respect to groundwater, priority pollutants, and maximum contaminant levels.

3. Describe three examples of step inputs and three examples of pulse inputs of pollutants to a groundwater system in your area.

4. Discuss the difficulties in installing a monitoring well with respect to obtaining a representative groundwater sample.

5. Describe the major ways of installing a groundwater monitoring well. List the advantages and disadvantages of each.

6. Describe each of the three ways of obtaining a K_d or K_p for a pollutant.

7. Select two pollutants from Table 2.6 and calculate K_p value for them using two of the predictive equations in this chapter. Compare the values between the two equations.

8. Explain the relationship between R and K.

9. Draw typical breakthrough curves (BTC) for pulse and step inputs of pollutant.

10. Use Figure 8.20 to explain the difficulties in estimating dispersion for an untested system.

11. Summarize the three ways of estimating/determining dispersion in a groundwater system.

Exercises

1. A storage tank breaks during an earthquake event, spilling gasoline treated with MTBE into the surrounding soil (glacial till) and contaminating the groundwater. Create plots of (1) concentration of MTBE versus distance (0–22 m) 10 years after the spill and (2) concentration of MTBE versus time (0–15 years) 10 m from the spill, using the following parameters (check your answers with Fate®):

Void volume: $11.5\,m^2$

Groundwater velocity: 5 m/yr

Retardation factor: 8.457

Initial concentration of MTBE resulting from spill: $35\,mg/m^3$

Half-life of MTBE: 1 year

Dispersion coefficient: $10\,m^2/yr$

Bulk density: $1.6\,g/cm^3$

K_d: 1.072 mL/g

Porosity: 0.23

Now, for your sensitivity analysis, prepare the previous plots using $1\,m^2/yr$ as your dispersion coefficient.

2. A tanker car from a mining operation leaks and spills 100 g HCN over a sand aquifer with an area of 20 m. The EPA asks you to assess the damage and to determine the probable concentration of HCN in the ground water. The velocity of the groundwater is 182.5 m/yr, the dispersion coefficient is $2\,m^2/yr$, and the retardation factor is 1.050. The half-life of HCN under the conditions at the site is 1.00 year. The bulk density of the sand is $1.5\,g/cm^3$ and the porosity is 0.3. The K_d for HCN is 0.01. Calculate the concentration of HCN 0.3, 0.9, and 2 years after the spill. Use Fate to create a graph of time versus concentration. Compare this scenario with one in which the pollutant travels more slowly (90 m/yr). In both cases measure 10 m away from the spill.

3. Arsenic in the ground contaminates an aquifer by a naturally occurring, continuous geological process. The sandy aquifer has a bulk density of $1.08\,g/cm^3$ and a solid density of $2.63\,g/cm^3$. The distribution coefficient (K_d) in this scenario is 0.027 mL/g. Consider that the initial concentration of arsenic at the source is 21 μg/L, the longitudinal dispersion is 0.1 m per day, and the linear water velocity is 100 m/yr. Using this information, determine how far and how quickly arsenic must travel through this natural aquifer in order to be completely dispersed (>0.50 μg/L). The data from this problem is organized in the table below.

Note: Arsenic does not have a rate constant as it is stable (does not degrade). However, you can use a very high value to graph this in Fate® (suggested value: 60,000).

Use Fate® to create a plot of the scenario and to check your results.

Bulk density:	$1.08\,g/cm^3$
Solids density:	$2.63\,g/cm^3$
K_d:	0.027 mL/g
Initial concentration:	21 μg/L
Dispersion:	0.1 m/day
Water velocity:	100 m/yr

4. Xylene is one of the major components of gasoline. It can be toxic if released into groundwater. Some of the harsher effects of xylene on the human body include brain hemorrhaging or death. One summer in 1979, before gas stations had started using double-layered underground storage tanks (USTs), a UST at a gas station began corroding. Gasoline had been spilling out of the UST for 10 years before any removal efforts were undertaken. Xylene has a half-life of 30 days. The ground beneath the UST is fairly porous: It has a bulk density of $1.44 \, g/cm^3$ and a solids density of $2.94 \, g/cm^3$. The initial concentration of xylene was 370 ppm. Below the UST the ground-water moved at a velocity of 70 ft/yr and dispersed longitudinally $5 \, m^2/yr$. The K_d of xylene is $1.224 \, cm^3/g$. For this system, what was the concentration of xylene 20 m from the UST 1 year, 2 years, 5 years, and 10 years after the spill began?

5. You are a chemist who has been hired to analyze the step release of a DNAPL (1,1,1-trichloromethane) into a coarse sand aquifer with a porosity of 30%. The average linear groundwater velocity for this aquifer is 10 m/yr with a longitudinal dispersivity of $19 \, m^2/yr$. The sand has a retardation factor of 42.6. The pollutant has a half-life of 56 days and is initially present at a concentration of 5.50 ppb. How much pollutant (what concentration) is present at distances of 5.0 m and 20.0 m from the release point at times of 3.0, 5.0, and 10.0 years after the accident?

Spreadsheet Exercise

Create a spreadsheet that performs the same calculations as Fate® for a pulse input of pollutant to a groundwater system. Construct your spreadsheet so that it is interactive (so that you can change numeric values for parameters and the plot automatically updates itself).

REFERENCES

Chiou, C. T., L. F. Peters, and V. H. Freed. A physical concent to soil–water equilibria for nonionic compounds. *Science* **206**, 831–832 (1979).

Christman, M. C., C. H. Benson, and T. G. B. Edil. Geophysical study of annular well seals. *Ground Water* **22**(3), 104–112 (2002).

Dunnivant, F. M. and N. E. Newman. Migration of Radionuclide Tracers through Fractured Media: Preliminary Modeling Results of Breakthrough Curves Obtained during the Large-Scale Aquifer Pumping and Infiltration Tests, U.S. Department of Energy, Idaho National Engineering Laboratory, INEL-95/288, ER-WAG7-84, November 1995.

Dunnivant, F. M., P. M. Jardine, D. L. Taylor, and J. F. McCarthy. Cotransport of cadmium and hexachlorbiphenyl by dissolved organic carbon through columns containing aquifer material. *Environ. Sci. Technol.* **26**(2), 360–368 (1992a).

Dunnivant, F. M., P. M. Jardine, D. L. Taylor, and J. F. McCarthy. Transport of naturally occurring dissolved organic matter in laboratory columns containing aquifer material. *Soil Science* **56**(2), 437–444 (1992b).

Dunnivant, F. M., M. E. Newman, C. W. Bishop, D. Burgess, J. R. Giles, B. D. Higgs, J. M. Hubbell, E. Neher, G. T. Norrell, M. C. Pfiefer, I. Porro, R. C. Starr, and A. H. Wylie. Water and radioactive tracer flow in a heterogeneous field-scale system. *Ground Water,* **36**(6), 949–958 (1998).

Dunnivant, F. M., I. Porro, C. Bishop, J. Hubbell, J. R. Giles, and M. E. Newman. Verifying the integrity of annular and back-filled seals for vadose-zone monitoring wells. *Ground Water* **35**(1), 140–148 (1997).

EPA, Guidance on Remedial Actions for Contaminated Ground Water at Superfund Sites, Office of Emergency and Remedial Response, EPA 540 G-88 003, OSWER Directive 9283, December 1–2, 1988.

Fetter, C. W. *Contaminant Hydrogeology*, Macmillan, New York, 1993.

Gelhar, L. W., C. Welty, K. R. Rehfeldt. A critical review of data on field-scale dispersion in aquifers. *Water Resour. Resear.* **28**(7), 1955–1974 (1992).

Johnson Filtration Systems, Inc. *Groundwater and Wells*, St. Paul, MN, 1986.

Karickhoff, S. W., D. S. Brown, and T. A. Scott. Sorption of hydrophobic pollutants on natural sediments. *Water Res.* **13**, 241–248 (1979).

Means, J. C., S. G. Wood, J. J. Hassett, and W. L. Banwart. Sorption of polynuclear aromatic hydrocarbons by sediments and soils. *Environ. Sci. Technol.* **14**, 1524–1528 (1980).

Wood, T. R. and G. T. Norrell. Integrated Large-Scale Aquifer Pumping and Infiltration Tests, Groundwater Report OU 7-06, Summary Report, U. S. Department of Energy, Idaho National Engineering Laboratory, INEL-96/0256, October 1996.

www.usgs.gov

www.epa.gov

CHAPTER 9

FATE AND TRANSPORT CONCEPTS IN ATMOSPHERIC SYSTEMS

CASE STUDY: THE ACCIDENT AT UNION CARBIDE—BHOPAL

One of the worst industrial accidents in history occurred overnight on December 2–3, 1984 at a chemical plant in Bhopal, India. The plant, owned and operated by Union Carbide India Limited (UCIL), was primarily a battery company but the owners also ventured into the manufacture of carbaryl, the active ingredient in the pesticide Sevin (Shrivastava, 1987, p. 42). Many reasons for the accident have been proposed, but cutbacks in operations and the fact that the refrigeration units for the chemical storage tanks were off-line weigh heavily in the cause of the accident (Bogard, 1989, p. 3). It has been estimated that as much as 42 tons of methyl isocyanate (MIC) escaped over a matter of minutes due to an uncontrolled reaction in storage tank 610 (Lapierre and Moro, 2002, p. 270). Thermal degradation of the MIC during the chemical reaction may have resulted in the formation of cyanide gas (Kurtzman, 1987, p. 101; Shrivastva, 1987, p. 70), but this is disputed by Union Carbide. MIC is almost twice as dense as air, so the gas blanketed the ground and did not readily mix with the surrounding air. The cloud, approximately 100 yards wide (Lapierre and Moro, 2002, p. 299), spread with the prevailing southerly wind into the impoverished neighborhoods around the plant. Death estimates vary widely, from 1754 to 15,000, while estimates of the number injured range from 200,000 to 300,000 (Shrivastva, 1987, p. 65; Bogard, 1989, p. 341).

In this chapter, we will learn about factors important in the fate and transport of airborne pollutants and how to model pollutant releases such as this one, using simple concepts included in the simulator Fate®. As in our other fate and transport models, we will use continuous (step) inputs and instantaneous (puff) inputs to the atmosphere. But as we will see in this chapter, atmospheric modeling requires knowledge of atmospheric and wind conditions. Also, transport and dilution of the pollution occur so quickly that we do not usually include degradation reactions in our model estimates.

After you finish reading this chapter, we will ask you to return to the Bhopal accident and model the event using Fate®. As input for your modeling approach, you will use the puff model in fate. Your input conditions are 42 tons of MIC (Moro, 2002, p. 270), a stack height of 0.00, a clear night with a temperature inversion, and a light wind (2–3 m/sec). For comparison purposes and concentration threshold limits, we will use a MIC threshold limit value of 0.02 parts per million (Kumar and Mukerjee, 1985, p. 131). Plot the concentration of MIC as a function of distance from the source at 20.0 minutes after the release. Note that this is an iterative approach, similar to that shown later in Figure 9.7, which will require you to calculate the concentration many times using Fate®.

9.1 INTRODUCTION

The atmosphere is the environmental medium where we live and breathe, and therefore we are very sensitive to inputs of pollution to the air. Modeling of atmospheric pollution can be used to determine human exposure to existing pollution sources and to predict future exposures from industrial accidents. There are many sources of atmospheric pollution, including industrial smokestacks, fugitive (or non-point) industrial emissions, gasoline stations, industrial accidents, automotive and railroad accidents, volcanoes, and forest fires. In this chapter, we will develop relatively simple models to predict the fate and transport of pollution released from these and similar sources. One major difference between the atmosphere and the other environmental media is that the atmospheric models must allow mixing in three dimensions since air and pollutants readily mix. Thus, in contrast to the one-dimensional modeling equations presented elsewhere in this textbook, we will be using a three-dimensional model for our atmospheric studies.

9.2 INPUT SOURCES

Input sources of pollution to the atmosphere are highly variable, and classification of these as either pulse or step depends on the time scale of observation. Pulse and step inputs are defined in the previous chapters as short-term and long-term, respectively. But many pollution events, both natural and man-made, do not neatly fit into these two categories. The complexities of the timing of these pollution events could be incorporated into a more complicated model, but this would require a numerical methods of analysis approach. Here, we want only a basic understanding of atmospheric modeling. Therefore, as we have for other systems, we will break the pollution events into pulse (referred to as puff in atmospheric scenarios) and step (plume) categories by selecting time scales that allow modeling by our two simple scenarios. For example, if a railroad tanker of acetone (a very volatile organic compound) derails and spills its contents, this could be considered a pulse release over a period of months if we are looking at long-range transport, but we would rarely be concerned with the downwind concentrations of acetone at these distances, because the concentrations would be extremely dilute. So, a more realistic modeling approach would be to model the release as a step release using the volatilization rate of acetone as our input source. An example that would fit into the pulse scenario for both the short and long-term time scales is the release of radiation from a catastrophic nuclear accident. For example, the release of radioactivity from Chernobyl could have been modeled as a pulse release in considering short-term radioactive dose rates for nearby countries. Scientists also monitored the spread of radiation around the world for the Chernobyl release, and this could also have been modeled as a pulse release but on a much longer time scale. So, in using the atmospheric models, the modeler must set the time duration (a boundary condition), which, in turn defines the source input function.

9.3 IMPORTANT FACTORS IN THE MODELING OF ATMOSPHERIC POLLUTION: CONCEPTUAL MODEL DEVELOPMENT

9.3.1 One- Versus Two- Versus Three-Dimensional Models

First, we will compare other fate and transport models to the general atmospheric model. The aquatic models in this textbook were given only for one dimension, the x or longitudinal direction. Streams and lakes usually can be adequately modeled using one-dimensional models since most of the dispersion is in the longitudinal direction, while groundwater systems can be modeled in one, two, or three dimensions (x, y, and z). Two dimensions are normally required for accurate modeling because the groundwater is not constrained by a river or lake bank, and dispersion can occur in all directions. Vertical dispersion, while important near a pollution source, becomes less important when the groundwater system is bounded by confining layers above and below the aquifer of interest, which is why we used the simpler one-dimensional model in considering both the instantaneous and pulse groundwater releases. However, deep aquifer systems require at least two-dimensional models. Virtually all atmospheric models, including the one used here and in Fate®, are three-dimensional.

9.3.2 Mixing and Dispersion in Atmospheric Systems

While the aquatic models may have seemed complicated, they are simple compared to most atmospheric models. To account for complex wind currents and mixing, atmospheric models have to incorporate three dimensions, which automatically makes the governing equations more complex. As usual, we make many assumptions so that our model will be more manageable. For example, the models given later in this chapter ignore buoyancy effects, and thus they are designed for gases with the same density as the atmosphere. As discussed above, the models distinguish between step and instantaneous sources, although many atmospheric pollution episodes can lie between these two extremes. Unlike the aquatic models that allow first-order decay processes, our atmospheric models do not allow degradation of pollutants. This assumption is justified for modeling of pollutants over relatively short distances (under 10 km or 7 miles). With a 5-mile-per-hour wind, it will take less than 1.5 hours for the pollutant to be dispersed this distance (our model limit, explained later), and most atmospheric photochemical reactions (except for the production of smog) require the pollutant to be in the atmosphere over a much longer time frame (hours to days). The dominant force resulting in the decrease in pollutant concentration is dispersion, which can rapidly dilute pollutant concentrations in the atmosphere. However, understanding and accounting for dispersion can be very complicated. First, we will look at the movement of atmospheric gases over the Earth's surface.

A profile of the wind's velocity with increasing height shows a steep increasing parabolic shape, with low velocity at the Earth's surface due to friction between the moving air and the ground. The surface wind velocity, however, is also subject to many complex variables. For example, the roughness of the Earth's surface can significantly impact the shape or steepness of the wind velocity-height profile. The wind velocity profile over an open grassland is illustrated on the right-hand side of Figure 9.1, showing that wind speed rapidly approaches its maximum as height above the surface increases. Compare this to an urban setting where tall buildings impede the path of the wind and slow its speed near the land surface. This expands the velocity–height gradient well above the Earth's surface. The resulting lower wind velocity could decrease the turbulence and subsequent dispersion and may result in stagnant pockets of the atmosphere that can contain clear or polluted air. The increase

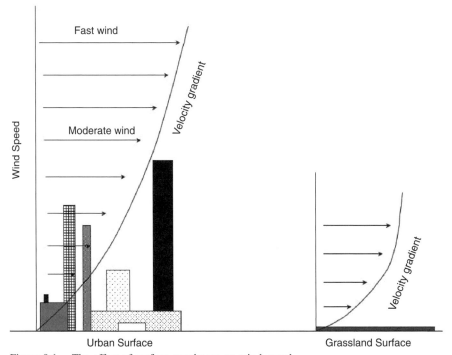

Figure 9.1.   The effect of surface roughness on wind speed.

in the surface's roughness due to the presence of buildings will greatly affect flow patterns and ground-level pollutant concentrations. The wind velocity and direction are also highly variable in urban settings. For example, in Chicago in winter, the cold wind may actually be blowing into your face every time you turn the corner of a building. Observations such as this demonstrate that atmospheric processes are too complicated even for our most sophisticated models. In our brief treatment here we will simplify our model by assuming that an average wind speed can be used and, in general, we will not account for differences in surface roughness or small-scale changes in wind direction.

While surface roughness can greatly affect turbulence and mixing, the magnitude of wind speed can also increase mixing. Recall that the movement of any fluid or gas is referred to as advection, and where gradients exist this movement results in the mixing of gases. We will refer to this mixing as dispersion, since the net result is a dilution of pollutant concentrations. Considering the combined effects of wind velocity and the atmospheric temperature as a function of height above the surface, we obtain the three basic turbulence scenarios shown in Figure 9.2. First, consider an isolated pocket of atmosphere at nighttime temperatures (shown in Figure 9.2a). This type of condition occurs where a thick cloud layer prevents the Earth from radiating its heat to space as it cools during the night. Under theses conditions, an emission from an industrial stack will take the shape of the plume shown in Figure 9.2a. The released gases will rise or sink until their density (temperature) matches that of the surrounding (diluting) atmospheric gases. Then the plume will take the shape of a thin layer.

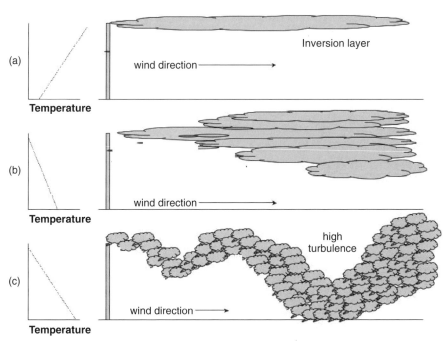

Figure 9.2.   Three basic turbulence scenarios for plumes.

Under daytime heating conditions, the temperature–height profile will be similar to the one shown in Figure 9.2b. The warmest point in the atmosphere is at the Earth's surface; as the heated air rises, it cools, thus producing the temperature profile shown in Figure 9.2b. In a steady wind, the plume will spread in all directions, but mostly in the longitudinal direction. With a lower temperature–height gradient and a higher wind velocity, extreme turbulence will be observed (Figure 9.2c). In order to attempt the modeling of these conditions, we must greatly simplify the temperature and wind relationships.

We begin our simplification process by attempting to combine the effects of wind velocity, temperature–height profiles, and cloud cover into a set of atmospheric stability categories. As we do this, remember that our goal is to come up with a way to characterize dispersion (mixing) of the pollutant with the atmospheric gases. Table 9.1 shows a qualitative approach to the combined effects of wind speed and cloud cover collected for rural settings in England. Cloud cover is a good reflector of heat back to the Earth. The categories range from strongly unstable (category A reflected in Figure 9.2c) to very stable (category G) and distinguish between day and night conditions.

Next, the somewhat qualitative categories in Table 9.1 are used to mathematically predict values for horizontal dispersion coefficients (Table 9.2), which are estimates of mixing in the x and y directions. We do not have a way to mathematically predict these values accurately, and the data in Tables 9.1 and 9.2 are empirical (based on experimental observations). We usually assume that dispersion (σ in Table 9.2) in the x and y directions is the same; thus Table 9.2 can be used to estimate σ_x and σ_y simultaneously. The equations given in Table 9.2 were used to draw the lines in Figure 9.3 for the six stability categories. Note that dispersion increases as you move away from the point source of pollution. This should be intuitive, since mixing

TABLE 9.1. Pasquill Stability Categories

Wind Speed (at 10-m elevation m/sec)	Day, Degree of Cloud Insulation			Night	
	Strong	Moderate	Slight	Thinly Overcast or Greater than 50% Low Clouds	Less than 50% Cloud Cover
<2	A	A–B	B	G	G
2–3	A–B	B	C	E	F
3–5	B	B–C	D	D	E
5–6	C	C–D	D	D	D
>6	C	D	D	D	D

Source: Turner (1994) and Pasquill (1961). Turner (1994) adds the following notes on selecting the appropriate category:

1. Strong insolation corresponds to sunny midday in midsummer in England; slight isolation to similar conditions in midwinter.
2. Night refers to the period from 1 h before sunset to 1 h after sunrise.
3. The neutral category D should also be used, regardless of wind speed, for overcast conditions during day or night and for any sky condition during the hour preceding or following night as defined in note 2.

**TABLE 9.2.   Pasquill–Gifford Horizontal Dispersion
Parameters (Turner, 1994)**

$$\sigma_y = 1000 \times \tan(T)/2.15$$

where x is the downwind distance (in km) from the point source
and T is one-half Pasquill's q in degrees. T, as a function of x, is
determined by each stability category from Table 9.1.

Stability	Equation for T
A	$T = 24.167 - 2.5334 \ln(x)$
B	$T = 18.333 - 1.8096 \ln(x)$
C	$T = 12.5 - 1.0857 \ln(x)$
D	$T = 8.333 - 0.7238 \ln(x)$
E	$T = 6.25 - 0.5429 \ln(x)$
F	$T = 4.167 - 0.3619 \ln(x)$

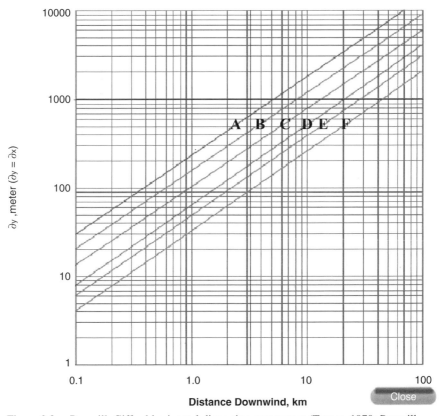

Figure 9.3.   Pasquill–Gifford horizontal dispersion parameters (Turner, 1970; Pasquill, 1961).

continues as you move away from the point source. So, for every pollutant concentration you attempt to estimate, you must select a distance from the point source. The limitation of this is that in Fate® you can only plot a slice of the concentration profile in the y and z plane. You therefore have to manually plot the concentration gradient in the x, or longitudinal, direction of wind flow in a spreadsheet.

Dispersion in the vertical (z) direction is somewhat more complicated to predict and again is based on experimental observations. We can estimate the vertical dispersion coefficient, σ_z, by using the same atmospheric stability categories from Table 9.1 but with a more precise treatment of the wind speed. The equation governing the estimate of vertical dispersion is

$$\sigma_z = ax^b$$

where x is the distance (in km) and a and b are fitting parameters obtained from Table 9.3.

Plots showing the dependence of vertical dispersion coefficients (in meters) on distance from the point source for the different stability categories are shown in Figure 9.4. We have been describing dispersion, but what exactly is it? As we have noted, dispersion is a physical process and is a function of the distance from the point source. Dispersion can be mathematically described as the mixing between the pollutant plume and the natural atmospheric gases. The values you read from the graph or calculate using the equations are given in meters or kilometers. Thus, the values given represent the width of the pollutant plume at the specified distance from the point source and reflect the amount of atmosphere the pollution has mixed with. You can visualize the spread of the plume downwind as a narrow funnel, flaring out as you move away from the point source and becoming more diluted in pollutant concentration.

9.4 MATHEMATICAL DEVELOPMENT OF MODEL

9.4.1 Step Input (Plume Model) of Pollutant

Using the many assumptions stated earlier and the estimated horizontal and vertical dispersion coefficients, a model [Eq. (9.1)] for the steady-state plume can be derived, using differential equation techniques to estimate the pollutant concentration at any point (x, y, and z) downwind from the continuous source. This is referred to as the steady-state plume model.

$$C(x,y,z) = \frac{Q_m}{2\pi\sigma_y\sigma_z u}\left(\exp\left[-\frac{1}{2}\left(\frac{y}{\sigma_y}\right)^2\right]\right)$$
$$\left(\exp\left[-\frac{1}{2}\left(\frac{z-H_r}{\sigma_z}\right)\right] + \exp\left[-\frac{1}{2}\left(\frac{z+H_r}{\sigma_z}\right)^2\right]\right) \tag{9.1}$$

where $C(x, y, z)$ is the concentration of pollutant in the plume as a function of x, y, and z (mass/length^3), x, y, and z are distances from the source (length) (see Figures

TABLE 9.3. Pasquill–Gifford Vertical Dispersion Parameter

Stability Boundary	Distance (km)	Vertical Dispersion Parameter: $\sigma_z = ax^b$, where x is in km		
		a	b	σ_z at upper
A	>3.11			5000
	0.5–3.11	453.85	2.1166	
	0.4–0.5	346.75	1.7283	104.7
	0.3–0.4	258.89	1.4094	71.2
	0.25–0.3	217.41	1.2644	47.4
	0.2–0.25	179.52	1.1262	37.7
	0.15–0.2	170.22	1.0932	29.3
	0.1–0.15	158.08	1.0542	21.4
	<0.1	122.8	0.9447	14.0
B	>0.35			5000
	0.4–35	109.30	1.0971	
	0.2–0.4	98.483	0.9833	40.0
	>0.2	90.673	0.93198	20.2
C	All values of x		61.141	0.91465
D	>30	44.053	0.51179	
	10–30	36.650	0.56589	251.2
	3–10	33.504	0.60486	134.9
	1–3	32.093	0.64403	65.1
	0.3–1	32.093	0.81066	32.1
	<0.3	34.459	0.86974	12.1
E	>40	47.618	0.29592	
	20–40	35.420	0.37615	141.9
	10–20	26.970	0.46713	109.3
	4–10	24.703	0.50527	79.1
	2–4	22.534	0.57154	49.8
	1–2	21.628	0.63077	33.5
	0.3–1	21.628	0.75660	21.6
	0.1–0.3	23.331	0.81956	8.7
	<0.1	24.260	0.83660	3.5
F	>60	34.219	0.21716	
	30–60	27.074	0.27436	83.3
	15–30	22.651	0.32681	68.8
	7–15	17.836	0.4150	54.9
	3–7	16.187	0.4649	40.0
	2–3	14.823	0.54503	27.0
	1–2	13.953	0.63227	21.6
	0.7–1	13.953	0.68465	14.0
	0.2–0.7	14.457	0.78407	10.9
	<0.2	15.209	0.81558	4.1

Source: Turner (1970) and Pasquill (1961).

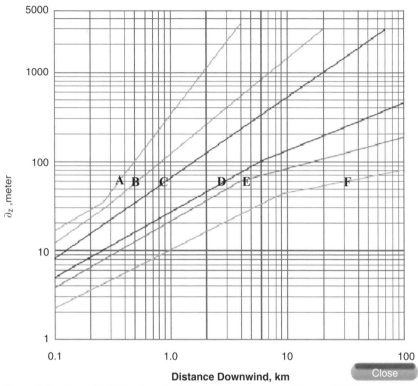

Figure 9.4. Pasquill–Gifford vertical dispersion parameters (Turner, 1970).

9.3 and 9.4), Q_m is the pollutant source (mass/time), $\sigma_x = \sigma_y$ is the horizontal dispersion coefficient (length), σ_z is the vertical dispersion coefficient (length), u is the wind velocity (length/time), and H_r is the height of the release (length).

Notice that the model is divided into two basic parts: the mass of pollutant released, represented by the first term in the equation, and dispersion, characterized by the wind speed, the y and z coordinates that yield estimates of dispersion (mixing), and the height of the release above the Earth's surface. Input values for all of these parameters are relatively simple to estimate or measure using the techniques described earlier. Note that the longitudinal distance, x, is a part of the equation since you must select the x distance to obtain estimates of σ_x and σ_z.

The equation can be simplified if the point of interest (at the receptor) is immediately downwind of the pollution source. For the concentration along this three-dimensional centerline of the plume ($z = 0$, and $H_r = 0$), we can use a simplification of Eq. (9.1):

$$C(x, y, 0) = \frac{Q_m}{\pi \sigma_y \sigma_z u} \left(\exp\left[-\frac{1}{2}\left(\frac{y}{\sigma_y} \right)^2 \right] \right) \tag{9.2}$$

A typical simulation of downwind pollutant concentration is shown in Figure 9.5 for a z value of 1.0 m (height above ground level), a y distance of 0.0 km (along the x–z axis), and an x value of 1.5 km (distance downwind). In Figure 9.5, the peak of the Gaussian-shaped plot is along the center x-axis ($y = 0$ m), and 1 m above ground or about nose level for a child. The pollutant concentration declines as you go to the left or right of the centerline (positive or negative y values). Note that the width of the main plume concentration covers a range of approximately 1200 m (from −600 m to the left to +600 m to the right) along the x-axis.

A similar output would be ·obtained by plotting a y value of 0.0 (along the center line), an x distance of 1.5 km, and calculating the pollutant concentration as you move up above the Earth's surface. This is illustrated in Figure 9.6. In this plot, as you go from left to right on the x-axis, you are moving up away from the Earth's surface.

A useful function of Fate® is to evaluate the pollutant concentration as a function of distance from the point source along the centerline of the wind flow. Fate® cannot plot this directly, since dispersion in the x, y, and z directions are functions of the distance from the point source. In order to accomplish this, we must repeatedly use Steps 5 and 6 in manual calculations or in the Fate® simulator. Systematically change the x distance, increasing it incrementally, and record the pollutant concentration given in Step 6 (in Fate®). A plot like the one shown in Figure 9.7 can

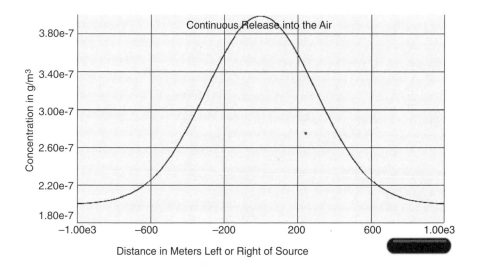

Figure 9.5. Output from Fate® for a continuous release (plume) of pollutant into the atmosphere at an x value of 1.5 km.

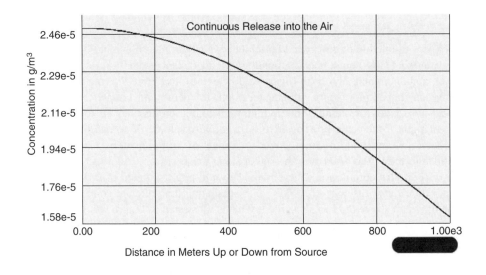

$$x = \boxed{1.5} \ \text{km} \qquad Q_m = \boxed{20} \ \text{g/sec} \qquad \partial_x = \boxed{298.2} \ \text{m}$$
$$z = \boxed{1.5} \ \text{m} \qquad \bar{u} = \boxed{0.8} \ \text{m/sec}$$
$$H_r = \boxed{30} \ \text{m} \qquad \partial_z = \boxed{1.071e3} \ \text{m}$$

Figure 9.6. Output from Fate® for a continuous release (plume) of pollutant into the atmosphere showing variations in plume concentration with changing vertical position in relation to the source ($x = 1.5$ km).

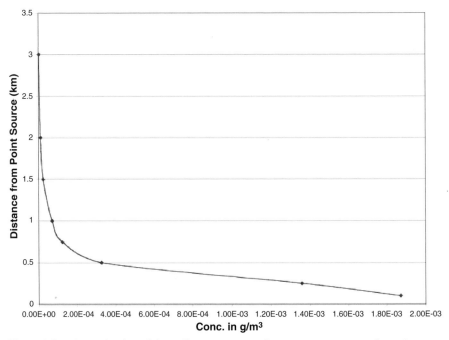

Figure 9.7. An evaluation of the pollutant concentration as you move away from the point source (plume model).

be obtained by summarizing the output data. Note that the pollutant concentration decreases, as expected, as you move away from the point source. This type of plot is similar to those used to illustrate the other fate and transport models in this text.

Example Problem.   For the following data (which would be measured at the source), manually calculate the downwind concentration profile over a distance of 1.5 km for (1) varying distance from the centerline (varying y coordinate) where z = 4, and (2) varying distance above the ground surface (z coordinate) where $y = 1.5$:

H_t: Height of pollutant release is 30.0 m.

u_s: Stack exit velocity of gases is 0.80 m/sec.

u: Wind speed is 0.80 m/sec.

d: Inside stack diameter is 0.80 m.

P: Atmospheric pressure is 1010 mbar.

T_s: Stack gas temperature is 285 K.

T_a: Atmospheric temperature is 280 K.

The source rate of emission is 20 g/sec.

Use an atmospheric stability condition of A (low wind speed, day time hours, and strong cloud cover).

Solution.   The calculated effective height of release is calculated by

$$\Delta H_r = \frac{\bar{u}_s d}{u}\left(1.5 + 2.68 \times 10^{-3}\, Pd\left(\frac{T_s - T_a}{T_s}\right)\right) \tag{9.3}$$

and results in an effective stack height of 31.2 m (the height the gaseous plume rises to). Note that temperature must be in degrees Kelvin.

From the equations for dispersion given in Table 9.2 and 9.3 or Figures 9.3 and 9.4, and an atmospheric stability in category A, the horizontal dispersion is 298.2 m and the vertical dispersion is 1.071×10^3 m.

$$C(x,y,z) = \frac{Q_m}{2\pi\sigma_y\sigma_z u}\left(\exp\left[-\frac{1}{2}\left(\frac{y}{\sigma_y}\right)^2\right]\right)\left(\exp\left[-\frac{1}{2}\left(\frac{z - H_r}{\sigma_z}\right)^2\right]\right.$$
$$\left. + \exp\left[-\frac{1}{2}\left(\frac{z + H_r}{\sigma_z}\right)^2\right]\right) \tag{9.4}$$

Using the continuous fate and transport equation [Eq. (9.4), the same as Eq. (9.1)], the concentration profile (in g/m³ versus m) for the positive and negative y directions is shown in Figure 9.8.

The concentration profile for the z (height) direction is shown in Figure 9.9.

9.4.2 · Instantaneous Input (Pulse or Puff Model) of Pollution

For a pulse rather than a step input, dispersion is handled a little differently. In the step (plume) model we can use either rural or urban dispersion estimates, whereas

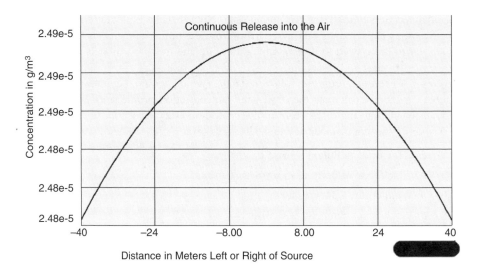

x = 1.5 km	Q_m = 20 g/sec	∂_x = 298.2 m
z = 4 m	$\bar{u}$ = 0.8 m/sec	
	H_r = 30 m	∂_z = 1.071e3 m

Figure 9.8. Concentration profile in the positive and negative y direction for the example problem (plume model).

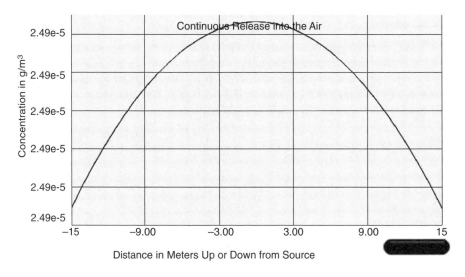

X = 1.5 km	Q_m = 20 g/sec	∂_x = 298.2 m
Y = 1.5 m	$\bar{u}$ = 0.8 m/sec	
	H_r = 30 m	∂_z = 1.071e3 m

Figure 9.9. Concentration profile in the positive and negative z direction for the example problem (plume model).

urban dispersion parameters are usually used for the pulse (puff) model. The "Puff" dispersion estimates are derived from experimental observations made by McElroy and Pooler (1968) near St. Louis and from Briggs (1972). Calculations for estimating the horizontal and vertical dispersion coefficients are shown in Tables 9.4 and 9.5. We will again assume that dispersion (σ) in the x and y directions is the same. Atmospheric stability categories are the same as those described in Table 9.1. Vertical and horizontal dispersion coefficients are shown in Figures 9.10 and 9.11.

Using the stability categories, wind speed, and the equations shown in Table 9.4, we can now estimate the atmospheric pollutant concentration downwind from an instantaneous source (also referred to as pulse or puff) by

$$C(x,y,z,t) = \frac{Q_m}{(2\pi)^{3/2}\sigma_x\sigma_y\sigma_z}\exp\left[-\frac{1}{2}\left(\frac{y}{\sigma_y}\right)^2\right]$$
$$\left(\exp\left[-\frac{1}{2}\left(\frac{z-H_r}{\sigma_z}\right)^2\right]+\exp\left[-\frac{1}{2}\left(\frac{z+H_r}{\sigma_z}\right)^2\right]\right) \quad (9.5)$$

where $C(x, y, z, t)$ is the concentration of pollutant in the plume as a function of x, y, and z (mass/length^3) and time; x, y, and z are distances from the source (length) (see Figures 9.10 and 9.11); t is time; Q_m is the pollutant source (mass/time); σ_x and σ_y are the horizontal dispersion coefficients (length); σ_z is the vertical dispersion coefficient (length); and H_r is the height of the release (length). Note that time is

TABLE 9.4. Urban Dispersion Parameters

Pasquill Type of Stability	σ_y (in meters)	σ_z (in meters)
A–B	$0.32/(0.0004x)^{-0.5}$	$0.24/(0.001x)^{0.5}$
C	$0.22/(0.0004x)^{-0.5}$	$0.20x$
D	$0.16/(0.0004x)^{-0.5}$	$0.14/(0.0003x)^{-0.5}$
E–F	$0.11/(0.0004x)^{-0.5}$	$0.08/(0.0015x)^{-0.5}$

For distances, x, between 100 and 10,000 m.

Source: Turner (1994), Briggs (1972), and McElroy and Pooler (1968).

TABLE 9.5. Open-Country Dispersion Parameters (not used in Fate°, but you may manually enter the calculated values)

Pasquill Type of Stability	σ_y (in meters)	σ_z (in meters)
A	$0.22x/(1+0.0001x)^{0.5}$	$0.20x$
B	$0.16x/(0.0001x)^{0.5}$	$0.12x$
C	$0.11x/(1+0.0001x)^{0.5}$	$0.08x/(1+0.0002x)^{0.5}$
D	$0.08x/(1+0.0001x)^{0.5}$	$0.06x/(1+0.0015x)^{0.5}$
E	$0.06x/(1+0.0001x)^{0.5}$	$0.03x/(1+0.0003x)$
F	$0.04x\,(1+0.0001x)^{0.5}$	$0.016x/(1+0.0003x)$

For distances x between 100 and 10000 m.

Source: Turner (1994), Briggs (1972), and McElroy and Pooler (1968).

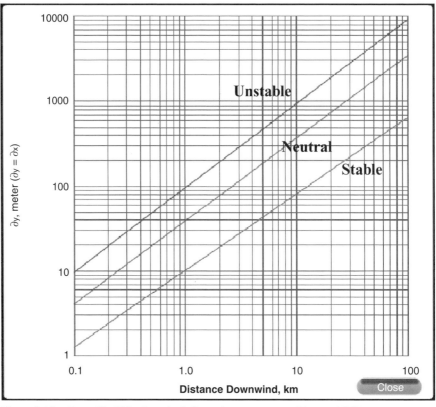

Figure 9.10. Pasquill–Gifford vertical dispersion parameters (Turner, 1970).

included here because the distance traveled (x) since the instantaneous pollutant release is a function of wind velocity (u) and time (t), where

$$x = ut$$

For the concentration along the centerline ($y = 0$, $z = 0$, and $H_r = 0$) we can use a simplification of Eq. (9.5), to yield

$$C(x, 0, 0, t) = C(ut, 0, 0, t) = \frac{Q_m}{\left(\sqrt{2\pi}\right)^{3/2} \sigma_x \sigma_y \sigma_z} \quad (9.6)$$

Simulation outputs [for Eq. (9.5)] from Fate® are shown in Figures 9.12 and 9.13 for viewing pollutant concentration along the y-axis and the z-axis, respectively.

Example Problem. For the following data (which would be measured on-site), manually calculate the downwind concentration profile for a distance of 10 km for (1) varying distance from the centerline ($y = 0$) where $z = 30$ and (2) varying distance above the ground surface ($z = 0$) where $y = 1.5$.

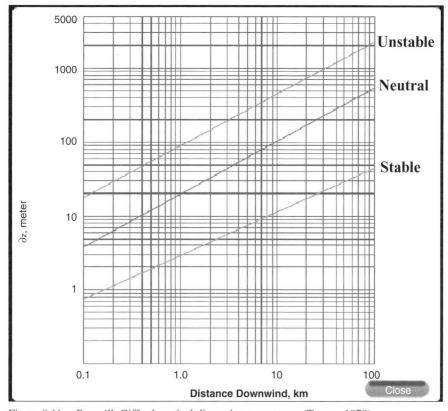

Figure 9.11.   Pasquill–Gifford vertical dispersion parameters (Turner, 1970).

Height of pollutant release is 30.0 m

Wind speed is 4.0 m/sec

Total source mass is 2000 g

The given wind speed results in an atmospheric stability of condition D for clear daytime conditions.

Solution.   Using the equations for dispersion or the figures, the horizontal dispersion is 0.7517 m and the vertical dispersion is 0.4990 m.

$$C(x,y,z,t) = \frac{Q_m}{(2\pi)^{3/2}\sigma_x\sigma_y\sigma_z} \exp\left[-\frac{1}{2}\left(\frac{y}{\sigma_y}\right)^2\right]$$

$$\left(\exp\left[-\frac{1}{2}\left(\frac{z-H_r}{\sigma_z}\right)\right] + \exp\left[-\frac{1}{2}\left(\frac{z+H_r}{\sigma_z}\right)^2\right]\right) \qquad (9.7)$$

Using Eq. (9.7), $z = 30$ m, and $y = 1.5$ m, the concentration profile (in g/m^3 versus m) for the positive and negative y directions is shown in Figure 9.14.

The concentration profile for the z (height) direction is shown in Figure 9.15.

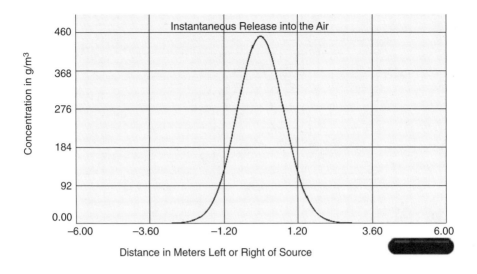

Figure 9.12.   Output from Fate® for a pulse release (puff) of pollutant into the atmosphere, with variation in horizontal distance from source.

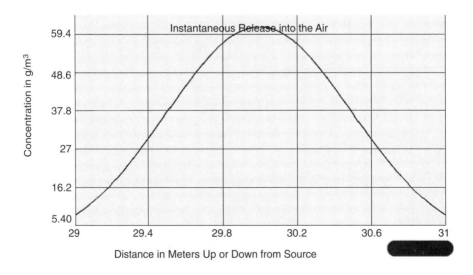

Figure 9.13.   Output from Fate® for a pulse release (puff) of pollutant into the atmosphere, with variation in vertical distance from source.

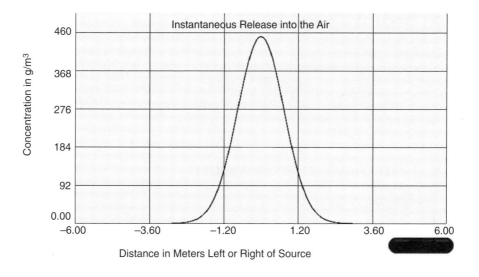

Figure 9.14.   Concentration profile for the positive and negative y direction for the example problem (puff model).

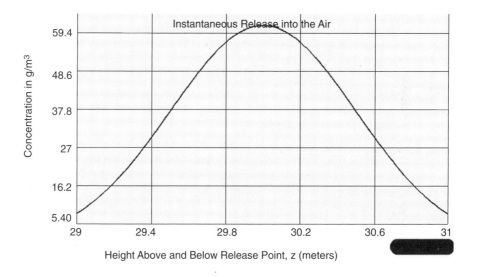

Figure 9.15.   Concentration profile for the positive and negative z direction for the example problem (puff model).

9.5   SENSITIVITY ANALYSIS

Each chapter has discussed the need to question the model and input parameters. Model parameters for a sensitivity analysis of the two atmospheric equations include the wind velocity flow (v) and dispersion (σ) in each direction. Thus, by systematically varying each of these, the modeler can tell how sensitive their model is to inaccurate or uncertain model parameters.

9.6   LIMITATIONS OF OUR MODEL

Chemistry. As noted earlier, one obvious parameter missing from the atmospheric modeling equations is chemical reactions. These can relatively easily included but are usually omitted because dilution of the pollution to insignificant concentrations usually occurs relatively quickly. This is due to the great dilution power of atmospheric mixing. Usually we are concerned with minutes to hours of travel time, which are considerably short as compared to the degradation rates for pollutants. Extremely reactive pollutants such as chlorine gas may require that a degradation rate be included.

Dispersion and Mixing. Dispersion is difficult to accurately quantify in any environmental media, and this is especially true in atmospheric systems. The difficulty in estimating this model parameter is due to the inconsistent surface of the ground. We have included two ways of estimating dispersion, and each is specific to ground surface setting (i.e., urban, rural, etc.). To further complicate our efforts, it should be obvious that wheat fields of Midwestern United States yield different mixing scenarios than forested lands of the Southeast. Similarly, urban setting with low buildings yield different mixing events than cities with skyscrapers. Professionals use site-specific dispersion parameters for the area of interest and a numerical methods modeling approach.

Wind Velocity. A highly variable parameter in determining mixing and pollutant transport is the wind velocity. Wind velocity is generally much more variable than is water velocity in a river or aquifer. The only way to handle large variations in wind velocity is, again, to use a numerical methods approach to model the system.

9.7   REMEDIATION

While, in theory, technologies such as carbon absorption exist for the remediation of atmospheric gases, the enormous scale of such atmospheric pollutant outputs has prevented any such attempt at remediation on a large scale. The one tried-and-true approach to atmospheric remediation is to eliminate the source of pollution and let nature take its course of dilution of the pollution in the prevailing wind. This may seem simplistic, but its efficacy was recently demonstrated during the electrical outage on the U.S. east coast and in Canada in 2003. Scientists, meteorologists, and residents all reported cleaner air after the event when coal-fired plants and industry were closed down for hours or days (Petkewich, 2004). Of course, given our current dependence on coal we cannot economically afford a long-term closure of these power-generating facilities. Still, it is interesting to consider the immediate impact of the removal of a large source of our air pollution. Again, the only cure for a polluted atmosphere is source reduction or elimination.

SUGGESTED READING

Kitman, J. L. The secret history of lead, *The Nation*, March 20, 2000. Also available at http://www.thenation.com/doc.mhtml?i=20000320&c=1&s=kitman

Concepts

1. Name three point sources of pollution to the atmosphere in your area.
2. Describe three step input scenarios of pollutants to the atmosphere.
3. Describe three pulse input scenarios of pollutants to the atmosphere.
4. Study the dispersion tables for the urban and rural settings. How does an increase in wind velocity affect dispersion?
5. Why did we not include chemical degradation in our models?
6. Both of the atmospheric models used in this chapter are three-dimensional. Would you expect the pollutant concentration to increase or decrease if we used a one-dimensional model with the same input parameters?

Exercises

1. Using the background example dataset in Fate®, conduct a sensitivity analysis on dispersion in each direction by varying wind speed and cloud cover conditions.
2. A new coal-fired power plant was constructed near the banks of a large river where it is fairly windy. Even with the new technology, there is still some mercury in the emissions from the plant. Use Fate® to construct an atmospheric dispersion graph assuming standard atmospheric pressure (approximately 1.0 atm) and the conditions listed below. Determine the concentration of mercury at a distance of 20 miles downwind of the stack.

   Height of initial release: 45 m

   Stack exit velocity: 1.2 m/sec

   Wind speed: 3.33 m/sec

   Inside stack diameter: 1.5 m

Stack gas temperature: 353

Air temperature: 60°F

Source flow: 0.0025 g Hg/sec

3. A tanker carrying chlorine gas crashes, releasing its contents into the surrounding area. Create a plot showing the concentration in g/m^3 versus distance in meters left or right of the source (−1.2 m to 1.2 m) using the following parameters:

Mass of gas spilled: 90,000,000 g

Wind speed: 12 m/sec

Height of release: 2 m

Atmospheric stability category: C

For a sensitivity analysis, change the wind speed to 1 m/sec and atmospheric stability at B. Check your answers with Fate®.

4. On a windy day, while working outside you can occasionally detect the smell of rotten eggs coming from the pulp mill east of your home. Since you know that hydrogen sulfide is highly toxic, and can kill at concentrations of only 50 ppm, you decide to calculate what concentrations could be reaching you from the mill about 6 km away. On the Internet, you find the concentration of H_2S emitted at the stack and other data you need to calculate the concentration downwind. Using data from the table below, calculate the concentration 6 km from the stack.

Effective Release Height: 28 m

Source flow: 2.18 g/sec

Wind speed: 6 m/sec

Atm stability: C

Distance downwind: 6 km

5. A train wreck has left a rusted gasoline tanker on its side and reeking of gasoline. The smell is emanating from a pool of gas on the ground next to the truck. The gasoline is evaporating at a rate of 1 g/sec, and it is being dispersed by a 2-m/sec wind during the middle of the day. Measurements are made at 4 m above ground, and the wind stability and radiation are described as type B in Fate®. At 0.1 and 4 km away, what are the concentrations of gasoline? Create a plot of gasoline concentrations from 0.02 to 5 km away.

Spreadsheet Exercise

Create a spreadsheet that performs the same calculations as Fate® for both the step and pulse equation. Construct your spreadsheet so that it is interactive (so that you can change numeric values for parameters and the plot automatically updates itself).

REFERENCES

Bogard, W. *The Bhopal Tragedy: Language, Logic, and Politics in the Production of a Hazard*, Westview Press, San Francisco, 1989.

Briggs, G. A. Discussion: Chimney plumes in neutral and stable surroundings. *Atmos. Environ.* **6**, 507–510 (1972).

Gifford, F. A. Turbulent diffusion typing schemes: A review. *Nuclear Safety* **17**, 1, 68–86 (1976).

Kumar, C. D. and S. K. Mukerjee. Methyl isocyanate: Profile of a killer gas, in *Bhopal: Industrial Genocide?* Arena Press, Hong Kong, 1985.

Kurtzman, D. *A Killing Wind: Inside Union Carbide and The Bhopal Catastrophe*, McGraw-Hill, New York, 1987.

Lapierre, D. and J. Moro. *Five Past Midnight in Bhopal*, Warner Books, An AOL Time Warner Company, New York, 2002.

McElroy, J. L. and F. Pooler. St. Louis Dispersion Study. U.S. Public Health Service, National Air Pollution Control Administration Report AP-53, 1968.

Pasquill, F. The estimation of the dispersion of windborne material. *Meterol. Mag.* **90**, 1063, 33–49 (1961).

Pasquill, F. *Atmospheric Dispersion Parameters in Gaussian Plume Modeling: Part II. Possible Requirements for Change in the Turner Workbook Values.* EPA-600/4-76-030b. U.S. Environmental Protection Agency, Research Triangle Park, NC, 1976.

Pethewich, R., *2003 blackout cleaned the air, Environ. Sci. Technol.,* Sept. 1, p. 320A, (2004).

Shrivastava, P. *Bhopal: Anatomy of a Crisis*, Ballinger Publishing Company, Cambridge, MA, 1987.

Turner, D. B. *Workbook of Atmospheric Dispersion Estimates*, U.S. Department of Health, Education, and Welfare, Cincinnati, OH, 1970.

Turner, D. B. *Workbook of Atmospheric Dispersion Estimates: An Introduction to Dispersion Modeling*, 2nd edition, Lewis Publishers, Ann Arbor, MI, 1994.

PART IV

RISK ASSESSMENT

"All things are poisons, for there is nothing without poisonous qualities. It is only the dose which makes the thing a poison."

—Paracelsus 1493–1541

CHAPTER 10

RISK AND THE CALCULATION OF HEALTH RISK FROM EXPOSURE TO POLLUTANTS

The EPA's mandated mission by Congress is to "protect human health and the environment." In the early days of EPA, it was felt that the agency regulated by decree—they defined a situation or practice as "bad" and ordered that it be stopped. Since the 1970s, the EPA's approach has been slowly transformed through the civil court process, so that they now have to more effectively substantiate why a practice is "bad" before they can regulate it. This prior litigation is probably the origin of risk assessment, which seeks to mathematically quantify how bad a situation is, and which is viewed by some environmentalists as an attempt by industry to undermine the power of EPA. Unfortunately, this dependence on a documented risk only serves half the mission of the EPA: the protection of human health. We cannot so easily and numerically quantify effects or place a monetary value on the environment.

The overall goal of this textbook was to develop ways to determine whether a polluted site warranted remediation or clean up, and ultimately to understand how the health effects of such a hazardous waste site or action are evaluated. We started by reviewing the chemistry of pollution and next presented basic fate and transport modeling procedures for predicting pollutant concentration as a function of chemical and physical conditions in the system. In this chapter, we will finally use all of the previous chapters to calculate the concentration of pollutant reaching a human (the receptor in our models). Based on this information, combined with the results of chemical exposure experiments involving animals, we will attempt to estimate the risk to humans. As you will see, our approaches are simple, and when you consider all of the possible sources of error or uncertainty, from the beginning of the process (pollutant source and mass characterization) through fate and transport modeling (with uncertainties in chemical degradation and physical dispersion), and now to risk assessment (where we extrapolate animal study data to human risk), we should recognize that our risk estimates may be orders of magnitude in error. Thus, risk assessment should be viewed as only one tool in our approach to environmental management. At the end of this chapter, we will attempt to summarize and evaluate our overall approach.

A Basic Introduction to Pollutant Fate and Transport, By Dunnivant and Anders

10.1 THE CONCEPT OF RISK

What is an acceptable risk? This is a major, and largely unanswerable, question. In order to attempt to answer this question, we will first look at commonly accepted risks that each of us exposes ourselves to every day. Table 10.1 shows the annual risk associated with a variety of actions. These are listed in decreasing degree of risk and show that smokers routinely accept an increased mortality risk of 3600 per million; that is, for every one million smokers, 3600 or 0.36% of them will die each year. As you can see, smoking is more risky than being a police officer or driving a car, yet millions of citizens willingly purchase cigarettes every day. Because this is a self-posed risk, it is perceived as more acceptable to most citizens than a risk that is imposed on them.

Another example of a popular perception of unknown risk can be seen in the question of drinking water treatment. Many citizens are becoming aware that the chlorination of our drinking water can result in the formation of cancer-causing agents, such as trihalomethanes (THMs). Some residents using treated drinking water purchase relatively costly water filters or boil their drinking water prior to use to remove the THMs. But when you look at the associated risk of drinking chlorinated water (as we will do in this chapter), you can clearly see that drinking chlorinated water is far safer than driving a car, working in the home, or even eating four tablespoons of peanut butter (refer to Table 10.1). But again, these are common practices we routinely accept—often, the main difference in terms of our risk perception is the distinction between a deliberate decision and an imposed situation.

TABLE 10.1. Some Commonplace Risks (Wilson and Crouch, 1987)

Action	Annual Risk (deaths per 1,000,000 persons at risk)
Cigarette smoking (1 pack/day)	3600
All cancers	2800
Mountaineering	600
Motor vehicle accident (total)	240
Police killed in line of duty	220
Air pollution	200
Home accidents	110
Frequent flying (professors)	50
Alcohol (light drinker)	20
Sea-level radiation	20
Eating 4 tbsp peanut butter/day (liver cancer from aflatoxin)	8
Electrocution	5.3
One-in-a-million chance of death per year	1
One-in-a-million chance of death per life	0.0014
Drinking water with EPA limit of chloroform (cancer over a 70-year period)	0.6
Drinking water with EPA limit of chloroform (cancer)	0.002

At the time of the writing of this textbook, the news media and the Center for Disease Control and Prevention (CDC) have been focusing on the problem of obesity in the United States population. This undoubtedly is linked to some of the leading causes of death in the United States, as seen in Table 10.2. The top three or four leading causes of death can be related to body weight. Such a consideration of the leading causes of death is one way of assessing risk.

Another way of looking at risk is to estimate what action will increase a person's chance of dying by one in a million. Table 10.3 lists a variety of activities that will increase your annual chances of death by one in a million. For example, it

TABLE 10.2. Leading Causes of Death in the United States for the Year 2000 (www.cdc.gov)

Rank	Cause of Death (all races, both sexes, and all ages)	Number	Percent of Total Deaths
	All causes	2,403,351	100.0
1	Diseases of the heart	710,760	29.6
2	Malignant neoplasms	553,091	23.0
3	Cerebrovascular diseases	167,661	7.0
4	Chronic lower respiratory diseases	122,009	5.1
5	Accidents (unintentional injuries)	97,900	4.1
6	Diabetes mellitus	69,301	2.9
7	Influenza and pneumonia	65,313	2.7
8	Alzheimer's disease	49,558	2.1
9	Nephritis, nephritic syndrome and nephrosis	37,251	1.5
10	Septicemia	31,224	1.3
	All other causes	499,283	20.8

TABLE 10.3. Activities that Increase Annual Mortality Risk by One in a Million (Wilson, 1979)

Activity	Type of Risk
Smoking 1.4 cigarettes	Cancer, heart disease
Living 2 months with a cigarette smoker	Cancer, heart disease
Eating 40 tablespoons of peanut butter	Liver cancer from aflatoxin
Eating 100 charcoal-broiled steaks	Cancer from benzo(a)pyrene
Flying 6000 miles by jet	Cancer by cosmic radiation
Living 2 summer months in Denver (at a mile elevation rather than at sea level)	Cancer by cosmic radiation
Traveling 300 miles by car	Accident
Flying 1000 miles by jet	Accident
Traveling 10 miles by bicycle	Accident
Traveling 6 minutes by canoe	Accident
Spending 1 hour in a coal mine	Black lung disease
Living 2 days in New York or Boston	Air pollution

has been estimated that only smoking 1.4 cigarettes per year, eating 100 charcoal-broiled steaks per year, or living in New York for a year will increase your risk of death, by the magic number of one in a million. These are risks that most of us are willing to assume, and all risk assessment calculations now consider a risk of one in a million to be acceptable, whatever the cause. Thus, if the risk associated with living near a chemical factory, drinking polluted water, or eating fruit containing a pesticide is less than one in a million, we (through our government) consider this activity, and the associated risk, acceptable. So, now we have a point of reference for our risk assessment calculations. We will accept a risk of one in a million.

In Chapters 5 through 9, we studied ways to estimate the concentration of a pollutant at various locations and times in river, lake, groundwater, and atmospheric systems. The modeling calculations estimated the concentration of pollutant that a receptor (human) consuming the water and air would be exposed to. First, we will use these concentrations to calculate the actual mass of a pollutant that a human would be exposed to under a set of "standard" conditions. The risk factors (Table 10.4) used in such dose calculations depend on certain standard conditions. For example, most calculations for adults use a body weight of 70 kg, while 15 kg is used for a child. Other suggested risk factors are the average water consumption at home and at the workplace and the average volume of air that we breathe. The use of recommended values makes comparisons of polluted waste sites and cancer clusters more feasible. In addition, EPA has compiled data from a variety of sources that can be used in risk assessment calculations. All of the data can be found at www.epa.gov/iris/, and data for specific pollutants are given in Tables 10.5–10.8. There is an extensive amount of data for each chemical in the IRIS database. For example, the printout for benzene is more than 20 pages long.

Now we will look at the process of estimating risks to human health. First, we calculate the dose rate, also referred to as the chronic daily intake (CDI), and later we use these values to estimate rates of carcinogenic and noncarcinogenic effects. Recall the general approach of environmental risk assessment outlined in Chapter 1: (1) Identify a potentially hazardous situation or polluted site, (2) identify the source of the pollutant and inputs to the systems, (3) estimate the concentration of pollutant reaching the receptor (human), use fate and transport modeling, and (4) evaluate health risk, the subject of this chapter. Finally, a remediation decision is made based on the results of all four steps and upon agreement between the liable party (the polluter), federal, state, and local governments, and the public.

10.2 DOSE RATES FROM VARIOUS SOURCES

There are a number of exposure routes by which humans can receive and uptake pollutants. These include not only drinking polluted water and breathing polluted air, but also ingestion during swimming activities, adsorption through the skin upon exposure to any contaminated material, direct intake of pollutants on soil, sediment, and dust, dermal adsorption from soil, sediment, and dust, and intake from foods. In the following pages, we will present calculations detailed by the EPA (1989a) that allow the estimate of pollutant intake from each of these sources. The overall process

TABLE 10.4. EPA-Recommended Exposure Factors for Risk Assessment (EPA, 1991)

Exposure Source	Exposure Pathway	Daily Intake	Exposure Frequency (days/year)	Exposure Duration (years)	Body Weight (kg)
Residential	Ingestion of potable water	2.00 L (adult) 1.00 L (child)	350	30	70 (adult) 15 (child)
	Ingestion of soil and dust	100 mg (adult) 2000 mg (child)	350	6 24	70 (adult) 15 (child)
	Inhalation of contaminants	20 m³ (adult) 12 m³ (child)	350	30	70 (adult) 15 (child)
Industrial and commercial	Ingestion of potable water	1.00 L	250	25	70 (adult)
	Ingestion of soil and dust	50 mg	250	25	70 (adult)
	Inhalation of contaminants	20 m³ (workday)	250	25	70 (adult)
Agricultural	Consumption of produce	42 g (fruit) 80 g (vegetable)	350	30	70 (adult) 15 (child)
Recreational	Consumption of fish	54 g	350	30	70 (adult) 15 (child)

TABLE 10.5. Oral RfDs for Selected Pollutants (http://www.epa.gov/iris/)

Compound	CASRN	Experimental Doses[a] (mg/kg · day, unless other wise noted)	UF	MF	RfD (mg/kd · day)	Critical Effect
Acetone	67-64-1	NOAEL: 900 LOAEL: 1700 BMDL: not determined	1,000	1	0.9	Nephropathy
Acrylamide	79-06-1	NOAEL: 0.2 LOAEL: 1	1,000	1	2E-4	Nerve damage
Alachlor	15972-60-8	NOAEL: 1 LOAEL: 3	100	1	1E-2	Hemosiderosis, hemolytic anemia
Aldicarb	116-06-3	NOAEL: 0.01 LOAEL: 0.025	10	1	1E-3	AChe inhibition
Aldrin	309-00-2	NOAEL: none LOAEL: 0.5 ppm diet 0.025	1,000	1	3E-5	Liver toxicity
Antimony	7440-36-0	NOEL: none LOAEL: 0.35	1,000	1	4E-4	Longevity, blood glucose, and cholesterol
Barium and compounds	7440-39-3	NOAEL: 65 LOAEL: 115	3	1	7E-2	Increased kidney weight
Benzene	71-43-2	BMDL = 1.2	300	1	4E-3	Decreased lymphocyte count
Beryllium and compounds	7440-41-7	BMD10: 0.46	300	1	2E-3	Small intestinal lesions
Bromoform	75-25-2	NOEL: 25	1,000	1	2E-2	Hepatic lesions
Cadmium	7440-43-9	NOAEL (water): 0.005	10	1	5E-4	Proteinuria
Carbon tetrachloride	56-23-5	NOAEL: 1 LOAEL: 10	1,000	1	7E-4	Liver lesions

Chemical	CAS					Effect
Chloroform	67-66-3	NOAEL: none LOAEL: 15	1,000	1	0.01	Fatty cyst formation
Chlordane (technical)	12789-03-6	NOAEL: 0.15 LOAEL: 0.75	300	1	5E–4	Hepatic necrosis
Chromium (VI)	18540-29-9	NOAEL: 25 (mg/L)	300	3	3E–3	None reported
2,4-Dichloro-phenoxyacetic acid (2,4-D)	94-75-7	NOAEL: 1.0 LOAEL: 5.0	100	1	1E–2	Hematologic, hepatic and renal toxicity
Dichlorvos	62-73-7	NOAEL: 0.05 LOAEL: 0.1	100	1	5E–4	Plasma and RBC ChE inhibition
Dieldrin	60-57-1	NOAEL: 0.1 (ppm)	100	1	5E–5	Liver lesions
Endrin	72-20-8	NOEL: 0.025 LOAEL: 0.05	100	1	3E–4	Lesions in liver occasional convulsions
Ethyl acetate	141-78-6	NOEL: 900 LOAEL: 3600	1,000	1	9E–1	Mortality body weight loss
Ethylbenzene	100-41-4	NOEL: 136 LOAEL: 408	1,000	1	1E–1	Liver and kidney toxicity
Formaldehyde	50-00-0	NOAEL: 15 LOAEL: 82	100	1	2E–1	Reduced weight gain histopathology
Heptachlor	76-44-8	NOEL: 3 (ppm diet) 0.15 LEL: 5 (ppm diet) 0.25	300	1	5E–4	Liver weight increases
Hexachlorobenzene	118-74-1	NOAEL: 1.6 (ppm diet) 0.08 LOAEL: 8.0 (ppm diet) 0.29	100	1	8E–4	Liver effects
Hexachloroethane	67-72-1	NOAEL: 1 LOAEL: 15	1,000	1	1E–3	Atrophy
Isophorone	78-59-1	NOEL: 150 LEL: None	1,000	1	2E–1	No observed effects

(*Continued*)

TABLE 10.5. Oral RfDs for Selected Pollutants (http://www.epa.gov/iris/) (continued)

Compound	CASRN	Experimental Doses[a] (mg/kg · day, unless other wise noted)	UF	MF	RfD (mg/kd · day)	Critical Effect
Methyl parathion	298-00-0	NOEL: 0.5 (ppm) 0.025 LEL: 5.0 (ppm) 0.25	100	1	2.5E–4	ChE inhibition
Methyl mercury (MeHg)	22967-92-6	BMDL05: 0.857 1.472 (μg/kg · day)	10	1	1E–4	Developmental neuropsychological impairment
Paraquat	1910-42-5	NOEL: 0.45 LEL: 0.93	100	1	4.5E–3	Chronic pneumonitis
Pentachlorobenzene	608-93-5	NOAEL: none LOAEL: 8.3	10,000	1	8E–4	Liver and kidney toxicity
Pentachlorophenol	87-86-5	NOAEL: 3 LOAEL: 10	100	1	3E–2	Liver and kidney pathology
Tetraethyl lead	78-00-2	NOAEL: none LOAEL: 1.7 (μg/kg · day) converted to 0.0012 mg/kg · day	10,000	1	1E–7	Histopathology of liver and thymus
Toluene	108-88-3	NOAEL: 312 mg/kg converted to 223 mg/kg · day LOAEL: 625 mg/kg converted to 446 mg/kg · day	1,000	1	2E–1	Changes in liver and kidney weights
Trifluralin	1582-09-8	NOEL: (30ppm) 0.75 LEL: (150ppm) 3.75 mg/kg · day	100	1	7.5E–3	Increased liver weights
Xylenes	1330-20-7	NOAEL: 250 LOAEL: 500	1,000	1	0.2	Decreased body weight, increased mortality

[a] Data from Louvar flouvar, 1998.

TABLE 10.6. Inhalation RfCs for Selected Pollutants (www.epa.gov/iris/)

Compound	CASRN	Experimental Doses[a] (mg/m³ unless otherwise noted)	UF	MF	RfD (mg/m³ unless otherwise noted)	Critical Effect
Acetonitrile	75-05-8	NOAEL: 336	100	10	6E-2	Mortality
Ammonia	7664-41-7	NOAEL: 6.4 (9.2 ppm)	30	1	1E-1	Pulmonary function or changes
Aniline	62-53-3	NOAEL: 19 (5 ppm) LOAEL: None	3000	1	1E-3	
Benzene	71-43-2	BMCL = 8.2	300	1	3E-2	Decreased lymphocyte count
Beryllium and compounds	7440-41-7	LOAEL(HEC): 0.55 µg/m³	10	1	2E-2 µg/m³	Beryllium sensitization
Chlordane (technical)	12789-03-6	NOAEL 1.0	1000	1	7E-4	Hepatic effects
Chromium (VI)	18540-29-9	NOAEL: none LOAEL: 2E-3	90	1	8E-6	Nasal septum atrophy
Dichlorvos	62-73-7	NOAEL: 0.05 LOAEL: 0.48	100	1	5E-4	Decreased brain colinesterase activity
Ethylbenzene	100-41-4	NOAEL: 434 (100ppm) LOAEL: 4,340 (1,000ppm)	300	1	1E+0	Developmental toxicity
Hydrogen sulfide	7783-06-4	NOAEL: 0.64 LOAEL: 41.7 (30ppm)	300	1	2E-3	Nasal lesions
Mercury, elemental	7439-97-6	NOAEL: None LOAEL: 0.025 (converted to LOAEL [ADJ] of 0.009)	30	1	3E-4	Hand tremor, memory disturbance
Methyl tert-butyl ether (MTBE)	1634-04-4	NOAEL: 1,453 (403 ppm) LOAEL: 10,899 (3,023 ppm)	100	1	3E+0	Increased absolute and relative liver and kidney weights
Toluene	108-88-3	NOAEL: None LOAEL: 332 (88 ppm)	300	1	4E-1	Neurological effects
Xylenes	1330-20-7	NOAEL(HEC): 39 LOAEL(HEC):	300	1	0.1	Impaired motor coordination

[a] Data from Louvar flouvar, 1998.

TABLE 10.7. Carcinogenicity Assessment: Carcinogenic Risk from Oral Exposure for Selected Pollutants (www.epa.gov/iris/)

Compound	CASRN	Classification	Oral Slope Factor per (mg/kg·day)	Drinking Water Unit Risk per (μg/L)
Acrylamide	79-06-1	B2	4.5	1.3E–4
Aldrin	309-00-2	B2	1.7E+1	4.9E–4
Aniline	62-53-3	B2	5.7E–3	1.6E–7
Azobenzene	103-33-3	B2	1.1E–1	3.1E–6
Benzene	71-43-2	A	1.5E–2 to 5.5E–2	4.4E–4 to 1.6E–3
Benzo [a] pyrene	50-32-8	B2	7.3E+0	2.1E–4
Bromoform	75-25-2	B2	7.9E–3	2.3E–7
Carbon tetrachloride	56-23-5	B2	1.3E–1	3.7E–6
Chloroform	67-66-3	B2	6.1E–3	—
Chlordane (technical)	12789-03-6	B2	3.5E–1	1E–5
3,3'-Dichloro-benzidine	91-94-1	B2	4.5E–1	1.3E–5
Dichlorvos	62-73-7	B2	2.9E–1	8.3E–6
Dieldrin	60-57-1	B2	1.6E+1	4.6E–4
2,4-/2,6-Dinitrotoluene mixture	NA	B2	6.8E–1	1.9E–5
Heptachlor	76-44-8	B2	4.5E+0	1.3E–4
Hexachlorobenzene	118-74-1	B2	1.6	4.6E–5
Hexachloroethane	67-72-1	C	1.4E–2	4.0E–7
Isophorone	78-59-1	C	9.5E–4	2.7E–8
Pentachlorophenol	87-86-5	B2	1.2E–1	3E–6
Trifluralin	1582-09-8	C	7.7E–3	2.2E–7

Note the water unit risk = risk per μg/L = slope factor $*$ 1/70 kg $*$ 2.00 L/day $\times$ 10^{-3}.

is referred to as a chronic daily intake (CDI), and it is expressed as the mass of pollutant taken up per unit body weight and unit time. We will address each common exposure route separately.

10.2.1 Ingestion of Pollutants from Drinking Water

Pollutant CDIs from drinking polluted water are calculated using

$$\text{Intake from drinking (mg/kg·day)} = \frac{\text{CW} \times \text{IR} \times \text{EF} \times \text{ED}}{\text{BW} \times \text{AT}} \quad (10.1)$$

where CW is the pollutant concentration in the drinking water (mg/L), IR is the ingestion rate of water (L/day; 2.00 L/day for an adult at the 90th percentile, 1.40 L/day for an average adult), EF is the exposure frequency (days/year; usually 365 days per year), ED is the exposure duration (number of years; 70.0 years for a conventional lifetime, 30.0 years, upper 90th percentile at one residence, 9.00 years, 50th percentile median time at one residence), BW is the body weight (kg; 70 kg is

TABLE 10.8. Carcinogenicity Assessment: Carcinogenic Risk from Inhalation Exposure for Selected Pollutants (www.epa.gov/iris/)

Compound	CASRN	Classification	Slope Factor[a] per (mg/kg·day)	Inhalation Unit Risk per (µg/m³)
Acrylamide	79-06-1	B2	45	1.3E−3
Aldrin	309-00-2	B2	17	4.9E−3
Benzene	71-43-2	A	2.9E−2	A range of 2.2E−6 to 7.8E−6 is the increase in the lifetime risk of an individual who is exposed for a lifetime to 1 µg/m³ benzene in air.
Beryllium and compounds	7440-41-7	B1	8.4	2.4E−3
Bromoform	75-25-2	B2	3.9E−3	1.1E−6
Cadmium	7440-43-9	B1	6.1	1.8E−3
Chloroform	67-66-3	B2	8.1E−2	2.3E−5
Chlordane (technical)	12789-03-6	B2	1.3	1E−4
Chromium (VI)	18540-29-9	A	41	1.2E−2
Dieldrin	60-57-1	B2	16	4.6E−3
Formaldehyde	50-00-0	B1	4.5E−2	1.3E−5
Heptachlor	76-44-8	B2	4.5	1.3E−3
Hexachloroethane	67-72-1	C	1.4E−2	4.0E−6
Nickel refinery dust	No CASRN	A	8.4E−1	2.4E−4

Note the air unit risk = risk per µg/m³ = slope factor * 1/70 kg * 20 m³/day × 10⁻³.

[a] Louvar, J. F. and B. D. Louvar. *Health and Environmental Risk Analysis: Fundamentals with Applications*, Prentice-Hall, Upper Saddle River, NJ, 1998.

used as an average), AT is the average time period of exposure (days; for a pathway-specific noncarcinogen, the value is usually ED × 365 days/year; for a carcinogen, use 70 years × 365 days/year). Specific values of each parameter can be adjusted as needed. For example, a child would drink less water each day, the receptor may be away from home most of the day or year, or the body weight of a specific receptor or group of receptors may differ from the average.

10.2.2 Ingestion of Water While Swimming

Exposure from ingestion of water during swimming can be calculated by

$$\text{Intake of water} \ (\text{mg}/\text{kg} \cdot \text{day}) = \frac{\text{CW} \times \text{CR} \times \text{ET} \times \text{EF} \times \text{ED}}{\text{BW} \times \text{AT}} \qquad (10.2)$$

where CW is the pollutant concentration in the water (mg/L), CR is the contact rate (0.050 L/hr; EPA, 1989b), ET is the exposure time (hours/swimming event), EF is the exposure frequency (swimming events/year, national average is 7 days/year), ED is the exposure duration [70 years (lifetime), 30 years (national upper-bound time (90th percentile) at one residence), 9 years (national median time (50th percentile) at one residence)], BW is the body weight [70 kg (average adult); for age-specific values, see EPA (1989b)], and AT is the averaging time [period over which the exposure is averaged, in days; for noncarcinogenic effects, use ED × 365 days; for carcinogenic effects use 70 years (lifetime) × 365 days/year].

10.2.3 Dermal Contact with Pollutants in Water While Swimming

The CDI of pollutants from dermal contact with pollutants in water can be calculated by

$$\text{Absorbed dose} \ (\text{mg}/\text{kg} \cdot \text{day}) = \frac{\text{CW} \times \text{SA} \times \text{PC} \times \text{ET} \times \text{EF} \times \text{ED} \times \text{CF}}{\text{BW} \times \text{AT}} \qquad (10.3)$$

where CW is the pollutant concentration in water (mg/L) and SA is the skin surface area available for contact (cm^3):

50th Percentile Total Body Surface Area (m^2)

Age (years)	Male	Female
3 < 6	0.728	0.711
6 < 9	0.931	0.919
9 < 12	1.16	1.16
12 < 15	1.49	1.48
15 < 18	1.75	1.60
Adult	1.94	1.69

50th Percentile Body-Part-Specific Surface Area for Males

Age (years)	Arms	Hands	Legs
3 < 4	0.096	0.040	0.18
6 < 7	0.11	0.041	0.24
9 < 10	0.13	0.057	0.31
Adult	0.23	0.082	0.55

PC is the chemical-specific dermal permeability constant (cm/hr). These must be obtained from the open literature or an average value can be used (8.4×10^{-4} cm/hr). ET is the exposure time (hours/day). The national average for swimming is 2.6 hr/day. EF is the exposure frequency (days/year). The national average for swimming is 7 days/year. ED is the exposure duration [years; 70 years for lifetime, 30 years (national upper-bound time (90th percentile) at one residence), and 9 years (national median time (50th percentile) at one residence)]. CF is the volumetric conversion factor for water ($1.00 \, \text{L}/1000 \, \text{cm}^3$). BW is the body weight [kg; 70.0 kg for an average adult; for age-specific values, see EPA (1989b)]. AT is the average time [period over which the exposure is averaged, days; for noncarcinogenic effects use $\text{ED} \times 365$ days/year, for carcinogenic effects, use 70 years $\times$ 365 days/year].

10.2.4 Ingestion of Pollutants in Soil

Ingestion of polluted soil can result from the ingestion of unwashed root crops and from children playing in dirt. The CDI of pollutants from the ingestion of polluted soil can be calculated from

$$\text{Intake} \, (\text{mg}/\text{kg} \cdot \text{day}) = \frac{\text{CS} \times \text{IR} \times \text{CF} \times \text{FI} \times \text{EF} \times \text{ED}}{\text{BW} \times \text{AT}} \tag{10.4}$$

where CS is the pollutant concentration in the soil (mg/kg), IR is the ingestion rate (mg soil/day) [suggested values (EPA, 1989c): 200 mg/day children 1–6 years old, 100 mg/day > 6 years old], CF is a conversion factor (10^{-6} kg/mg), FI is the fraction of soil ingested from the polluted site (versus soil ingested from a nonpolluted site or playground; this value varies based on population activity patterns), EF is the exposure frequency (365 days/year), ED is the exposure duration [years; 70 years for a lifetime exposure, 30 years (national upper-bound time (90th percentile) at one residence), 9 years (national median time (50th percentile) at one residence)], BW is the body weight [70 kg for an average adult, 16 kg for children aged 1 to 6 years (50th percentile)], and AT is the averaging time [period over which the exposure is averaged, in days (for noncarcinogenic effects, this is equal to $\text{ED} \times 365$ days/year; for carcinogenic effects, use 70 years $\times$ 365 days/year)].

10.2.5 Intake from Dermal Contact with Pollutants in Soil

Humans can also obtain a pollutant dose from working with polluted soil, via absorption through the skin. This dose can be calculated by

$$\text{Absorbed dose (mg/kg} \cdot \text{day)} = \frac{CS \times CF \times SA \times AF \times ABS \times EF \times ED}{BW \times AT} \quad (10.5)$$

where CS is the pollutant concentration in the soil (mg/kg), CF is a conversion factor (10^{-6} kg/mg), and SA is the skin surface area available for contact (cm^2/event). Suggested values for SA are:

50th Percentile Total Body Surface Area (m^2)

Age (years)	Male	Female
3 < 6	0.728	0.711
6 < 9	0.931	0.919
9 < 12	1.16	1.16
12 < 15	1.49	1.48
15 < 18	1.75	1.60
Adult	1.94	1.69

50th Percentile Body Part-Specific Areas for Males (m^2)

Age (years)	Arms	Hands	Legs
3 < 4	0.096	0.040	0.18
6 < 7	0.11	0.041	0.24
9 < 10	0.13	0.057	0.31
Adult	0.23	0.082	0.55

AF is the soil to skin adherence factor (mg/cm^2; 1.45 mg/cm^2 for commercial potting soil for hands, 2.77 mg/cm^2 for kaolin clay for hands), ABS is the absorption factor (unitless) which accounts for desorption of the pollutant from the soil matrix and absorption of the pollutant across the skin (literature data for this parameter are limited), EF is the exposure frequency (events/year), ED is the exposure duration [for a lifetime use 70 years, use 30 years (the national upper bound (90th percentile) at one residence), use 9 years (the national medium time (50th percentile) at one residence)], BW is the average body weight (use 70 kg for an average adult), and AT is the averaging time [period over which the exposure is averaged, days (for noncarcinogenic effects use ED × 365 days/year, for carcinogenic effects use 70 years × 365 days/year)].

10.2.6 Inhalation of Airborne (Vapor Phase) Pollutants

Chapter 9 discussed ways of estimating the downwind pollutant concentration of a pollutant from a chemical spill. Other atmospheric exposure routes include vapors from household products, gasoline fumes from automobile filling operations, and volatilization of pollutants from household water. Another common inhalation route is breathing vapors while showering with polluted water. Estimates of CDI from vapors can be calculated by

$$\text{Intake}\left(\text{mg}/\text{kg} \cdot \text{day}\right) = \frac{\text{CA} \times \text{IR} \times \text{ET} \times \text{EF} \times \text{ED}}{\text{BW} \times \text{AT}} \tag{10.6}$$

where CA is the pollutant concentration in the air (mg/m^3), IR is the inhalation rate (m^3/day) [suggested values are: adult (upper-bound) 30 m^3/day, adult (average) 20 m^3/day; for hourly rates see EPA (1989b); use 0.6 m^3/hr for showering events], ET is the exposure time (days) [use 12 minutes converted to days for upper-bound shower duration (90th percentile), use 7 minutes converted to days for average shower duration (50th percentile)], EF is exposure frequency (times/day), ED is the exposure duration (years) [use 70 years for a lifetime exposure, use 30 years at one residence (national upper-bound time (90th percentile)), use 9 years at one residence (national median time (50th percentile))], BW is the body weight (kg) [use 70 kg for an average adult, age-specific values are given in EPA (1989b)], and AT is the averaging time (time period over which the exposure is averaged, days) [for non-carcinogenic effects upper-bound shower duration use ED × 365 days/year, for carcinogenic effects use 70 years × 375 days/year].

10.2.7 Ingestion of Contaminated Fish and Shellfish

We not only have to be concerned with the intake of pollutants from water, but also with foods that come from polluted water sources. Aquatic species often accumulate pollutants into their muscle and organ tissue, which we use for food. Thus, significant doses of pollutants can result from eating these polluted organisms. A recently discovered example is the presence of mercury in several species, especially those at the top of the food chain. The dose of a pollutant can be calculated by

$$\text{Intake}\left(\text{mg}/\text{kg} \cdot \text{day}\right) = \frac{\text{CF} \times \text{IR} \times \text{FI} \times \text{EF} \times \text{ED}}{\text{BW} \times \text{AT}} \tag{10.7}$$

where CF is the pollutant concentration in the fish or shellfish (mg/kg), IR is the ingestion rate (kg/meal) [suggested values: 0.284 kg/meal (90th percentile for fin fish; Pao et al., 1982), 0.113 kg/meal (50th percentile for fin fish; Pao et al., 1982), 132 g/day (95th percentile daily intakes averaged over 3 days for consumers of fin fish; Pao et al., 1982), 38 g/day (50th percentile daily intake, averaged over 3 days for consumers of fin fish; Pao et al., 1982), 6.5 g/day (daily intake averaged over a year; EPA 1989b); other specific values for age, sex, race, region, and other fish species are available in EPA (1989b)], FI is the fraction of the daily fish intake from the polluted source (a case-specific value), EF is the exposure frequency (meals/year) [general value: 48 days/year (average per capita for fish and shellfish; EPA, 1989d)], ED is the exposure duration [for lifetime exposure use 70 years, for 30 years at one residence (national upper-bound time (90th percentile); EPA, 1989b), for 9 years at one residence (national median time (50th percentile); EPA, 1989b)], BW is the body weight (kg) [70 kg for an average adult (EPA, 1989b); for age-specific values see EPA (1989b)], and AT is the period of exposure (for noncarcinogenic effects use ED × 365 days/year, for carcinogenic effects use 70 years × 365 days/year).

10.2.8 Ingestion of Contaminated Fruits and Vegetables

Some plants grown on polluted soil will take up pollutants through their root systems, and these pollutants may concentrate in the plants' fruits and vegetables. We can calculate the CDI from eating contaminated produce by

$$\text{Intake (mg/kg} \cdot \text{day)} = \frac{CF \times IR \times FI \times EF \times ED}{BW \times AT} \tag{10.8}$$

where CF is the pollutant concentration in the fish (mg/kg-day), IR is the ingestion rate (kg/meal) [food-specific values can be found in Pao et al. (1982)], FI is the fraction of food ingested from the contaminated source (as opposed to food from a noncontaminated source; unitless), EF is the exposure frequency (meals/year) (region-specific values must be used), ED is the exposure duration (years) [for lifetime exposure use 70 years, use 30 years for living at one residence (national upperbound time (90th percentile; EPA, 1989b), use 9 years for living at one residence (national median time (50th percentile: EPA, 1989b)], BW is the body weight [for an average adult use 70 kg, for age-specific values see EPA (1989b)], and AT is the averaging time (period over which the exposure is averaged, days) [for noncarcinogenic effects use ED × 365 days/year, for carcinogenic effects use 70 years × 365 days/year].

10.2.9 Ingestion of Contaminated Meat, Eggs, and Dairy Products

Contaminated food can result from a variety of sources. A more general equation for calculating CDIs from food is

$$\text{Intake (mg/kg} \cdot \text{day)} = \frac{CF \times IR \times FI \times EF \times ED}{BW \times AT} \tag{10.9}$$

where CF is the pollutant concentration in the food (mg/kg), IR is the ingestion rate (kg/meal) [suggested values: 0.28 kg/meal for beef (95th percentile; Pao et al., 1982), 0.112 kg/meal for beef (50th percentile; Pao et al., 1982), for specific values for other meats see Pao et al. (1982); 0.150 kg/meal for eggs (95th percentile; Pao et al., 1982), 0.064 kg/meal for eggs (50th percentile; Pao et al., 1982); for specific values for milk, cheese, and other dairy products see Pao et al. (1982)], FI is the fraction of food ingested from the contaminated source (as opposed to food from a noncontaminated source; unitless), EF is the exposure frequency (meals/year) (region-specific values must be used), ED is the exposure duration (years) [for lifetime exposure use 70 years, use 30 years for living at one residence (national upperbound time (90th percentile; EPA, 1989b), use 9 years for living at one residence (national median time (50th percentile: EPA, 1989b)], BW is the body weight [for an average adult use 70 kg; for age-specific values see EPA (1989b)], and AT is the averaging time (period over which the exposure is averaged, days) [for noncarcinogenic effects use ED × 365 days/year, for carcinogenic effects use 70 years × 365 days/year].

10.3   HEALTH RISK CALCULATIONS FOR CARCINOGENS

Perhaps the most disputed, but very important, aspect of determining the effects of a pollutant on human health is our attempt to estimate the increase in cancer risk in a population exposed to the pollutant at a specific concentration. We start with the standard approach of developing a dose–response curve, such as the one shown in Figure 10.1, by exposing an animal to increasing concentrations of pollutant. After a given time period—days, months, or even a lifetime—the animal is sacrificed and examined for signs of cancer development. In rare cases, human data are available, not from experiments, of course, but from legal, illegal, or in many cases accidental exposures to pollutants. The obvious problem with animal experiments is determining how nonhuman responses in exposure studies relate to human responses to the same pollutant. In most cases, we do not know but rather approximate the relationship. To complicate the situation further, dose–response experiments are conducted at relatively high dose rates (CDIs), so that we can easily see a percentage level response (cancer rate). Clearly, a percentage response rate in humans is unacceptable given our general accepted level of one in a million. Thus, we must severely extrapolate the experimental animal data not only from species to species but also from intense exposures down to very small doses and response rates.

We begin our extrapolation process by assuming that the dose–response curve is linear at low pollutant concentrations (the extreme lower left-hand section of Figure 10.1). Note that for cancer risk assessment we assume that there is no lower

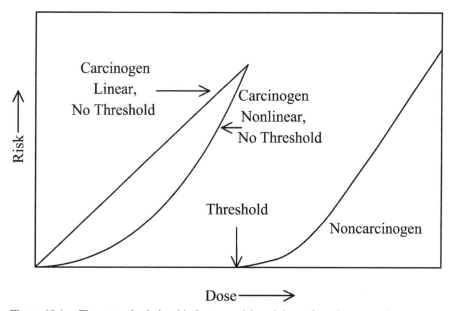

Figure 10.1.   The general relationship between risk and dose of carcinogen and noncarcinogen pollutants.

threshold where the pollutant concentration does not have an effect. As a result of our linearization assumption, we will have a straight line with the slope of y over x, or risk per dose. The units of risk are the fraction of the population **developing** cancer (not dying from cancer). Risk is a unitless fraction, while the dose has units of mg/kg·day. Thus, the slope of the plot has units of $(\text{mg/kg·day})^{-1}$.

This slope of the plot, referred to as the slope factor or potency factor, is the basis of our risk assessment calculations. There are a number of models used to estimate slope factors, most of which produce conservative estimates, conservative in the sense that they maximize the likelihood of detecting cancer development risk. Usually, the 95th percentile value is given. In addition, a weight-of-evidence classification is given to each slope factor. A summary of the classification system is given in Table 10.9 and is based on the conclusiveness (certainty) of the data (i.e., human test data versus animal test data, correlation of results, etc.).

The slope factor is used, along with the CDI calculations, to estimate the risk of developing cancer, usually over a 70-year exposure period. We will work several example problems to illustrate the use of the slope factor.

Example Problem. Most municipalities in the United States treat their water with some form of chlorine. Since almost all waters have some natural organic matter present, a by-product of the chlorine treatment is the formation of chlorinated organics referred to as trichloromethanes (THMs). Calculate the cancer risk of drinking water containing THMs (in the form of chloroform, $CHCl_3$). Assume that an adult drinks 2.00 L/day for 70 years with the chloroform at the THM drinking water limit (the governmentally regulated pollution concentration limit) of 0.10 mg/L.

To solve this problem, we would first use Eq. (10.1) to calculate the chronic daily intake (CDI).

Intake from drinking $(\text{mg/kg} \cdot \text{day})$

$$= \frac{CW \times IR \times EF \times ED}{BW \times AT} = \frac{(0.10\text{ mg/L}) \times (2.00\text{ L/day}) \times 365\text{ day/year} \times 70\text{ years}}{70\text{ kg} \times (70\text{ years} \times 365\text{ days/year})}$$

$$= 2.86 \times 10^{-3}\text{ mg/kg} \cdot \text{day} \tag{10.1}$$

TABLE 10.9. EPA Weight of Evidence Classification System for Carcinogenicity

Group	Description
A	Human carcinogen
B1 or B2	Probable human carcinogen
	B1 indicates that limited human data are available.
	B2 indicates sufficient evidence in animals and inadequate or no evidence in humans.
C	Possible human carcinogen
D	Not classifiable as to human carcinogenicity
E	Evidence of noncarcinogenicity for humans

From Table 10.7, the slope factor for chloroform is 6.1×10^{-3} (mg/kg·day)$^{-1}$ for the oral route, thus the risk of cancer is

$$\text{Risk} = \text{CDI} \times \text{slope factor}$$

$$= 2.86 \times 10^{-3} \text{ mg/kg} \cdot \text{day} \times 6.1 \times 10^{-3} \left(\text{mg/kg} \cdot \text{day}\right)^{-1}$$

$$= 1.74 \times 10^{-5} \text{ or } 17.4 \text{ per one million people}$$

Recall that the slope factors given by the U.S. EPA represent the upper-bound (95th percentile) of developing cancer, not dying from cancer. This means that over a 70-year period, we can estimate that 17.4 more cases of cancer per million people will develop in the population drinking the water.

Suppose the city supplying the water has a population of 3.5 million. How many actual cases of cancer per year will result from drinking the water over a 70-year period?

$$3,500,000 \text{ people} \left(\frac{17.4 \text{ cancer patients}}{1,000,000 \text{ people}}\right) \frac{1}{70 \text{ years}} = 0.87 \text{ cancers per year}$$

The average annual cancer death rate in the United States is 193 per 100,000 people. How does the calculated cancer development rate compare in our city of citizens that drink the chlorinated water?

$$3,500,000 \frac{193 \text{ cancer/year}}{100,000 \text{ people}} = 6755 \text{ cancer deaths per year}$$

Given that 6755 people living in the city will die from cancer each year, it is unlikely that we will statistically notice the additional 0.87 cases of cancer. This last type of calculation, comparing the added risk of cancer development to the existing rate of cancer deaths, should be used in risk assessment for all carcinogens.

For a thorough risk assessment, we must also consider what the rate of death would be if we did not disinfect our drinking water and sewage waste, by chlorination or any other treatment. Numbers are difficult to accurately estimate, but for undeveloped countries where waterborne disease is common, we find that approximately 4 million people die each year of diarrheal diseases that infect approximately 2 billion people per year (UNEP, 1993). This translates into a death rate of 2000 per million citizens. This clearly illustrates the benefit of disinfecting our drinking water and sewage waste (17.4 cases of cancer per million people as opposed to 2000 deaths per million people from not treating the water).

Slope factors are also used to estimate when an unacceptable risk will develop. In our case, the risk is in terms of the number of cancer cases and, as noted earlier, it is becoming commonly accepted to accept any risk below one in a million. When the risk equals or becomes greater than one in a million, governments usually impose pollutant exposure limits, referred to as drinking water equivalent levels (DWEL), at the concentration that causes this highest acceptable risk. We will next use the slope factor to illustrate how to estimate this pollutant concentration.

Example Problem. A well-known, cancer-causing chemical is benzene. Table 10.7 quotes a slope factor range from 1.5×10^{-2} to 5.5×10^{-2} (mg/kg·day)$^{-1}$ for benzene.

Calculate the DWEL that would result in a one-in-a-million risk of developing cancer over a 70-year time span.

$$CDI = \frac{\text{Risk}}{\text{Slope factor}} = \frac{10^{-6}}{5.5 \times 10^{-2} \, (mg/kg \cdot day)^{-1}}$$

$$= 1.82 \times 10^{-5} \, mg/kg \cdot day$$

For a 70-kg person drinking 2.00 L/day, the DWEL would be

$$CDI = 1.82 \times 10^{-5} \, mg/kg \cdot day = \frac{X \, mg/L \times 2.00 \, L/day}{70 \, kg}$$

$$X \, mg/L = 6.36 \times 10^{-4} \, mg/L \text{ or } 0.636 \, \mu g/L \text{ of benzene}$$

Thus, benzene concentrations greater than 0.636 µg/L should be avoided to keep our cancer risk below one in a million. Similar approaches can be used for any of the CDI estimation equations in Section 10.2. We will next work an example of cancer risk assessment for a worker in an agricultural setting.

Example Problem. A farm worker is exposed to Trifluralin, a commonly used pesticide, while working in the fields. Calculate the risk of the worker exposed to $0.01 \, mg/m^3$ who works 8 hours per day, 7 days a week, for 4.0 months a year, and for 25 years. Assume a body weight of 70.0 kg and an inhalation volume of $20 \, m^3$ per 8 hours. The slope factor for Trifluralin is $8.4 \times 10^{-1} \, (mg/kg \cdot day)^{-1}$.

$$\text{Intake by inhalation} \, (mg/kg \cdot day) = \frac{CA \times IR \times ET \times EF \times ED}{BW \times AT}$$

$$= \frac{0.010 \, mg/m^3 \times 20 \, m^3/day \times (29 \, days/month)(4 \, months/years)(25 \, years)}{70 \, kg \times 25 \, years \times 365 \, days/years}$$

$$= 9.08 \times 10^{-4} \, mg/kg \cdot day \tag{10.6}$$

$$\text{Risk} = CDI \times \text{Slope factor} = 9.08 \times 10^{-4} \, mg/kg \cdot day \times 8.4 \times 10^{-1} \, (mg/kg \cdot day)^{-1}$$

$$= 7.6 \times 10^{-4} \text{ or } 763 \text{ cancer cases per million}$$

$$= 7.6 \times 10^{-4} \times 10^6 = 763 \text{ cancer cases per million}$$

Obviously, this is a significant risk and adequate protection of the workers, such as a breathing apparatus, should be required.

There is one final note concerning the risk of developing cancer from exposure to a pollutant. Cancer risks are assumed to be cumulative, which means the total cancer risk is the sum of all individual cancer risks. For example, if you have a cancer risk of 0.00004 from pollutant exposure route one, a cancer risk of 0.00008 from pollutant route two, and 0.000005 from exposure route three, then your total cancer risk would be the sum of these, or 0.000125 (a cancer risk of 125 cases per million people).

10.4 HEALTH RISK CALCULATIONS FOR NONCARCINOGENS

In considering the carcinogenic effects of chemicals, there is no threshold dose of a chemical to which an animal species can be exposed safely. In risk calculations for noncarcinogens, though, it is accepted that there is a threshold chemical concentration, below which the chemical is not harmful. These values are calculated by taking a population of animals, exposing them to increasing doses of the chemical, and observing health effects. A generic plot of such an experiment is shown in Figure 10.2, where the response is shown on the y-axis and the chemical concentration is on the x-axis. Note the presence of the threshold, which is essentially just an estimate. There are several important points on the plot that need to be mentioned. The first is the lowest-observed-effect level (LOEL; not shown in Figure 10.2) which is lowest dose administered to the animal that results in a response (death or health effect). The next point is the no-observed-effect level (NOEL; not shown in Figure 10.2) or the highest dose that does not produce a health effect or response. The LOEL and NOEL can be refined further by noting whether the health effect is an adverse effect. This changes the names to the no-observed-adverse-effect level (NOAEL) and the lowest-observed-adverse-effect level (LOAEL). These latter two levels are the ones commonly reported by researchers. A list of these doses and effects for selected pollutants are summarized in Tables 10.5 and 10.6.

The final dose level given in Figure 10.2 is the reference dose (RfD). The RfD listed by EPA is considered the maximum acceptable daily intake (ADI) of the pollutant. Concentrations or masses below this dose are considered acceptable and should not cause adverse health effects in humans. But how do we arrive at the RfD?

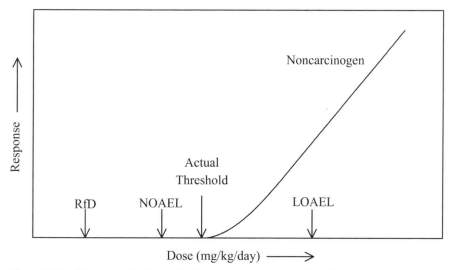

Figure 10.2.   The general relationship between response and dose for a noncarcinogen pollutant illustrating the concept of threshold level.

The RfD is obtained by dividing the NOAEL by several safety factors. The safety factors are also referred to as uncertainty factors because they reflect the uncertainties associated with extrapolating the results between individuals in a population and from animal testing to humans. The first uncertainly factor divides the NOAEL by a factor of 10 to account for sensitivity differences in the exposed human population. This accounts for variations between individuals, including pregnant women, infants, the elderly, and normal, healthy adults. A second adjustment (the magnitude depends on the situation) occurs if the NOAEL results from animal testing data that are extrapolated to humans. A final factor of 10 is used when no data are available on human responses to the pollutant. Values of these latter two adjustment factors range from one to ten. Thus, the RfD will usually be only a fraction of the NOAEL. Note the difference between the NOAEL and RfD given in Tables 10.5 and 10.6.

In many ways, the predicted response from noncarcinogenic risk assessment is less quantitative than the series of calculations involved in cancer risk assessment. For example, we cannot exactly tell the magnitude of the observed effect but only whether one should occur. To do this, we use a hazard quotient:

$$\text{Hazard quotient} = \frac{\text{Average daily dose during the exposure period } (\text{mg}/\text{kg} \cdot \text{day})}{\text{RfD}}$$

$$(10.10)$$

Hazard quotient values of 1.00 or greater indicate that an adverse health effect or risk of toxicity should occur. Values below 1.00 are considered safe. Another major difference between cancer and noncarcinogenic risk assessment is that, for noncarcinogens, we use the specific exposure time, not the average exposure over a 70-year period. As in cancer risk assessment, risks are additive.

Example Problem. A contaminated drinking water contains the following concentrations of gasoline products: benzene at 0.010 mg/L, tetraethyl lead at 0.015 mg/L, toluene at 0.050 mg/L, and xylene at 0.050 mg/L. Calculate the cumulative hazard quotient for a 70.0-kg person who drinks the water at a rate of 2.00 L/day. The RfD values in mg/kg-day are 4.0×10^{-3} for benzene, 1.00×10^{-7} for tetraethyl lead, 2×10^{-1} for toluene, and 0.2 for xylene.

First calculate the average daily doses (ADD) for each pollutant:

$$\text{ADD}_{\text{Benzene}} = \frac{0.010 \text{ mg}/\text{L} \times 2.00 \text{ L}/\text{day}}{70.0 \text{ kg}} = 2.86 \times 10^{-4} \text{ mg}/\text{kg} \cdot \text{day}$$

$$\text{ADD}_{\text{Tetraethyl lead}} = \frac{0.000015 \text{ mg}/\text{L} \times 2.00 \text{ L}/\text{day}}{70.0 \text{ kg}} = 4.29 \times 10^{-7} \text{ mg}/\text{kg} \cdot \text{day}$$

$$\text{ADD}_{\text{Toluene}} = \frac{0.050 \text{ mg}/\text{L} \times 2.00 \text{ L}/\text{day}}{70.0 \text{ kg}} = 1.43 \times 10^{-3} \text{ mg}/\text{kg} \cdot \text{day}$$

$$\text{ADD}_{\text{Xylene}} = \frac{0.050 \text{ mg}/\text{L} \times 2.00 \text{ L}/\text{day}}{70.0 \text{ kg}} = 1.43 \times 10^{-3} \text{ mg}/\text{kg} \cdot \text{day}$$

Next, calculate the hazard quotient (HQ) for each pollutant:

$$HQ_{Benzene} = \frac{2.86 \times 10^{-4} \text{ mg/kg} \cdot \text{day}}{4.0 \times 10^{-3} \text{ mg/kg} \cdot \text{day}} = 0.0715$$

$$HQ_{Tetraethyl\ lead} = \frac{4.29 \times 10^{-7} \text{ mg/kg} \cdot \text{day}}{1.00 \times 10^{-7} \text{ mg/kg} \cdot \text{day}} = 4.3$$

$$HQ_{Toluene} = \frac{1.43 \times 10^{-3} \text{ mg/kg} \cdot \text{day}}{2.0 \times 10^{-1} \text{ mg/kg} \cdot \text{day}} = 0.00715$$

$$HQ_{Xylene} = \frac{1.43 \times 10^{-3} \text{ mg/kg} \cdot \text{day}}{0.2 \text{ mg/kg} \cdot \text{day}} = 0.00715$$

The sum of the HQ is 4.3, where tetraethyl lead is responsible for essentially all of the risk.

10.5 BIOCONCENTRATION CALCULATIONS

In Chapter 2 and 3, we discussed bioconcentration and looked at DDT data from a food chain on Long Island, NY (Table 2.3). The data from Table 2.3 are reproduced in Table 10.10, where we used it to calculate bioconcentration factors (BCFs) for each animal species. Each BCF is calculated by dividing the DDT concentration in each animal by the water concentration. As you can clearly see, DDT is increasingly bioconcentrated as we move up the food chain, from 800 in the first tropic level to 528,000 in birds that eat fish.

Bioconcentration factors, given in Table 10.11, can be used to estimate cancer risk for a person eating contaminated fish, as is illustrated in the following problem.

Example Problem.   EPA estimates that an average person might eat 54 g of fish per day. Calculate the cancer risk of a 70.0-kg person who eats 54 g of fish per day, 350

TABLE 10.10. Bioconcentration Factor for DDT in a Long Island Food Web (United States)

Organism	DDT Residues (ppm)	BFC (L/kg)
Water	0.00005	—
Plankton	0.04	800
Silverside minnows	0.23	4,600
Sheephead minnows	0.94	19,200
Pickerel (predatory)	1.33	26,600
Needlefish (predatory)	2.07	41,400
Heron (feeds on small aquatic animals)	3.57	71,400
Herring gull (scavenger)	6.00	120,000
Osprey egg	13.8	276,000
Merganser (fish-eating duck)	22.8	456,000
Cormorant (feeds on large fish)	26.4	528,000

TABLE 10.11.   Bioconcentration Factors (BFCs) for Selected Pollutants (U.S. EPA, 1986)

Compound	BCF for Fish (L/kg)	Compound	BCF for Fish (L/kg)
Acenaphthene	242	Fluorine	1,300
Acrylonitirle	48	Heptachlor	15,700
Aldrin	28	Heptachlor epoxide	14,400
Antimony	1	Hexachlorobenzene	8,690
Arsenic	44	Hexachlorobutadiene	2.8
Benzene	5.2	Hexachlorocyclopentadiene	4.3
Benzidine	87.5	α-Hexachlorocyclohexane (HCCH)	130
Beryllium	19		
Bis(2-chloroethyl)ether	6.9	β-HCCH	130
Bis(chloroethyl)ether	0.63	γ-HCCH (Lindane)	130
Cadmium	81	Hexachloroethane	87
Carbon tetrachloride	19	Kepone	8,400
Chlordane	14,000	Lead	49
Chlorobenzene	10	Mercury (alkyl)	3,750
Chloroform	3.75	Mercury (inorganic)	5,500
Chromium III	16	Methyl parathion	45
Chromium VI	16	Nickel	47
Copper	200	Pentachlorobenzene	2,125
DDE	51,000	Pentachlorophenol	770
DDT	54,000	Phenanthrene	2,630
1,2-Dichlorobenzene	56	Phenol	1.4
1,3-Dichlorobenzene	56	Polychlorinated biphenyls (PCBs)	100,000
1,4-Dichlorobenzene	56		
3,3′-Dichlorobenzidine	312	Selenium	16
1,2-Dichloroethane	1.2	Silver	3,080
1,1-Dichloroethylene	5.6	1,2,4,5-Tetrachlorobenzene	1,125
1,2-Dichloroethylene	1.6	2,3,7,8-TCDD (Dioxin)	5,000
1,2-Dichloroethylene (cis)	1.6	1,1,2,2-Tetrachlorothethane	42
1,2-Dichloroethylene (trans)	1.6	Tetrachloroethylene	31
Dichloromethane	5	2,3,5,6-Tetrachlorophenol	240
2,4-Dichlorophenol	41	Toluene	10.7
1,3-Dichloropropene	1.9	Toxaphene	13,100
Dieldrin	4,760	1,2,4-Trichlorobenzene	2,800
Diethyl phalate	117	1,1,1-Trichloroethane	5.6
2,3-Dinitrotoluene	3.8	1,1,2-Trichloroethane	5
2,4-Dinitrotoluene	3.8	Trichloroethylene	10.6
2,5-Dinitrotoluene	3.8	2,4,5-Trichlorophenol	110
2,6-Dinitrotoluene	3.8	2,4,6-Trichlorophenol	150
3,4-Dinitrotoluene	3.8	tris(2,3-Dibromopropyl) phosphate	2.7
N,N-Diphenylamine	30		
1,2-Diphenylhydrazine	25	Vinyl chloride	1.17
Ethylbenzene	37.5	Zinc	47
Fluoranthene	1,150		

Exhibit A-1 in Superfund Public Health Evaluation Manual, Office of Emergency and Remedial Response, Washington, DC).

days each year for 30 years. Assume that the water the fish was taken from has a Chlordane concentration of 0.0100 ppb (µg/L). The oral slope factor for Chlordane is 3.5×10^{-1}.

First we must calculate the concentration in the fish.

$$\text{Chlordane concentration in fish} = \text{pollutant concentration in fish} \times \text{BCF}$$
$$= 0.0000100 \text{ mg/L} \times 14{,}000 \text{ L/kg}$$
$$= 0.14 \text{ mg Chlordane/kg fish}$$

Using Eq. (10.7), we can then calculate the CDI:

$$\text{Intake (mg/kg} \cdot \text{day)} = \frac{\text{CF} \times \text{IR} \times \text{FI} \times \text{EF} \times \text{ED}}{\text{BW} \times \text{AT}}$$

$$= \frac{0.14 \text{ mg/kg} \times 0.054 \text{ kg/day} \times 1.00 \times 350 \text{ days/year} \times 30 \text{ years}}{70 \text{ kg} \times 30 \text{ years} \times 365 \text{ days/year}}$$

$$= 1.04 \times 10^{-4} \text{ mg/kg} \cdot \text{day} \tag{10.7}$$

$$\text{Risk} = \text{CDI} \times \text{Slope factor} = 1.04 \times 10^{-4} \text{ mg/kg} \cdot \text{day} \times 3.5 \times 10^{-1} \text{ (mg/kg} \cdot \text{day)}^{-1}$$
$$= 3.62 \times 10^{-5} \text{ or 36 cancer cases per million people eating the fish}$$

Thus, 36 cancer cases should develop among one million people consuming the contaminated fish.

10.6 PUTTING IT ALL TOGETHER: MARGIN OF ERROR (UNCERTAINTY) OF THE ENTIRE ESTIMATION PROCESS

We are finally at the culmination of a long and complicated discussion of source characterization, chemistry, fate and transport modeling, and now risk assessment. But how good are our risk assessment values? We have discussed a number of precautions in every chapter. The major ones include the following:

1. Every number we generate in our modeling efforts depends on an accurate characterization of the waste site or polluted system with respect to pollutant mass dumped or contained in the system. These values are usually based on company records, employee memory, or varying amounts of sampling and analysis, and these may not be the most reliable sources. Thus, there may be considerable error with our source estimate.

2. In Chapters 2 and 3 we discussed the chemical aspects that are important in successful fate and transport modeling. As we clearly saw, many degradation rate constants depend on pH, E_H, the number of viable microorganisms, the concentration of degrading chemical (such as natural organic matter), and the temperature. Sorption phenomena can be important in lake, river, and especially groundwater systems. Finally, chemical speciation will also influence toxicity and therefore risk. Chemical data, produced by complex experimen-

tation and analysis, are often used in our fate and transport modeling equations, and the accuracy of our modeling efforts is closely linked to the accuracy of this chemical data.

3. A major factor that is difficult to estimate in fate and transport modeling, but one that greatly influences pollutant concentration, is mixing or dispersion in the system. As we saw for groundwater systems, estimates of the dispersion can vary by four orders of magnitude. Errors in such estimates will again significantly impact pollutant concentration, and therefore risk.

4. One fact that has become distinctly clear in collecting risk assessment data for this textbook is the lack of human-specific toxicity data on many pollutants. We must therefore attempt to account for this paucity of data by including uncertainty factors in the RfD, RfC, and slope factors.

So, are our efforts adequate, and do they provide an adequate margin of safety given the uncertainty in our source, fate and transport, and toxicity data? In general, we feel that the answer to this question is yes and that our accuracy will improve as more toxicity data become available. At every point in our assumptions, we tend to make *conservative estimates* of parameters that will affect pollutant concentrations and risk. For example, it is always best to assume *maximum* pollutant input to a system, *minimal* chemical and biological degradation, and *minimal* dispersion, since this will result in calculation of the *maximum pollutant concentration*. To this end, we also include relatively large uncertainty factors in risk assessment. And perhaps the most conservative of all of our approaches is declaring that the highest acceptable risk is one in a million, which in most cases is beyond our abilities to detect given our day-to-day life practices.

We should also keep in mind that risk assessment calculations targeted toward human populations, such as those considered here, should only be one tool in our approach to determining acceptable risk and especially determining whether or not a site should be remediated. For example, although a toxic metal mining operation in the middle of proverbial nowhere may pose no risk to any human, the mission of EPA also includes protecting the environment. Both factors must be given weight in our decision-making processes. Because no formal risk assessment calculations address environmental protection, it must often be built into the dialogue in other ways.

To summarize, the errors and uncertainties are very large, but by making conservative assumptions we can provide meaningful information to assist in making policies to protect the (human) public health.

Concepts

1. Using Tables 10.1–10.3, justify the acceptability of the one-in-a-million approach used by EPA with regard to risk assessment.

2. Draw a typical dose–response curve for a carcinogenic pollutant.

3. Draw a typical dose–response curve for a noncarcinogenic pollutant and label appropriate dose levels.

4. Define RfD, NOAEL, threshold, and LOAEL.

Exercises

1. Mercury is released into the atmosphere by burning coal. When it rains, mercury enters the ocean, where it is bioconcentrated in marine life forms. Calculate the risk of cancer for a 65-kg person who eats 45 g of tuna, 325 days per year for 35 years. Assume that mercury is present in the water at 2.5 ppb. Mercury's oral slope factor is 3×10^{-4} and it has a bioconcentration factor of 5500 L/kg.

2. It has been found that exposure to high levels of aniline results in cancerous effects on the lungs, as well as upper respiratory tract irritation and congestion. Aniline has been classified as carcinogenic and is very toxic to humans. It may be produced during the burning of plastics or from burning tobacco. Assume that an adult drinks 2.00 L of water per day for 70 years with an aniline concentration of 0.60 mg/L. Calculate the CDI and then calculate the risk. The slope factor is 5.7×10^{-3} (mg/kg·day)$^{-1}$. Next, suppose that the city supplies water to 5 million people. How many actual cases of cancer are predicted to result from drinking the water over a 70-year period?

3. Studies indicate that the long-term intake of bromoform in drinking water can cause cancer in animals and humans. The slope factor for bromoform is 7.9×10^{-3} (mg/kg·day)$^{-1}$. Calculate the DWEL that would result in a one-in-a-million risk of developing cancer over a 70-year time span.

4. A worker works 8 hours a day, 5 days a week, 10 months a year and is exposed to 0.01 mg/m^3 of nickel refinery dust for 21 years. Assume that his body weight is 70.0 kg and that he inhales a volume of 20 m^3/day. The slope factor of nickel refinery dust is 8.4×10^{-1} (mg/kg·day)$^{-1}$. What is the risk of developing cancer?

5. Contaminated drinking water near a farm contains the following concentrations of pesticide products: Aldrin at 0.0025 mg/L, Dieldrin at 0.0003 mg/L, and Endrin at 0.0007 mg/L. Calculate the cumulative hazard quotient for a local 70.0-kg person that drinks 2.0 L/day of the water. The RfD values in mg/kg·day at 3×10^{-5} for Aldrin, 5×10^{-5} for Dieldrin, and 3×10^{-4} for Endrin.

6. Calculate the risk of a 70.0-kg person who eats fish 300 days a year for 50 years. The water that the fish was taken from has a 3,3'-dichlorbenzidine concentration of 1.1 ppb. Assume that the person eats 54 g of fish per day. The oral slope factor for the pollutant is 4.5×10^{-1}.

7. In the past, before labor laws and intensive study on the health effects of pesticides were completed, fumigation planes were guided by people on the ground with brightly colored flags. Pesticides were sprayed directly onto the flaggers. One common pesticide applied in this manner was chlordane. What would the chronic daily intake (CDI) be for a person who lived near a fumigated field for 30 years, if the pollutant concentration was 0.02 mg/m^3 for 12 minutes once every week?

8. Calculate the risk of developing cancer for an average adult that inhales 0.003 mg/m^3 of gaseous carbon tetrachloride per day from a nearby leaking

dry cleaning facility. The person is only exposed immediately after they return home from work for a total of 23 minutes each day, for 5 days a week. The person lives by the facility for 9 years. Assume an inhalation rate of $0.6\,m^3/hour$. The slope factor for chloroform is 1.3×10^{-1}.

9. Determine the average daily doses (ADD) for the inhalation of elemental mercury for an average adult living only a few miles away from the coal-fired plant. Each day an average person breathes in $20\,m^3$ of air. Every day the person inhales an average of $0.04\,mg/m^3$ of mercury. What is the hazard quotient?

10. Polychlorinated biphenyls (PCBs) are no longer produced in the United States, but they are still found in relatively high concentrations in animals around the world. Many people enjoy consuming fish on a regular basis. What is the cancer risk for a 70-kg person what eats 6 oz of swordfish steak on Mondays and an 8 oz tuna steak on Thursdays? This pattern of fish consumption is constant for 25 years. (There are 28.35 grams in an ounce.) The bioconcentration factor for PCBs is 100,000 L/kg. The water the swordfish was taken from has a PCB concentration of 5.0 ppb, while the tuna was raised in water containing 3.5 ppb. The slope factor for PCBs is 2.0 (www.epa.gov/iris/).

11. Azobenzene is a commonly used herbicide. An average adult weighing 70 kg lives a few miles from a spraying operation that creates an average risk of 840 cancers per million. What is the chronic daily intake for this adult? The slope factor is 0.11.

12. A painter has painted houses for 45 years. There are many chemicals in paint that, upon volatilization, pose a risk to human health. Two chemicals found in paint are benzene (a carcinogen) and toluene (a noncarcinogen). The painter weighs 70 kg, breathes $22\,m^3$ per day of air, and works 5 days a week. Calculate the painter's lifetime intake of each of the chemicals assuming an average air concentration of $0.010\,mg/m^3$. Using this calculated intake, calculate the cancer risk associated with benzene (slope factor = $2.9 \times 10^{-2}\,mg/kg\cdot day^{-1}$) and the hazard quotient of toluene (RfD = $4.00 \times 10^{-1}\,mg/m^3$).

13. You are a researcher who is constantly exposed to appendages and organs preserved in formaldehyde. You inhale an average dose of $0.02\,mg/kg\cdot day$. The RfD for inhaled formaldehyde is $2.0 \times 10^{-1}\,mg/kg\cdot day$. Calculate the hazard quotient for the exposure. Are you at risk of developing a problem?

14. MTBE, an additive in gasoline, has recently been in the media as a drinking water contaminant due to leaking gasoline storage tanks. Calculate the cancer risk from drinking water that has an MTBE concentration of 40.0 ppb (considered safe according to the EPA website). Assume that a 70-kg adult drinks 2.0 L/day for 70 years. Finally, calculate the DWEL that would result in a one-in-a-million risk of developing cancer over a 70-year time span.

15. Drinking water contains the following concentrations of contaminants: Aldrin at 0.0400 mg/L, toluene at 0.060 mg/L, benzene at 0.010 mg/L, and

trifluralin at 0.05 mg/L. Calculate the cumulative hazard quotient for a 70.0-kg person who drinks the water at a rate of 2.00 L/day. The RfD values in mg/kg·day are 7.5×10^{-3} for trifluralin, 4.0×10^{-3} for benzene, 2×10^{-1} for toluene, and 3×10^{-5} for Aldrin.

16. Dichlorvos is a contact insecticide that is used as a household and public health insecticide. As the active ingredient of Nuvan 500 EC, Dichlorvos was used extensively in salmon farming to control the salmon louse *Lepeoph-theirus salmonis*. The principal, direct routes of entry for dichlorvos into waters include industrial effluents and accidental discharges (e.g., from pesticide manufacturing plants, formulation plants and marketing outlets), use in salmon fisheries, disposal of unused insecticide and the cleaning of application and mixing equipment. Dichlorvos may also indirectly enter the aquatic environment via spray drift during application and in land runoff. Dichlorvos has a potential for bioaccumulation into marine life, such as in the salmon consumed by humans. Calculate the Dichlorvos-associated cancer risk of a 70.0-kg person who eats 54 g of fish for 350 days out of the year for 30 years. Assume that the water the fish was taken from has a dichlorvos concentration of 2.00 ppb. The oral slope factor for dichlorvos is $2.9 \times 10^{-1} (\text{mg/kg·day})^{-1}$.

17. A secretary reviewing records at a local hospital notices that an unusually high number of patients have had liver lesions in the past few years and alerts the health officials. Finding no significant correlation between these patients, investigators decide to test the drinking water. They find high concentrations of carbon tetrachloride, a compound known to cause liver lesions, far above the LOAEL, and link the pollutant to discharges from an upstream industrial site. If the measured concentration of CCl_4 is 55 ppb, what is the average CDI of those people drinking the water? How does this concentration compare to the maximum acceptable contaminant level (MCL) of CCl_4 (0.005 mg/L)? Assume an intake of 2.00 L water per day, an exposure frequency of 350 days per year, exposure duration of 5 years, a body weight of 70 kg, and an average time of exposure of 5 years (at 365 days per year. Finally, from the calculated daily intake (CDI) you just determined, find the risk of people drinking the contaminated water if the slope factor is 0.13 kg·day/mg. Judging by this estimated risk, how many cases, on average, would you expect to see each year in a population of 150,000 people?

18. Suppose that the polluting industry in problem number 21 also discharged other pollutants, such as 1,2,4,5-tetrachlorobenzene. If this compound were discharged into a lake, it could enter the food chain by being adsorbed by or sorbing to microorganisms in the water. If the concentration of tetra-chlorobenzene is the lake is 0.30 ppm, what is the concentration in local fish if the bioconcentration factor is 1125 L/kg? Suppose that local fisherman in the area consume one-third of the EPA average fish consumption (54 g). What is their intake of tetrachlorobenzene in mg/kg-day?

19. A worker is exposed to chromium (VI) every day as a by-product from welding. Assuming that she works 8 hours per day, 5 days a week for 10

years, find the risk of the welder exposed to 0.0001 mg/m^3. Assume an inhalation rate of 20 m^3/day.

20. A contaminated water source contains many hazardous materials. The water contains 0.025 mg/L acrylamide, 0.010 mg/L chloroform, 0.050 mg/L ethyl acetate, and 0.030 mg/L Trifluralin. Find the cumulative hazard quotient for a 70-kg person who drinks 2.00 L/day. RfD values are listed in Tables 10.5–10.8.

21. Some shellfish have recently been discovered to contain Dieldrin, a pesticide. If a person eats 54 g of shellfish in one sitting, find the cancer risk of a 70-kg person that eats shellfish 200 days out of the year for 40 years. Assume that half the fish eaten is contaminated and that half the fish eaten in not contaminated. The water that the shellfish was taken from contains 0.100 ppb Dieldrin.

22. A twofold excess risk of lung cancer was observed in cadmium smelter workers. The study group consisted of 602 white males, with an average body weight of 70 kg, who have been employed during the years 1940–1969. The researchers were able to ascertain that the increase in lung cancer was probably not due to the presence of arsenic or to smoking. The men are assumed to have worked 8-hour days, 365 days a year, in order to assess a maximum exposure value. The average concentration of cadmium dust in the air was 5 mg/m^3. The inhalation rate for an average adult is 20 m^3/day. Use these data to confirm or dispute the stated twofold excess risk of lung cancer, assuming that the risk of cancer was 2500 per million persons at risk. The slope factor for cadmium is 6.1 (mg/kg·day)$^{-1}$. How many actual cases of cancer are predicted to occur among this worker group?

23. Sixty-four production workers exposed to ammonia in a soda ash facility agree to participate in a study to assess the cumulative hazard for workers at this site. The mean age of the workers was 38.9 years, with an average weight of 70 kg and a mean duration of exposure of 12.2 years. The mean ammonia exposure was 6.4 mg/m^3. The RfD is 1×10^{-1} mg/m^3. Calculate the hazard quotient.

REFERENCES

Environmental Protection Agency (EPA). *Superfund Public Health Evaluation Manual*, Office of Emergency & Remedial Response, Exhitit A-1, Washington, D.C., 1986.

Environmental Protection Agency (EPA). *Risk Assessment Guidance for Superfund*, Vol. 1, Human Health Evaluation Manual (Part A), Interim Final. Office of Emergency and Remedial Response, EPA/540/1-89/002, 1989a.

Environmental Protection Agency (EPA). *Exposure Factors Handbook*. Office of Health and Environmental Assessment, EPA/600/8-89/043, 1989b.

Environmental Protection Agency (EPA). *Interim Final Guidance for Soil Ingestion Rates*. Office of Solid Waste and Emergency Response. (OSWER Directive 9850.4), 1989c.

Environmental Protection Agency (EPA). *Guidance Manual for Assessing Human Health Risks form Chemically Contaminated Fish and Shellfish*. Office of Marine and Estuarine Protection. EPA/503/8-89/002, 1989d.

Environmental Protection Agency (EPA). Human Health Evaluation Manual, Supplemental Guidance: Standard Default Exposure factors, OSWER Directive 9285.6-03, Washington, D.C., 1991.

Louvar, J. F. and B. D. Louvar. *Health and Environmental Risk Analysis: Fundamentals with Applications*, Prentice-Hall, Upper Saddle River, NJ, 1998.

Pao, E. M., K. H. Fleming, P. M. Gueuther, and S. J. Mickle. Food Commonly Eaten by Individuals: Amount per Day and per Eating Occasion. U.S. Department of Agriculture, 1982.

UNEP. Environmental Data Report, 1993–1994, United Nations Environment Programme, Blackwell, Oxford, 1993.

Wilson, R. Analyzing the daily risks of life. *Technol. Rev.* **81**(4), 41–46 (1979).

Wilson, R. and E. A. C. Crouch. Risk assessment and comparisons: An introduction. *Science* **April 17**, 267–270 (1987).

PART V

ENVIRONMENTAL LEGISLATION IN THE UNITED STATES AND EUROPE

"A law is valuable not because it is law, but because there is right in it."
—Henry Ward Beecher

A Basic Introduction to Pollutant Fate and Transport, By Dunnivant and Anders
Copyright © 2006 by John Wiley & Sons, Inc.

CHAPTER 11

ENVIRONMENTAL LAWS

Frank Dunnivant, Lance DeMuth, Savanna Ferguson,
Rose Kormanyos, Sarah McConnell, Loren Sackett,
and Jill Schulte

Discussions of environmental laws can become burdened by technicalities. Thus, I have decided to take a different approach to this chapter. The contents of this chapter are the result of presentations and papers written by many students in Environmental Chemistry during the spring of 2004 and Chemistry of the Natural World during the spring of 2003 and 2004. The students listed as authors are those whose papers are incorporated in one form or another in this chapter. I have added my own interpretations and literature research to the students' work. We feel this approach, in the words of students, provides a more interesting presentation of this relatively dry, but important material.

11.1   ENVIRONMENTAL MOVEMENTS IN THE UNITED STATES

A myriad of social and historical factors may have contributed to the environmental movement and to the eventual formation of the Environmental Protection Agency. Some argue that environmentalism has been an American value since the United States' earliest history. The pilgrims and colonialists monitored depletion of their natural resources: "as early as 1626 the members of the Plymouth Colony passed ordinances regulating the cutting and sale of timber on colony lands" (Switzer, 1994, p. 3). Following is a timeline outlining the important eras and turning points of the modern environmental movement.

11.2.1   Timeline of the U.S. Environmental Movement

1800–1900: The nascent environmental debate is split into two opposing sides: the conservationists and the preservationists. The conservationists "believed sustainable exploitation of resources was possible," while the preservationists "sought to preserve wilderness areas from all but recreational and educational use" (Switzer, 1994, p. 7).

A Basic Introduction to Pollutant Fate and Transport, By Dunnivant and Anders
Copyright © 2006 by John Wiley & Sons, Inc.

In literature, Henry David Thoreau, Ralph Waldo Emerson, and John Muir appeal to peoples' natural longing for the good life, by portraying nature as the ultimate utopia, offering freedom, escape, and fulfillment for every person. Writing at the close of the nineteenth century, Muir uses poetic language to portray the destruction of nature as the ultimate evil. He hopes people will identify with the need for wildness. Many are affected by his essays, including then-President Theodore Roosevelt.

1900–1920: Progressive Era, marked by the mobilization of government and private organizations such as the 1919 National Parks Association and the 1905 National Audubon Society.

1901–1909: Under President Theodore Roosevelt, conservationism gains dominance. Roosevelt sets aside millions of acres for national forests (Barton, 2002, p. 102).

May 1908: White House holds a Conference on Resource Management, which concludes with the creation of a National Conservation Commission to monitor and take an inventory of natural resources.

1912: National Conservation Congress focuses its session on "the conservation of human life" (Switzer, 1994, p. 8), foreshadowing a new debate, that between human welfare and the welfare of the environment.

1930s: Out of the dust bowl and the beginning of the Great Depression, environmental consciousness emerged as part of a new, holistic and communal way of thinking about life and the world. People became more concerned with "maintaining the whole community of life in stable equilibrium with its habitat" (Barton, 2002, p. 70), and began to include the protection of the environment as one of their new priorities.

1935–1945: World War II and the New Deal shape the next development in environmental philosophy, as people move from the cities to the suburbs and "upwardly mobile white collar workers left crowded cities for localities with clean air, gardens, and grass," while "at the same time some rural folk watched with dismay as their small towns became urbanized" (Landy et al., 1994, p. 22). This dramatic environmental change opens the public eye to concrete examples of deforestation, development, and natural resource issues. Meanwhile, the "population was becoming younger, more financially secure, and better educated" (Landy et al., 1994, p. 22). The population's youth coupled with the economic prosperity of the time, allowed "affluence, leisure, mobility, and a greater understanding of physical and biological science [to combine] to create a new awareness of, and interest in, the natural world" (Landy et al., 1994, p. 22). The luxury of environmental awareness, which would have been mocked during the Great Depression, found itself embraced in this new society, a society which would learn from the New Deal that "government could be used to achieve social goals," a realization that the youth of the 1960s eagerly took to heart (Landy et al., 1994, p. 22).

1962: Rachel Carson's *Silent Spring* begins to raise awareness of the gravity of environmental issues and the danger that our society might essentially be welcoming future generations into silent, desolate, and bleak "uncontrolled environmental decay" (Quarles, 1976, p. 11). The work "attracted immediate attention and wound up causing a revolution in public opinion" (Lewis, 1985).

1969: The moon landing offers a patriotic and poetic metaphor for environmentalism: "when astronauts turned their cameras homeward, capturing the image of a delicate blue planet, the world looked upon itself with fresh understanding," although "their photographs showed clouds of pollution hovering over North America (EPA, Dec. 1, 1995; Quarles, 1976, p. 11).

1965–1970: The momentum of the anti-Vietnam War protest clears the way for the environmental movement. The movement towards environmental awareness "has been one of American's most effective protests. It gained power from the rebellion of youth, which dominated the civil rights and antiwar movements of the 1960s" (Quarles, 1976, p. xiv). Popularly, the environmental movement was carried on the shoulders of America's youth and activists. The author of an editorial in the *New York Times* in 1969 observed, "call it conservation, the environment, ecological balance, or what you will, it is a cause more permanent, more far-reaching, than any issue of our era—Vietnam and Black Power included" (Lewis, 1985, p. 2).

1970: The first Earth Day enjoys mass participation. Its widespread popularity influences Congress to pass legislation protecting the environment, giving the movement added momentum. Earth Day also results in "environmentalism" becoming a household word. Not until after Earth Day did surveyed members of the public rank environmentalism among the most important problems facing the nation—in fact, it ranked second in a May 1970 Gallup poll (Switzer, 1994, p. 15).

1975: Edward Abbey's *The Monkeywrench Gang* portrays radical environmentalists as heroes sabotaging the attempt to destroy nature.

1978: At Love Canal, where the Hooker Chemical Company had buried toxic material in a landfill that was later covered over and developed with housing, the pollution begins to seep into groundwater and basements, poisoning residents of the community built on the land. The event "had the dramatic elements of a great story . . . irresponsible corporations, indifferent bureaucrats, and arrogant scientists." The Love Canal crisis receives extensive media coverage, causing people nationwide to sympathize with the affected residents (Taylor, 1995, p. 38). During the 1960s and 1970s, television networks had begun to realize the tremendous profitability of the news, given that "the news is substantially cheaper to produce than entertainment programs, and widely viewed" (Landy et al., 1994, p. 23). And environmental stories held particular appeal: "Oil covered birds, belching smokestacks, rusting storage drums, and inspection crews in 'moonsuits' are all visually compelling. Environmental stories have the particular advantage that often the crew can set up at its leisure, obtain the desired angles, and have plenty of vivid footage to show at the six and the eleven PM new cases." While the media surely could not plant concern for an issue into the mind of an apathetic individual, the dramatics of the television media likely catalyzed the environmental movement and improved the force it would carry.

These are just a few of the events that helped shape and force environmental policy development in the United States, and as we will see in this chapter they helped form comprehensive and effective laws for the protection of human health and to some extent helped protect the environment.

11.2 THE HISTORY OF THE ENVIRONMENTAL PROTECTION AGENCY (U.S. EPA): ADMINISTRATORS

As we saw in the last section, the idealism, activism, and momentum of the environmental movement coalesced in the first celebration of Earth Day on April 22, 1970. Earth Day "lives in popular memory to this day as a joyous and life-affirming moment in American history" (Lewis, 1985, p. 3). Protestors delivered strong messages to the community, the president, and the government as "oil-coated ducks were dumped on the doorstep of the Department of the Interior . . . A student disguised as the Grim Reaper stalked a General Electric Company stockholder's meeting . . . Demonstrators dragged a net filled with dead fish down Fifth Avenue, and shouted to passers-by 'this could be you!'" (Lewis, 1985, p. 3). The government heard their message loud and clear, and as early as July 9, Nixon sent his Reorganization Plan No. 3, which would become the EPA, to Congress. The intention of the agency was to "establish and enforce environmental protection standards, conduct environmental research, provide assistance to others combating environmental pollution, and assist the Council of Environmental Quality (CEQ) in developing and recommending to the President new policies for environmental protection" (Lewis, 1985, p. 4).

Combining offices and responsibilities from the Department of the Interior, the Department of Agriculture, the Department of Health, Education, and Welfare, the Department of Atomic Energy Commission, the Federal Radiation Council, and the Council on Environmental Quality, the EPA began strongly, setting the following mission statement (www.epa.gov):

The mission of the U.S. Environmental Protection Agency is to protect human health and to safeguard the natural environment—air, water, and land—upon which life depends.

EPA's purpose is to ensure that:

- All Americans are proected from significant risks to human health and the environment where they live, learn, and work.
- National efforts to reduce environmental risk are based on the best available scientific information.
- Federal laws protecting human health and the environment are enforced fairly and effectively.
- Environmental protection is an integral consideration in U.S. policies concerning natural resources, human health, economic growth, energy, transportation, agriculture, industry, and international trade, and these factors are similarly considered in establishing environmental policy.

- All parts of society—communities, individuals, businesses, state and local governments, tribal governments—have access to accurate information sufficient to effectively participate in managing human health and environmental risks.

- Environmental protection contributes to making our communities and ecosystems diverse, sustainable and economically productive.

- The United States plays a leadership role in working with other nations to protect the global environment.

With these goals, the EPA certainly had its work cut out for it. Following are short bios of all the EPA administrators since the administration's creation. As you read these, note who they are, where they come from, and what environmental experience (if any) they bring to the EPA.

William D. Ruckelshaus

Tenure: 1970–1973, 1983–1985

Education: Harvard Law School

Previous Public Service: Deputy Attorney General in Indiana, minority and then majority leader in the Indiana House of Representatives.

Achievements/Philosophy: "concentrated on developing the new agency's organizational structure; enforcement actions against severely polluted cities and industrial polluters; setting health-based standards for pollutants and standards for automobile emissions; requiring states to submit new air quality plans; and the banning of the general use of pesticide DDT" (www.epa.gov/history/admin/agency/index.htm, accessed February 5, 2003).

Later career: Vice president, Weyerhaecser Company, 1976–1973; Chief Executive Officer, Browning-Ferris Industries, Inc., 1988–1995.

Russell E. Train

Tenure: 1973–1977

Education: Columbia Law School

Previous Public Service: Attorney on the staff of the Joint Committee on Internal Revenue Taxation of the U.S. Congress; Chief Counsel and Minority Advisor of the Ways and Means Committee of the U.S. House of Representatives; assistant to the Secretary and Head of the Legal Advisory Staff of the Treasury Department; judge in U.S. Tax Court.

Previous Environmental Experience: Helped found the African Wildlife Leadership Foundation; president of the Conservation Foundation; co-chair, Task Force on Environment (under Nixon); chairman, Council on Environmental Quality; named National Wildlife Federation Conservationist of the Year, 1975.

Achievements/Philosophy: "supported EPA's expansion of interest in international affairs; the approval of the catalytic converter to achieve Clean Air Act automobile emission reductions; the implementation of the Toxic Substances Control Act (TSCA) and the National Pollution Discharge Elimination System (NPDES); and EPA's work to balance the demands of the energy crisis with environmental issues" (www.epa.gov/history/admin/agency/index.htm, accessed February 5, 2003).

Later career: President, World Wildlife Fund; co-chairman, Conservationists for Bush.

Douglas M. Costle

Tenure: 1977–1981

Education: University of Chicago Law School

Previous Public Service: Trial attorney in the Civil Rights Division of the U.S. Department of Justice; attorney for the Economic Development Administration of the U.S. Department of Commerce; U.S. Army Reserves military intelligence.

Previous Environmental Experience: Senior staff associate of environmental and natural resources for President's Advisory Council on Executive Organization (led the very study that later advocated the formation of the EPA).

Anne M. Gorsuch

Tenure: 1981–1983

Education: University of Colorado Law School; Fulbright Scholarship in Jaipur, India

Previous Public Service: Hearing officer for the Real Estate Commission and State Board of Cosmetology, Optometric Examiners, Professional Nursing, and Veterinary Medicine; member of Colorado legislature: vice-chairman of the Judiciary Committee, member of the Finance Committee and Appropriations committee, author and prime sponsor of Colorado Presumptive Sentencing Law (sponsored 21 successful bills, voted Outstanding Freshman Legislator); member of the U.S. House of Representatives: Chairman of the House State Affairs Committee, of the House-Senate Legal Services Committee, and on the Interim Committee on Hazardous Waste (sponsored Air-Pollution-Control Inspection and Maintenance Legislation in the House).

Other Previous Experience: Assistant Trust Administrator, First National Bank of Denver, Colorado.

Lee M. Thomas

Tenure: 1985–1989

Previous Public Service: At the EPA: acting assistant administrator for Solid Waste (1983) (directed the Superfund hazardous waste program); prior to the EPA: Federal Emergency Management Agency, executive director, associate director for state and local programs and support; director of the Division of Public Safety Programs for the Governor (South Carolina).

Other Previous Experience: Independent criminal justice planning and managing consultant; chair, National Criminal Justice Association (two consecutive terms).

William K. Reilly

Tenure: 1989–1993

Education: Yale (B.A.), Columbia (M.A. in urban planning)

Previous Public Service: Captain, U.S. Army (two years); senior staff member, Center for Environmental Quality (CEQ).

Previous Environmental Experience: President, World Wildlife Fund and The Conservation Foundation.

Achievements/Philosophy: Interested in "protecting the health of not just people and wildlife, but of the biospheres in which we live" and "committed to continuing that quest for a sounder balance between never-to-be-neglected human health goals and the long-term challenge of preserving for future generations both the ecospheres and our natural resources" (Lewis, 1985).

Carol M. Browner

Tenure: 1993–2001 (longest serving administrator in the history of the agency)

Education: University of Florida Law School

Previous Public Service: General Counsel for the Florida House of Representatives Government Operation Committee,

Other Previous Experience: Worked for Citizen Action in Washington, DC.

Achievements/Philosophy: Grassroots idealism: "the environment and the economy can go hand in hand. We can set tough standards to protect the environment and public health—but do so in ways that promote innovation, flexibility, and American competitiveness" (www.epa.gov/history/admin/agency/index.htm, accessed February 5, 2003).

Christie Whitman

Tenure: 2001–2003

Education: Wheaton College

Previous Public Service: 50th Governor of New Jersey; previously, led the New Jersey Board of Public Utilities and the Somerset County Board of Freeholders.

Achievements/Philosophy: As Governor, and also as a preservationist, achieved noticeable environmental results in New Jersey and "under her environmental leadership, New Jersey's air became significantly clearer [. . . and] the state is on target to reduce greenhouse gas emissions below 1990 levels" (www.epa.gov/history/admin/agency/index.htm, accessed February 5, 2003). Her preservationist efforts have also ensured that "by 2010, New Jersey will have permanently preserved 40 percent of it total landmass" (www.epa.gov/history/admin/agency/index.htm, accessed February 5, 2003).

Michael Leavitt

Tenure: 2003–2005

Previous Public Service: Utah's 14th Governor, leader on homeland security, welfare reform and environmental management.

Achievements/Philosophy: As governor, helped clean up air over the Grand Canyon that resulted in 70 recommendations to improve visibility on the Colorado Plateau.

Stephen L. Johnson

Tenure: 2005–

Education: Taylor University and George Washington University

Previous Public Service: Stephen L. Johnson was sworn in as the 11th Administrator of the U.S. Environmental Protection Agency on May 2, 2005. Prior to becoming Administrator, Johnson had served as the Acting Administrator (since January 2005), Deputy Administrator (from August 2004 to January 2005) and Acting Deputy Administrator of the Agency (from July 2003 to August 2004) and has been a part of the EPA for 24 years. He was Assistant Administrator of EPA's Office of Prevention, Pesticides, and Toxic Substances (OPPTS) from June 2001 to July 2003. Johnson had been OPPTS Acting Assistant Administrator since January 2001, and had held top leadership positions in that office since January 1999, first serving as Acting Deputy Assistant Administrator. He was named Deputy Assistant Administrator in April 2000, and then was reassigned as Principal Deputy Assistant Administrator. He had also served as Deputy Director of the Office of Pesticide Programs (OPP) since May 1997. Other senior level positions held by Mr. Johnson at the EPA include: Director of OPP's Field Operations Division, Deputy Director of OPP's Hazard Evaluation Division and Executive Secretary of the Scientific Advisory Panel for the Federal Insecticide, Fungicide, and Rodenticide Act.

11.3 MAJOR U.S. ENVIRONMENTAL LAWS

As we saw in the previous section, there were many factors that led to legislative environmental action by the Federal government. Twelve major environmental laws were passed from the 1960s to 1990. These are listed in Table 11.1 by year of passage and include names that should be familiar to anyone with environmental interests. But how does an environmental law come about? The following gives a relative simply path for the formation of a law. Generally, a member of Congress proposes a bill, and then the bill is discussed, debated, and, if necessary, revised. If both houses of Congress approve the bill, then the President has the option to approve or veto it. If approved, the new law is called an act, and the text passed by Congress is known as a public statute. Once an act is passed and signed by the President, the House of Representatives standardizes the text which it publishes in the U.S. Federal Register (www.epa.gov/epahome/laws.htm).

Yet the laws themselves do not, and indeed cannot, contain all the details required for their implementation. Laws give a Federal agency the power to regulate, and then a specified agency must then develop regulations that set specific rules about what is legal and what is not legal. For example, the Clean Air Act gave EPA the power to regulate sulfur dioxide, but EPA scientists had to determine what level (emission concentration) was safe. This latter process, the defining of what is safe and how to monitor it, is a monumental task and in many cases it takes years to actually enact a regulation after the law has been passed.

Many interested parties provide information to lawmakers during the early and later stages of a law and regulation. Industry certainly has vested economic inter-

TABLE 11.1. Summary and Timeline of the Major Twelve Environmental Laws

1947	Federal Insecticide, Fungicide, and Rodenticide Act (FIFRA)
1955	Clean Air Act (CAA)
1969	National Environmental Policy Act (NEPA)
1965	Solid Waste Disposal Act
1970	Clean Air Act (CAA)
1970	Occupational Safety and Health Act (OSHA)
1972	Federal Insecticide, Fungicide, and Rodenticide Act (FIFRA)
1972	Federal Water Pollution Control Act (FWPCA)
1974	Safe Drinking Water Act (SDWA)
1976	Toxic Substances Control Act (TSCA)
1976	Resource Conservation and Recovery Act (RCRA)
1976	Solid Waste Disposal Act (SWDA)
1977	Clean Air Act Amendments
1977	Clean Water Act Amendments
1980	Comprehensive Environmental Response, Compensation, and Liability Act (Superfund)
1984	Hazardous and Solid Waste Amendments (HSWA)
1986	Safe Drinking Water Act Amendments
1986	Superfund Amendments and Reauthorization Act (SARA)
1987	Clean Water Act Amendments
1990	Oil Pollution Act
1990	Pollution Prevention Act
1990	Clean Air Act Amendments

ests in any environmental regulation and can contribute significantly to the formation or blockage of a law and the setting of standards of a regulation. This industry involvement is often a source of controversy. The most famous recent case is the Energy Task Force formed by Vice President Dick Cheney, for which the list of members was kept from the public. Some claim that this represents an attempt to hide industry influence on the task force, and this has resulted in a major legal fight for Freedom of Information seekers. Likewise, environmental organizations, who see themselves as politically defending the health of the public and the environment, attempt to influence the formation, structure, and passage of laws and regulations. Some of the most influential environmental organizations are shown in Table 11.2.

Laws in the United States can essentially be formed by two avenues: either (a) as statutes and regulations formed by Congress and the President or (b) by case law, under which industry, citizens, or groups file suit in order to question or attempt to clarify a law. In case law, a plaintiff who feels a private or civil wrong or injury has been committed may claim that a "tort" has been committed and ask the court to provide a remedy in the form of an action for damages. The alternative to a tort is a "nuisance claim," for cases of alleged unreasonable or unlawful use by an industry or person of their property or injury to the rights of another person or group. Examples of judicial clarification include court cases from both sides (industry and environmental organizations). Industry can sue, stating that a regulation by EPA is

TABLE 11.2. Major U.S. Environmental Organizations

Organization	Founded
Sierra Club	1892
National Audubon Society	1905
National Parks and Conservation Association	1919
Izaak Walton League	1922
Wilderness Society	1935
National Wildlife Federation	1936
Environmental Defense Fund	1967
Friends of the Earth	1970
National Resources Defense Council	1970
Greenpeace (worldwide)	1971

outside the intent of the law passed by Congress, and citizens or groups can sue EPA or an industry, stating that they are not following established regulations. Both types of action, statute and case law, have been instrumental in the formation of our environmental regulations.

In the remainder of this section, we will look at brief summaries of major environmental laws, with some interpretations as needed. Summaries for some of the relatively simpler laws are very short, while others deserve more discussion to attain a true appreciation of their intent. We will start in chronological order for the laws shown in Table 11.1.

11.3.1 The Federal Insecticide, Fungicide, and Rodenticide Act (FIFRA)

The passing of the FIFRA in 1947 was the first attempt to regulate the manufacture and use of pesticides across the nation. Although the enforcement efforts, and thus the act itself, were weak, the act was a starting point for a major movement. The act was amended in 1972; in 1988, to regulate all phases of pesticide sale, use, handling, and disposal; and in 1996, under the Food Quality Protection Act (FQPA), to establish tolerances for pesticide residues in food. The heart of FIFRA is the pesticide registration program. Before a pesticide can be manufactured, distributed, or imported, it must have approval from the EPA (the EPA assumed this responsibility upon its creation). In determining whether to register a pesticide, EPA considers its economic, social, and environmental costs and benefits.

FIFRA also defined the terms pesticide and pest. A pesticide is any substance, or mixture of substances, intended for a pesticidal purpose (for preventing, destroying, repelling, or mitigating any pest or for use as a regulator, defoliant, or desiccant). A pest is defined as an insect, rodent, nematode, fungus, weed, terrestrial and aquatic plant, virus, bacteria, or any other living organism that the U.S. EPA designates as a pest.

Some of the most significant aspects of FIFRA include the following:

1. Pesticides and their containers must be labeled with ingredients, instructions for use, EPA registration number, and all necessary warnings or restrictions.

2. Pesticides are divided into two categories:

   a. General-use pesticides may be applied by anyone; no permit is required.

   b. Restricted-use pesticides may only be applied by (1) private applicators in an agricultural setting on their own land, (2) commercial applicators who apply pesticides to other people's lands for a fee, and (3) experimental use applicators (manufacturers or researchers).

3. States were given the right to regulate the sale and use of pesticides within their borders, so long as the regulations are not less stringent that those specified in the FIFRA regulations.

4. The 1972 amendments stated that a balanced approach to considering the benefits and risk of a pesticide should be used in licensing. It also prohibited the use of cancer-causing agents, but defined an acceptable risk as less than a one-in-a-million chance of causing cancer (This one-in-a-million concept was developed and discussed in Chapter 10).

5. The 1988 amendments sped up the process of registering a pesticide and shifted the costs of canceled pesticides from the EPA (and the taxpayer) to the manufacturer (costs such as removal from retail store shelves and disposal).

6. The 1988 amendment called for a nine-year toxicology testing period of the 700–900 active ingredients found in pesticides approved prior to 1984. (The same negligible risk of one cancer in a million was applied in these analyses.)

7. The 1996 amendments required that there must be a reasonable certainty of no harm from aggregate exposure (exposure to more than one pesticide at one time). These amendments also required that existing tolerances (acceptable levels of exposure) be reassessed under new standards, and they required EPA to make an explicit determination of acceptable residue levels in foods for infants and children.

8. FIFRA established penalties for misuse of a pesticide. These include the following: (1) a $5000 per-offense fine for commercial applicators, wholesalers, dealers, retailers, or distributors who violate FIFRA or their permit (violators are also subject to a $25,000 fine and one-year imprisonment for knowingly violating a permit for FIFRA), (2) producers of pesticides who knowingly violate FIFRA are subject to criminal penalties for up to $50,000 and one-year imprisonment, and (3) private applicators are subject to a $1000 fine after a written warning. Due to the complexity of environmental laws and over-lapping laws, violators of FIFRA can also be subject to penalties under the Food Quality Protection Act; the Federal Food, Drug, and Cosmetic Act; the Resource Conservation and Recovery Act; and the Occupational Safety and Health Act.

11.3.2 The Air Quality Act, The Clean Air Act, and Amendments

The Clean Air Act is probably the most amended environmental act in U.S. history. While we would like to think an act is amended because of increased knowledge or tightening of pollutant emissions, this has not necessarily been the case with the Clean Air Act. Although, in general, air quality has improved since the introduction of clean air legislation in 1955, some critics of the process feel we have regressed in our efforts by deregulating what we are having trouble controlling or what we find too expensive to control. Two examples of these problematic pollutants are acid rain and carbon dioxide. But strides have been made in improving the air quality in most cities.

Given the high number of amendments to the Clean Air Act, and in order to give a more clear explanation of the Act, we will present only the general goals of the Act and amendments.

In the early years, the goal of the Act was to regulate air emissions from area, stationary, and mobile sources. This was attempted by setting National Ambient Air Quality Standards (NAAQS) that set (specified) the maximum amount of pollutant allowed in the air anywhere in the nation. The responsibility to accomplish or meet the NAAQS was given to the individual states, but if a state did not meet the NAAQS, the EPA would take control. Individual states developed State Implementation Plans (SIPs) applicable to their particular industrial pollutant sources. There are two basic types of sources defined by EPA: (a) stationary sources such as power plants and factories and (b) mobile sources such as cars, trucks, and other motor vehicles. Geographical areas meeting primary and secondary standards are referred to as attainment areas, while those not meeting the standards in Table 11.3 are referred to as nonattainment areas.

The amendments in the 1970s, especially the 1977 amendment, set new goals (dates) for achieving the NAAQS, since many areas of the nation had failed to meet the earlier deadlines. The 1977 amendments and previous NAAQS focused on a set of common air pollutants referred to as Criteria Air Pollutants. These include ozone, volatile organic compounds (VOCs), nitrogen dioxide, carbon monoxide, particulate matter, sulfur dioxide, and lead. Health effects, major sources, and primary and secondary standards for these pollutants (established in 1977) are shown in Table 11.3. VOCs, nitrogen dioxide, and ozone are contributors to the formation of smog, the brown haze that exists over large cities such as Los Angeles and Atlanta. Nitrogen dioxide and sulfur dioxide are precursors to acid rain and are especially important over the Midwest, where the majority of the U.S. coal-fired power plants are located. Much of the acid precipitation (rain, snow, and dry precipitation) then falls over the northeastern United States and Canada.

The 1990 amendments established a permit system so that each state issues a permit to polluters, stating which pollutants are being released, how much pollutant may be released, and what kinds of steps the source's operator is taking to reduce pollution, including plans to monitor (measure) the pollution. Operators must pay a fee for the permits, and the money from the fee helps pay for state air pollution control activities. The 1990 amendments also gave new enforcement powers to the EPA, allowing the EPA to fine violators for exceeding the limits of their permits. As with most Clean Air Act amendments, these established new deadlines for the EPA and states to meet and reduce air pollution. "Market-based approaches" were also introduced in the 1990 amendments, enabling owners of permits to trade pollution allowances. For example, emissions of the precursors to acid rain from coal-fired power plants are openly traded on the market.

TABLE 11.3. Criteria Air Pollutants, Effects, Sources, and Primary and Secondary Standards

Pollutant	Effects	Sources	Primary Standard	Secondary Standard
Ozone	Respiratory tract problems (difficulty breathing, reduced lung function, possible premature aging of lung tissue, and asthma), eye irritation, nasal congestion, reduced resistance to infection, damage to trees and crops	Chemical reaction of pollutants, volatile organic compounds and nitrogen oxides	0.12 ppm (160 μg/m³) for 1 hr	Same as primary standard
Particulate matter	Eye and throat irritation, bronchitis, lung damage, impaired visibility, cancer	Burning of wood, diesel, and other fuels; industrial plants; agriculture; unpaved roads	260 μg/m³ for 24 hr	150 μg/m³ for 24 hr
Carbon monoxide	Impairment of bloods ability to carry oxygen; damage to cardiovascular, nervous and pulmonary systems	Burning of gasoline natural gas, coal, and oil	9 ppm (10 mg/m³ for 8 hr; 35 ppm (40 mg/m³ for 1 hr	Same as primary standard
Sulfur dioxide	Respiratory tract problems (diminished lung capacity and permanent damage to lung tissue); primary component of "killer fogs"; precursor to acid rain which damages trees, vegetation, and aquatic life	Burning of coal and oil, especially high-sulfur coal from the Eastern United States; industrial processes	0.14 ppm (365 μg/m³) for 24 hr	None
Nitrogen dioxide	Respiratory illnesses and lung damage; damage to immune system; precursor to acid rain	Burning of gasoline, natural gas, coal, and oil	0.05 ppm (100 μg/m³) arithmetic mean	Same as primary standard
Nonmethane hydrocarbons (NMHC or VOCs)	Cancer; precursor to acid rain	Incomplete combustion of hydrocarbons (gasoline); industrial processes	0.24 ppm (160 μg/m³) for 3 hr	Same as primary standard
Lead	Brain damage and retardation, especially in children	Leaded gasoline, paint, smelters, manufacture of lead storage batteries	1.5 μg/m³ arithmetic mean	Same as primary standard

Sources: Bryner, 1995; Lave and Omnenn, 1981; www.epa.gov/oar/oaqps/peg_caa/pegcaa02.html.

349

While ozone is considered a pollutant at ground level, its concentration in the stratosphere is essential for life on Earth, because it absorbs (blocks) the entry of harmful high-energy radiation (ultraviolet radiation). Certain pollutants have been shown to destroy this essential stratospheric ozone, and the 1990 Clean Air Act set a schedule for ending the production and use of these chemicals. This schedule is shown in Table 11.4. To give one example of ozone-destroying chemicals, chlorofluorocarbons (CFCs), common propellants in aerosol cans in the twentieth century, are not readily degraded and thus are able to migrate upward in the atmosphere. Once in the stratosphere, they destroy ozone through a series of complex reactions. The removal of ozone allows more harmful radiation to enter the lower atmosphere, potentially increasing the incidence of skin cancer and cataracts and harming the basis of our food chain (photosynthesis-based organisms).

The most recent act dealing with air quality is the Clear Skies Initiative recently signed by President George W. Bush. This initiative has received intense criticism from environmentalists and has even been dubbed the "Clean Air Sham" and "Clean Lies." The long-term effect of this Act is yet to be determined, but the Bush Administration claims that the initiative will slash three major pollutants by 70% over the next 15 years. Environmentalists' predictions range widely, and they include among the worst cases a 50% increase in sulfur dioxide, a 190% increase in mercury, and a 36% increase in nitrogen dioxide.

Two major pollutants have yet to be addressed by the Clean Air Act amendments, not due to lack of science or health risk but due to a lack of leadership in Washington. These pollutants are mercury and carbon dioxide. The major source of mercury to the atmosphere is from the burning of coal across the world, which accounts for 52.7 million grams of mercury per year (Seignecr et al., 2004). Waste incineration produces another 32.2 million grams of mercury per year, while chloralkali facilities, in manufacturing chlorine bleach, generate another 6.8 million grams per year (Seignecr et al., 2004). Technologies exist today that would eliminate the vast majority of these emissions, but implementation of such technologies are often fought by industry in the name of economics. Still, mercury is one of the major hazards facing us today.

TABLE 11.4.   Ozone-Destroying Chemicals and Proposed Dates for Ending Production

Chemical Name	Use	When U.S. Production Ends
CFCs (chlorofluorocarbons)	Solvents, aerosol sprays, foaming agents in plastic manufacture	January 1, 1996
Halons	Fire extinguishers	January 1, 1994
Carbon tetrachloride	Solvents, chemical manufacture,	January 1, 1996
Methyl chloroform (1,1,1-trichloroethene)	Widely used solvent; in the work place, in consumer solvents, used in auto repair and maintenance products	January 1, 1996
HCFCs (hydro CFCs)	CFC substitutes, chemicals slightly different from CFCs	January 1, 2003

Source: www.epa.org.

Most scientists find it amazing that carbon dioxide has not been regulated. We regulate nitrogen dioxide, a minor contaminant leaving the exhaust of our automobiles, while the major component, carbon dioxide, is not yet considered a pollutant by the EPA. Carbon dioxide, of course, is the major culprit in global warming, a topic that is beyond the scope of this textbook. These two major policy failings must be addressed in future amendments to the CAA.

11.3.3 The National Environmental Policy Act

The National Environmental Policy Act (NEPA) is the Magna Carta of environmental law. Yet there are few today, even among environmentalists, who know what the law contains or understand the potential it once had, and may still have, to shape the U.S. federal policy toward the environment. The statute has been misunderstood ever since it was signed into law by President Nixon on January 1, 1970.

There are four main purposes to the Act (NEPA, 1969): (1) "to declare a national policy which will encourage productive and enjoyable harmony between man and his environment," (2) "to promote efforts which will prevent or eliminate damage to the environment and biosphere and stimulate the health and welfare of man," (3) "to enrich the understanding of the ecological systems and natural resources important to the Nation," and (4) "to establish a Council on Environmental Quality." Thus, NEPA seeks to provide guidance to federal agencies not through budgeting or strict regulations, but through an ethical code (Lindstrom and Smith, 2001, p. 8). Historically, environmental laws have focused on specific environmental concerns; this is not the case with NEPA. *It is a broad, umbrella law that seeks to incorporate ecological thinking and systems theory into the political process itself.* The Act was an attempt by Congress to institute a law that would prevent future environmental destruction rather than perpetuate the crisis mentality so common in environmental legislation, which focuses on reclamation. NEPA, as envisioned by its authors, had the potential to be the most powerful and most important environmental law ever enacted in the United States. If implemented correctly, it is capable of improving the quality of both human and nonhuman life. Especially in the context of the political process, where progress typically is approached incrementally with constant compromise, the National Environmental Policy Act is a monumental piece of legislation. However, this also makes it difficult to incorporate the necessary science behind the act into effective policy.

Public misunderstanding of the law's intent was evident from its inception. The *New York Times* and *Washington Post* articles regarding NEPA upon its passage referred to it as Nixon's new anti-pollution law rather than what its authors intended it to be: a revolutionary, ecological approach to governmental interaction with the environment (Caldwell, 2001). It has subsequently been degraded so thoroughly by judicial misinterpretation and executive neglect that it has degenerated into one of the least effective and certainly one of the least well known of all environmental laws. We will now look at some of the features of the Act in detail.

Section 101 of Title I outlines a powerful and all-encompassing but well-defined policy for federal interaction with the environment. There are six points in this section, declaring that it is the intention of the federal government "to improve and coordinate" federal action and resources in order that the nation (NEPA, SEC. 101 [42USC $ 4331, b]) (1) fulfill its responsibility as "trustee of the environment" for future generations, (2) assure for all Americans a healthy environment, including aesthetic and cultural components, (3) discover the maximum number of "beneficial uses of the environment" without degrading the environment, or causing unintended or undesirable outcomes, (4) preserve national heritage and maintain an environment that is diverse and offers a variety of "individual choice," (5) "achieve a balance between population and resource use that will permit high standards of living and a wide sharing of life's amenities," and (6) enhance renewable resources and provide the highest level of recycling for nonrenewable resources.

The second major section, Section 102 of Title II, of the Act has come to be known as the action forcing mechanism. The first requirement set out in Section 102 is that all government agencies use a systematic and interdisciplinary methodology in decision making in order to ensure that the environment and social factors are considered in agency proposals. It is further required that federal agencies develop a system to be used in decision making, with the aid of the Council on Environmental Quality (established in Title II of the Act), which will guarantee that unquantifiable environmental factors are given consideration alongside technical and economic factors. The third requirement, (C), has five parts and has been manifested in the ubiquitous Environmental Impact Statement (EIS), the single greatest legacy of NEPA: "In every recommendation or report on proposals for legislation and other major federal actions significantly affecting the quality of the human environment, a detailed statement by the responsible office on (1) the environmental impact of the proposed action, (2) any adverse environmental effect that cannot be avoided should the proposal be implemented, (3) alternatives to the proposed action, (4) the relationship between local short-term used of man's environment and the maintenance and enhancement of long-term productivity, and (5) any irreversible and irretrievable commitments of resources which would be involved in the proposed action should it be implemented" (NEPA, SEC. 102 [42USC $ 4332] (C)).

Furthermore, prior to making any detailed statement, the responsible federal official shall consult with and obtain comments of any federal agency that has jurisdiction by law or special expertise with respect to any environmental impact involved. Copies of such statements and the comments and views of the appropriate federal, state, and local agencies, which are authorized to develop and enforce environmental standards, shall be made available to the President, the Council on Environmental Quality and to the public (NEPA, SEC. 102 [42USC $ 4332], (C)).

The EIS was meant to force agencies to consider, and take steps to minimize, their effect on the environment and be punished if they did not. The EIS is not only requisite for the ongoing activities of regular federal agencies, but applies to any major actions made by independent agencies that are licensed by the government.

Another important section of the Act is Title II, Section 202, which achieves the fourth purpose of NEPA by establishing the Council on Environmental Quality (CEQ). The Council is to have three members who are appointed by the President with the advice and consent of the Senate. One of these members serves as chairman. The major portion of the section outlines the rigorous guidelines for the selection of a chairman who shall be "exceptionally well qualified to analyze and interpret environmental trends and information of all kinds." It was the intention of the authors that the CEQ members be environmental experts, free of political ties and constraints—that is, individuals who would not be prone to administration favoritism. The specific duties of the Council include (1) assisting and advising the President with the annual Environmental Quality Report (discussed below), (2) gathering information on environmental trends and referring it to the President in a timely manner, (3) reviewing the policies and programs of the federal government with respect to NEPA, (4) developing policies for the President that promote environmental quality, (5) researching and analyzing ecological systems and the environment, (6) documenting changes in the environment and collecting data that aids in understanding the causes of such changes, (7) reporting annually to the President on the condition of the environment, and (8) creating and supplying to the President studies, reports, and recommendations for policy and legislation that the President may request.

Section 102, again of Title II, establishes that the President shall submit an annual Environmental Quality Report to Congress on the environmental condition of the United States. In this report, the President must not only relate the environmental health of all of the varying biomes and urban communities of the United States, but must also report current and foreseeable trends in environmental quality and use, and predict how these trends will affect the nation.

In total, the Supreme Court has ruled twelve times against the full enforcement of NEPA and never ruled for it (Lindstrom and Smith, 2001, p. 120). The Supreme Court's downsizing of NEPA has led to a weakened enforcement of the Act by the lower courts, which had previously been NEPA's greatest chance for success.

Diminishment of the Act in the courts has led to a similar degradation of NEPA among government agencies. Even the creation of the EPA and subsequent environmental laws have weakened NEPA by granting exemptions from the EIS. These exemptions are frequently given to actions that are condoned by the EPA, an agency that is prone to administration preferencing. All of the presidents since NEPA's enactment have sought to abolish the CEQ or lower its budget. Even the EIS, the only remaining stronghold of NEPA, has been undermined by the lack of support for the Act among high-ranking government officials. Apparent compliance with the EIS often serves to improve the appearance of an agency's consideration of environmental effect while, in reality, affecting the decision-making process not at all. After working for the Department of Energy, one of the authors of this text (Dunnivant) can strongly vouch for the truth of this statement.

To date, NEPA has resulted in over 1000 lawsuits (Lindstrom and Smith, 2001, p. 100). Due to the judicial downsizing of the Act, environmental organizations have turned their efforts toward projects that may be applicable to other, better-enforced statutes. Indeed, even the impact statements that are prepared today are frequently researched and compiled not by federal officials, but by special environmental research firms that are hired to conduct EIS and Environmental Assessment (EA) studies. While individuals within such organizations may be far better qualified to research the environmental effects of federal projects, they are also legal mercenaries, inclined to create reports that will ensure the approval of their client's proposal. Even if the EIS determines that the proposal will do irreparable harm to the environment, and even if the EIS contains the comments of thousands of outraged citizens and agency officials, the action may still proceed. Having fulfilled the procedural requirements of the Act, the agency will have satisfied the level of NEPA compliance required by the courts. To restore NEPA to its original potential, there must be a complete reinterpretation and enforcement of the Act. NEPA in its simplicity is nonetheless radical, and despite the growing wave of environmentalism, a law like NEPA would never pass Congress today.

11.3.4   The Solid Waste Disposal Act, Resource Conservation and Recovery Act (RCRA), and Amendments

Disposal of solid wastes produced by human activities has posed problems since humans gave up the nomadic lifestyle. As humans settled in larger and larger groups in isolated areas, wastes reached sufficient concentrations such that they were no longer naturally removed and waste management plans had to be created. Continued development of sessile and ever larger communities exacerbated the problem. Congress began to address an impending crisis after World War II, as the United States faced dealing with the solid and hazardous waste generated during the war effort and as cities continued to deal with growing solid waste burdens. As the problem grew during the 1950s and 1960s, it led to the passage of the Solid Waste Disposal Act of 1965. The Act has subsequently been amended several times and restated in the Resource Conservation and Recovery Act (RCRA) of 1976 and the Hazardous and Solid Waste of Amendments of 1984. Note that the original Act was passed prior to the creation of the EPA in 1970. However, amendments to this act have set up an Office of Solid Waste within the EPA, which is responsible for establishing a regulatory program for waste that includes methods for identifying solid and hazardous wastes, determining the degree of hazard, and establishing recycling programs. The amendments also created the "cradle to grave" concept of legal responsibility for the waste, and they required the Office of Solid Waste to dictate how a waste is tracked and monitored between the source and its ultimate destination, whether a recycling plant or disposal site.

We will address each of these aspects of the Act and its successors and amendments further, but first, how did the RCRA define solid and hazardous wastes (both of which it addresses)? "Solid waste means any garbage, refuse, sludge from a waste treatment plant, water treatment plant, or air pollution control facility and other discarded material, including solid, liquid, semi-solid, or contained gaseous material from industrial, commercial, mining, and agricultural operations" (Section 1004 of RCRA). It does not include discharges from permits issued under the Federal Water Pollution Control Act or nuclear waste as defined by the Atomic Energy Act of 1954.

Hazardous waste, in turn, "means a solid waste or combination of solid wastes, which because of its quantity, concentration, or physical, chemical, or infectious characteristics may

(a) cause or significantly contribute to an increase in mortality or an increase in serious irreversible, or incapacitating reversible, illness; or

(b) pose a substantial present or potential hazard to human health or the environment when improperly treated, stored, transported, or disposed of, or other managed . . ." (Section 1004 of RCRA).

Although the RCRA thus relatively clearly defined solid and hazardous wastes, it did not specify exact standards for their disposal. As we saw earlier in Section 2.1 of this chapter, these types of decisions are left up to the EPA to research and set appropriate standards.

The EPA established three ways that a solid waste can be classified as hazardous: (1) if it is specifically listed in EPA regulations as such (for example, waste halogenated solvents or pesticides), (2) if it is tested by exact procedures and meets one of the four characteristics defined by EPA, including ignitability, corrosiveness, reactivity, or toxicity, and (3) if it is declared to be hazardous by the generator (the manufacturer or source of the waste). Procedures were developed by the EPA for conducting each of the tests mentioned in (2) above. Toxicity testing was the most complicated and involved subjecting the waste to two standardized procedures, the Extraction Procedure Toxicity Test (EP Toxicity Test) and the Toxicity Characteristic Leaching Procedure (TCLP).

A procedure for tracking wastes from source to disposal was also developed, hinging on a manifest system. A manifest form is shown in Figure 11.1. The manifest form is initiated at the source of the waste, and the form must be present with the waste as it travels to recycling, disposal, or destruction facilities. If an inspection is conducted by the EPA at the site of origin, during transport, temporary storage, or recycling, or at the disposal or destruction site, the manifest must be present or fines and imprisonment can result. The purpose of this manifest tracking system is to record who is responsible for the waste at each step of the process, as well as the volumes of waste generated and disposed, in order to provide a means of establishing liability if a future problem occurs. This obviously leads to the concept of "cradle to grave" liability, which states in simple terms that anyone dealing with the waste can be held accountable for future cleanup costs or lawsuits. Thus, it is in the best interest of the creator of the waste to hire reputable firms at each step of the handling, recycling, distribution, and disposal process.

RCRA also established detailed requirements for the construction and operation of solid and hazardous waste landfills, recycling facilities, and destruction facilities, including transportation and labeling of containers involved in transport. Specific design requirements of interest for landfills include (1) double liners in the bottom of the landfill, (2) leachate collection systems, (3) groundwater monitoring around the site, and (4) leak detection systems. Incinerators were required to have removal efficiencies of 99.99% for principal organic hazardous constituents and 99.9999% for acutely hazardous wastes (such as dioxin). Landfills and incinerators were licensed for specific types of waste, and some specific pollutants were banned from these facilities (a common example being polychlorinated biphenyls). Later amendments banned the disposal of bulk liquid hazardous waste in landfills or impoundments.

One other important program set up under RCRA required the oversight and management of storage tanks, specifically underground storage tanks (USTs). One of the most common uses of USTs in the 1970s (when RCRA was passed), as well

Please print or type   (Form designed for use on elite (12 - pitch) typewriter)   Form Approved. OMB No. 2050 - 0039 Expires 9 - 30 - 91

UNIFORM HAZARDOUS WASTE MANIFEST	1 Generator's US EPA ID No.	Manifest Document No.	2. Page 1 of	Information in the shaded areas is not required by Federal law

3. Generator's Name and Mailing Address	A. State Manifest Document Number
	B. State Generator's ID
4. Generator's Phone ( )	

5. Transporter 1 Company Name	6. US EPA ID Number	C. State Transporter's ID
		D. Transporter's Phone
7. Transporter 2 Company Name	8. US EPA ID Number	E. State Transporter's ID
		F. Transporter's Phone
9. Designated Facility Name and Site Address	10. US EPA ID Number	G. State Facility's ID
		H. Facility's Phone

11. US DOT Description (Including Proper Shipping Name, Hazard Class, and ID Number)	12. Containers No.	Type	13. Total Quantity	14. Unit Wt/Vol	I. Waste No.
a.					
b.					
c.					
d.					

G
E
N
E
R
A
T
O
R

J. Additional Descriptions for Materials Listed Above	K. Handling Codes for Wastes Listed Above

15. Special Handling Instructions and Additional Information

16. **GENERATOR'S CERTIFICATION:** I hereby declare that the contents of this consignment are fully and accurately described above by proper shipping name and are classified, packed, marked, and labeled, and are in all respects in proper condition for transport by highway according to applicable international and national government regulations.

If I am a large quantity generator, I certify that I have a program in place to reduce the volume and toxicity of waste generated to the degree I have determined to be economically practicable and that I have selected the practicable method of treatment, storage, or disposal currently available to me which minimizes the present and future threat to human health and the environment; **OR**, if I am a small quantity generator, I have made a good faith effort to minimize my waste generation and select the best waste management method that is available to me and that I can afford.

Printed/Typed Name	Signature	Month	Day	Year

T
R
A
N
S
P
O
R
T
E
R

17. Transporter 1 Acknowledgement of Receipt of Materials				
Printed/Typed Name	Signature	Month	Day	Year

18. Transporter 2 Acknowledgement of Receipt of Materials				
Printed/Typed Name	Signature	Month	Day	Year

F
A
C
I
L
I
T
Y

19. Discrepancy Indication Space

20. Facility Owner or Operator: Certification of receipt of hazardous materials covered by this manifest except as noted in item 19.				
Printed/Typed Name	Signature	Month	Day	Year

EPA Form 8700 - 22 (Rev. 9 - 88) Previous editions are obsolete.

Figure 11.1. The Hazardous Waste Manifest Form required by EPA to ship hazardous waste.

as today, is at gasoline filling stations. These tanks were originally made of metal that eventually rusted when buried in wet soil. Thus, numerous tanks did leak gasoline into the surrounding soil and groundwater. Cleanup and prevention of such leaks required a major effort by the EPA, but an end is in sight, and the program has been extremely successful in eliminating or minimizing releases from gasoline storage tanks. Operators were required to monitor for leaks through testing and through inventory assessments. When a leak was detected, the tank had to be dug up and replaced, and thousands of tanks have been replaced since the UST program began.

One very important distinction should be made. RCRA was responsible for controlling the generation, transport, and disposal or destruction of new and future solid and hazardous waste, and initially it was responsible for cleaning up dangerous waste sites through the cradle to grave liability actions. Yet RCRA did very little to address the issue of waste already disposed of or released to the environment. Such waste, in abandoned or illegal waste sites, had to be addressed by a later act, the Comprehensive Environmental Response, Compensation, and Liability Act of 1980, which will be discussed in Section 11.3.8.

11.3.5   Occupational Safety and Health Act (OSH Act)

While the OSH Act is not specifically considered an environmental legislative act, it is the act most responsible for safety in the workplace, where humans are often at risk of being exposed to pollutants. Thus, we will include a brief description of the act here. Prior to 1970, studies strongly indicated to Congress that excessive economic costs were being imposed on workers, in the from of lost productivity, wage loss, medical expenses and disability compensation payments, due to personal injuries and illnesses arising from poor work conditions. While all of the injuries were not due to chemical or pollutant exposure, many were, and these are the focus of our concern. The OSH Act created the National Institute for Occupational Safety and Health (NIOSH) as the research institution for the Occupational Safety and Health Administration (OSHA), with the purpose of researching and setting standards for workplace health and safety. These standards include chemical exposure limits and restrictions on conditions that can harm workers. For example, time-weighted exposure limits now exist for many chemicals, specifying the acceptable level (concentration) of each chemical that a worker can be exposed to over a specified period of time (usually based on an eight-hour workday). It is widely recognized that the creation and enforcement of these standards have been instrumental in reducing illness and cancer in industrial workers.

11.3.6   The Federal Water Pollution Control Act, the Clean Water Restoration Act, the Safe Drinking Water Act, and Amendments

Today we are concerned with both the availability (or scarcity) of water and the quality of the water. These concerns are best summarized by Charles C. Johnson, Former Assistant Surgeon General of the United States who said (www.epa.gov/history/topics/fwpca/05.htm)

Our water resources, more perhaps than any other, illustrate the interaction of all parts of the environment and particularly, the recycling process that characterizes every resource of the ecosystem. . . . Everything that man himself injects into the biosphere—chemical, biological or physical—can ultimately find its way into the Earth's water. And these contaminants must be removed, by nature or by man, before that water is again potable.

This statement also encompasses the need for maintaining the quality of our dwindling water resources. But this is not a recent development. Water treatment to improve the taste and odor were recorded as early as 4000 B.C. Ancient Sanskrit and Greek writings indicated the use of water treatment methods such as filtering dirty water through charcoal, exposing the water to sunlight, boiling by placing a hot metal rod in the water, and straining. Chemical treatment, such as adding alum to remove suspended particles, was used as by Egyptians as early as 1500 B.C. (EPA, 2000) Indications of poor quality such as visible cloudiness (turbidity), taste, and odor were the driving forces behind early water treatment, since bacteria, the source of disease, had not been discovered.

Large-scale filtration was introduced in the 1700s as an effective way to remove particles from water, with sand filtration being regularly used in Europe in the 1800s. The epidemiologist Dr. John Snow proved in 1855 that cholera was a waterborne disease by linking an outbreak in London to a single public well contaminated by sewage (EPA, 2000). In the 1880s, Louis Pasteur developed the "germ theory." During the late 1800s and early 1900s, water treatment focused more on disease-causing microbes or pathogens in public water supplies. Research at the time showed not only that turbidity was an aesthetic problem but also that particles harbored pathogens. In 1908, chlorine was first used as a primary disinfectant in the Jersey City (New Jersey) drinking water plant.

The government first stepped in in 1914, when the U.S. Public Health Service set standards for the bacteriological quality of drinking water. These standards were subsequently revised in 1925, 1946, and 1962. In 1962, 28 substances were regulated in drinking water. The first major legislation was the River and Harbor Act of 1886, which was basically an act to protect waterways from substances or material that would interfere with shipping. The Federal Water Pollution Control Act (FWPCA) of 1948 was enacted to "enhance the quality and value of our water resources and to establish a national policy for the prevention, control, and abatement of water pollution" (www.epa.gov/history/topics/fwpca/05.htm). The Water Pollution Control Act of 1956 strengthened enforcement provisions of the 1948 Act, and the federal role in regulating states was further expanded in 1965 with the enactment of interstate water quality standards. The Clean Water Restoration Act of 1966 imposed fines on a polluter who failed to submit a required report. Additional acts and amendments on water quality include the FWPCA Amendments of 1972, the Safe Drinking Water Act of 1974, Clean Water Act Amendments of 1977, and the Safe Drinking Water Act Amendments of 1986 and 1996. As you can see, there has been considerable legislative activity with respect to water, which emphasizes the importance of water to human life and stable civilization.

So what have these Acts actually done to improve water quality? They defined sources of pollutants in two ways. Point sources are broadly defined to include any "discernible, confined, and discrete conveyance" including any pipe, ditch, channel, tunnel, conduit, well, or container from which pollutants are or may be discharged. Point sources are usually subdivided into industrial and municipal sources. Nonpoint sources include a range of inputs to surface waters, including runoff from agricultural fields, feedlots, paved streets and parking areas, mining sites, forestry operations, and atmospheric deposition. Pollutants or contaminants of concern specified in these Acts include (1) conventional pollutants such as human wastes, food from sink disposals, laundry and bath waters, and waters containing fecal coliforms and oil and grease, (2) toxic pollutants such as organics (pesticides, solvents, polychlorinated biphenyls, and dioxins) and metals, and (3) nonconventional pollutants such as nutrients containing nitrogen and phosphorus. The Acts also established a list of water quality criteria that originally included 115 "priority pollutants" and set maximum contaminant levels (MCLs) for particular pollutants or contaminants in drinking water. Thus, these Acts gave EPA the power to (1) identify pollutants or contaminants and (2) determine the maximum contaminant level.

One of the most important elements of these Acts established the National Pollutant Discharge Elimination System (NPDES), which stated that all individual, government, and industrial point sources must have a permit to discharge waste into any waterway. The permit states what pollutant can be emitted and at what level (mass loading) the pollutant can be released. Violations of the permit can result in fines and imprisonment.

Government funding to build sewage and water treatment plants for municipalities was provided under the FWPCA of 1948, the Clean Water Act Amendments of 1956, and additional funding in 1961 and 1971. The construction and operation of these plants were instrumental in cleaning up sewage and associated bacteria from our waterways.

A chronological account of the evolution of recent water quality legislation follows.

1. The original Safe Drinking Water Act of 1974
    (a) federally mandated the regulation of contaminants that pose a health risk to the public based on frequency of the chemical and contaminant nature,
    (b) established a maximum contaminant level goal (MCLG), which is the limit below which there is no known or expected threat to public health,
    (c) designated that EPA would specify a maximum contaminant level (MCL) that is as close to the MCLG as deemed feasible through technology and cost effectiveness, and
    (d) provided that, in the event that no MCL can be determined, the EPA set a required treatment technique that specifies a way to treat the water to remove contaminants.

2. In 1979, the presence and concentrations of six synthetic organic chemicals, 10 inorganic chemicals, turbidity, total coliform bacteria, radium-226, radium-228, and total trihalomethanes were added to the control of EPA. Also added was the provision that in the event of a violation, the established system (state or federal government) must notify the public.

3. The Amendments of 1986
    (a) established regulations for 83 contaminants,
    (b) required disinfection of all public water supplies,
    (c) specified filtration requirements for nearly all water systems that draw water from surface sources,

    (d) developed additional requirements to protect groundwater supplies (Well-head Protection programs and Sole Source Aquifer Program)

    (e) established monitoring requirements on nonregulated contaminants every five years so the EPA could decide if these should be regulated in the future,

    (f) implemented a new ban on lead-based solder, pipe and flux materials in water distribution systems, and

    (g) specified the "best available technology" for the major drinking water contaminant groups (pathogens, organic and inorganic contaminants, and disinfectant by-products).

4. The Amendments of 1996

    (a) required consumer confidence reports,

    (b) required source water assessments,

    (c) specified State capacity development strategies,

    (d) gave water plant certification revisions,

    (e) required public notification improvements,

    (f) provided new publicly accessible drinking water contaminant databases,

    (g) required annual compliance reporting,

    (h) required health care provider outreach and education, and

    (i) developed the Drinking Water State Revolving Fund (DWSRF) for treatment facility upgrades.

To summarize, the Clean Water Acts (1) gave EPA the authority to implement pollution control programs such as the setting of water quality standards, (2) made it unlawful for any person to discharge any pollutant from a point source into navigable waters without a permit, (3) funded the construction of water and sewage treatment plants, and (4) recognized the need for planning to address the critical problems posed by non-point-source pollution.

11.3.7 The Toxic Substances Control Act

The Toxic Substances Control Act was signed by President Carter in 1977 and gave the EPA the authority to track and regulate the more than 75,000 industrial chemicals produced or imported into the United States. TSCA authorized the EPA to obtain information on all existing and newly created chemical substances, including chemicals that were determined to cause a risk to public health or the environment. TSCA was considerably different from earlier laws; previously the EPA only had the authority to intervene after damage became evident. TSCA gave the EPA the authority to regulate before a chemical was commercially manufactured.

In controlling and testing chemicals, TSCA gave the EPA certain powers. These include requiring a "premanufacture notice," so that chemical manufactures had to apply to EPA to commercially produce a substance. Prior to the manufacture of a chemical, the producer or importer had to conduct testing as defined by the EPA to ensure the safety of the chemical. Based on the results of the testing, the EPA had the right to ban the chemical from being manufactured or imported if it imposed an unreasonable risk to human health or the environment. The EPA was also given authority to act in cases of imminent health hazards and to seize a chemical substance, mixture, or article containing a specified chemical. TSCA also required the EPA to create the Toxic Substances Release Inventory (TSRL), which contains information on the safety and toxicity of specific chemicals and how much of each compound can safely be released to the environment.

TSCA was divided into four titles:

Title 1: Control of Toxic Substances. This included provisions for chemical substances and mixtures; manufacturing and processing notices; regulating hazardous chemicals, substances and mixtures; managing imminent hazards; and reporting and retaining information.

Title II: Asbestos Hazard Emergency Response (enacted by Congress in 1986). This authorized EPA to amend its TSCA regulations to impose more requirements on asbestos abatement in schools. It also required EPA to determine the extent of the danger to human health posed by asbestos in public and commercial buildings.

Title III: Indoor Radon Abatement. This was added in 1988 to respond to the human health threat posed by exposure to radon. It required EPA to publish an updated citizen's guide to radon health risk and to perform studies on the radon levels in schools and federal buildings.

Title IV: Lead Exposure Reduction. This was added in 1992 in an attempt to reduce environmental lead contamination and prevent adverse health effects as a result of lead exposure, particularly to children, from lead-based paint hazards and lead contamination in paint and toys.

11.3.8 The Comprehensive Environmental Response, Compensation, and Liability Act

The Resource Conservation and Recovery Act (RCRA) and its amendments established a much needed and extensive program to manage newly created hazardous waste, and it also set out rules for its proper disposal. However, RCRA did little to nothing for the hundreds to thousands of abandoned waste sites from previous poor and irresponsible disposal practices. Examples of these include Love Canal, Times Beach, and the Valley of the Drums, as well as the numerous contaminated sites resulting from the improper disposal of hazardous wastes generated during the Cold War by the U.S. government. The need to clean up these sites led Congress to pass the Comprehensive Environmental Response, Compensation, and Liability Act (CERCLA) of 1980, which was amended in 1986 by the Superfund Amendments and Reauthorization Act (SARA). Collectively, the resulting legislation is referred to as "Superfund." Again, the primary purpose of Superfund was to clean up unsafe and in many cases abandoned hazardous waste sites.

We will start by looking at the differences between CERCLA and SARA. CERCLA established a $1.6 billion fund, derived mostly from feedstock taxes on the chemical industry, to clean up the waste sites over a five-year period. We now see, over 20 years later, that this was a naive and optimistic effort, since there are hundreds of sites yet to be cleaned up. CERCLA also established prohibitions and requirements concerning closed and abandoned hazardous waste sites, provided for liability of persons responsible for the releases of hazardous waste at these sites, and established a trust fund to provide for cleanup when no responsible party could be identified. CERCLA also authorized two types of responses: (a) short-term removals for sites requiring prompt response and (b) long-term remedial actions to reduce dangers that are serious but are not immediately life-threatening.

SARA made several important changes and additions to the program. SARA (1) stressed the importance of permanent remedies, (2) required Superfund actions to consider the standards and requirements found in other state and federal environmental laws and regulations, (3) provided new enforcement authorities and settlement tools, (4) increased state involvement in Superfund, (5) increased the focus on human health, (6) revised the Hazardous Ranking System (HRS) to ensure that it accurately assessed the relative degree of risk to human health and to the environment, (7) encouraged citizen participation in deciding how best to deal with a site, (8) created the Emergency Planning and Community Right-to-Know Act (EPCRA) to give the public access to information concerning a site, and (9) increased the size of the cleanup fund from $1.5 billion to $8.5 billion.

So, how does the Superfund process work? This is outlined in Figure 11.2. First, a site is identified in any of a number of ways and evaluated using criteria specified in the Hazard Ranking System (HRS), including proximity to a large population, nature of the contaminants, and potential exposure pathways. The result is an HRS score related to the relative risk, and if the score exceeds a threshold value, the site is placed on the National Priorities List (NPL). The site then undergoes a remedial investigation (RI) that describes the history of the contamination, the current concentration of pollutant in the air, water, soil, and any waste, and a detailed risk assessment. Next a feasibility study (FS) is conducted to evaluate remediation (cleanup) options for the site. The FS contains a variety of remediation options, from no action (in which the site is eliminated from public access but not actually remediated) to extensive remediation. A risk assessment is conducted for each remediation approach. Nine criteria are used to evaluate each remediation option: (1) short-term effectiveness, (2) long-term effectiveness, (3) implementability, (4) reduction in toxicity, mobility, and volume, (5) cost-effectiveness, (6) compliance with established standards, (7) human health protection, (8) state concurrence, and (9) local acceptance. In the end, a record of decision (ROD) is agreed upon and issued by EPA, specifying the selected remedy and time scale for the remediation.

One major aspect of Superfund is that it created a method of determining liability, and thus determining who should pay for the remediation efforts. The Superfund process establishes potentially responsible parties (PRPs) who can be held accountable for these costs, which follows the "cradle-to-grave" concept created under RCRA. Any party who has generated the waste, owned a facility associated with the waste, stored the waste, transported the waste, or disposed of the waste can be held accountable for remediation costs. It is well known that government proj-

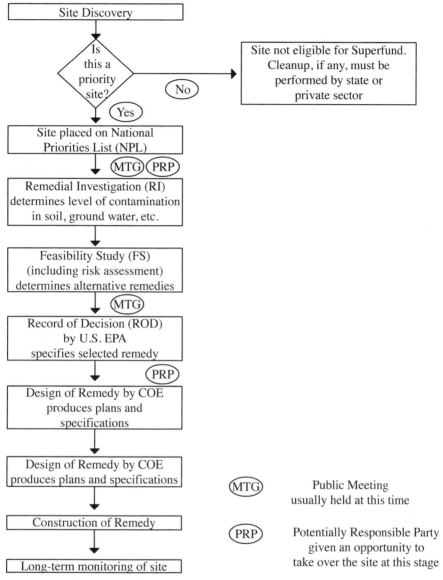

Figure 11.2.   An outline of the Superfund process.

ects are usually more costly than a private approach to the same problem, so EPA will allow a PRP to conduct the remediation effort, but under the strict oversight of EPA or the state where the site is located.

At the time of this writing, Superfund has completed work at 40 sites across the country, for a total of 886 cleanups or 59% of the sites listed on the NPL. It has conducted 699 ongoing cleanups at 436 projects, but the NPL increases in number by about 10 sites per year. Controls are now in place at 82% of the NPL sites to prevent unacceptable human exposures.

11.3.9 The Oil Pollution Act

On March 24, 1989, the *Exxon Valdez* struck the Blight Reef in Prince William Sound in Alaska, releasing 11 million gallons of crude oil. This accident sparked public and legislative action to attempt to prevent future accidents and establish strict liability for such accidents. The Oil Prevention Act of 1990 was Congress's response. This Act (1) strengthened EPA's ability to prevent and respond to future catastrophic oil spills, (2) established a trust fund, financed by a tax on oil, to clean up future spills when the responsible party is incapable or unwilling to do so, (3) required oil storage facilities and shipping vessels to submit to the federal government plans detailing how they will respond to large oil discharges, and (4) required the development of Area Contingency Plans to prepare and plan for oil spill response on a regional scale.

Other major provisions of the Act include

(1) the establishment of legally responsible parties for the vessel or facility from which oil is discharged and determination of liability for specified damages resulting from the discharged oil and removal costs and

(2) the establishment of new liability determination procedures by

(a) increasing the liability for tank vessels larger than 3000 tons to $1200 per gross ton or $10 million, whichever is greater,

(b) making responsible parties at onshore facilities and deepwater ports liable for up to $350 million per spill, and

(c) making holders of leases or permits for offshore facilities liable for up to $75 million per spill, plus removal costs.

11.3.10 The Pollution Prevention Act

Much as the National Environmental Policy Act was designed to change the way we think in approaching projects with environmental impacts, the Pollution Prevention Act (PPA) of 1990 was designed to promote a change in attitude among polluters and reduce pollutant volumes through voluntary source reduction and recycling. The PPA focused on industry and government in laying out a plan to reduce the amount of pollution through cost-effective changes in production, operation, and raw material use. The act was designed to (1) reduce the amount of hazardous substance, pollutant, or contaminant entering any waste stream or released into the environment prior to recycling, treatment, and disposal and (2) reduce the hazards to public health and the environment related to the release of these chemicals.

Note the focus on reduction of pollution prior to treatment or disposal. This distinct focus is in contrast to the more common focus on compliance and proper treatment and disposal, which often failed to reduce the amount of pollution generated. The PPA focused on source reduction, increasing the efficiency of use of energy, water, or other natural resources, and protecting our resource base through conservation. To achieve these goals, the PPA fosters pollution prevention through the collection and dissemination of data and technical assistance. These reductions are tracked by requiring the owner or operator of a facility that uses chemicals or produces pollution to file a toxic source reduction and recycling report. The regulations do not impose additional regulator obligations on the chemical industry other than increased reporting of their efforts to reduce the creation of pollution. Thus, the act is essentially designed to promote attitude change.

11.3.11 The Endangered Species Act of 1966 and Amendments

The Endangered Species Act (ESA) and its Amendments provide a program for the conservation of threatened and endangered plants and animals and the habitats where they live. Species of interest include birds, insects, fish, reptiles, mammals, crustaceans, flowers, grasses, and trees. The U.S. Fish and Wildlife Service and the Department of the Interior maintain the Threatened and Endangered Species System (TESS), which includes 632 endangered species (326 plants) and 190 threatened species (78 plants). These include well-known species such as the bald eagle and the grizzly bear.

While this Act is not directly involved with EPA, the EPA has historically been required to consult with the Fish and Wildlife Service and the Department of Interior when reviewing a pesticide or chemical, in order to evaluate its adverse effects on endangered species as well as the environmental fate of the chemical. Under FIFRA, the EPA could issue emergency suspension of certain pesticides to cancel or restrict the chemical's use if an endangered species will be adversely affected. This all changed in August 2004, when the Bush Administration developed new rules for EPA to approve pesticides without consulting the federal wildlife agencies about the potential harm to endangered species (Hileman, 2004). Thus, while the Endangered Species Act is an important environmental law, it no longer affects chemical regulations. This diminishes the Act's holistic, habitat-based approach to management of endangered species.

11.3.12 Marine Protection, Research, and Sanctuaries Act (MPRSA) of 1972

This act is commonly known as the Ocean Dumping Act. The purpose of this law is to regulate the dumping of all types of materials into ocean water and to prevent or limit the dumping of any material that would adversely affect (a) human health or welfare or (b) the marine environment. The Act specifically prohibits (1) transportation of material from the United States for the purpose of ocean dumping, (2) transportation of material from anywhere for the purpose of ocean dumping by U.S. agencies of U.S. flagged vessels, and (3) the dumping of material transported from outside the United States into the U.S. territorial zone (12 nautical miles). Penalties from breaking laws established under the Act include (1) a civil penalty of up to $50,000 against persons in violation of MPRSA or a permit, (2) a fine of up to $125,000 for the illegal dumping of medical waste, and (3) criminal penalties including a fine and up to five years in jail for knowingly violating MPRSA, its regulation, or an MPRSA permit. In addition, each day of violation is considered as a separate offense. The MPRSA is enforced by the EPA working with the U.S. Coast Guard to monitor and conduct surveillance operations on shipping vessels.

11.4 EPA's RECORD

So, how has the EPA done in controlling or reducing environmental pollution since its creation in 1970? There are mixed views on the answer to this question, but most of the comments are positive. A list of major environmental events is given in Table 11.5. Don't just casually dismiss this table, but look closely at each year and highlight events that you remember, that you have heard of from the past, or that directly affect you or someone in your family. Although a bit dated now, two feature articles that evaluate the EPA's performance are "Reform or Reaction: EPA at a Crossroads," in *Environmental Science & Technology* (Andrews, 1995) and a series of short articles in *Chemical and Engineering News* (Ember, 1995a; Ember, 1995b; Kirschner, 1995; Hileman; 1995; Rawls, 1995; Lepowski, 1995).

There are basically three important groups that judge EPA: the regulated industries, the public affected by the pollution, and the scientists that monitor and study the effects of the pollution. Almost every member of these groups note the following undisputable accomplishments by EPA (Hileman, 1995):

- Sewage, oil, grease, and visible industrial pollutants have largely disappeared from rivers, streams, and lakes in the United States.
- The number of fishable and swimmable rivers has doubled since the passage and initial enforcement of the Clean Water Act.
- Urban air quality in most cities has dramatically improved, but more work needs to be done in this area.
- Our efforts toward controlling ozone-depleting chemicals are unprecedented.
- The enforcement of the Toxic Release Inventory and the Community-Right-to-Know provision of the Superfund Amendment and Reauthorization Act of 1986 have greatly improved the safety and knowledge, and decreased the toxin exposure, of the general public.
- Coastal cities no longer simply dump garbage and sewage into the oceans.
- Domestic and hazardous waste landfills are now lined to inhibit leaching into surrounding soil and groundwater.
- Raptors, or birds of prey, have made a dramatic comeback since the banning of DDT and other chlorinated pesticides.

These are major accomplishments by any measure, and most of the world is still catching up to our environmental living standards, while a few countries are admittedly slightly ahead (most of Europe). To paraphrase Boris Yeltsin, if Russia had the environmental problems of the United States, it would have no problems. The United States is currently attempting to address the environmental disasters that resulted from our side of the Cold War, but Russia has not even begun to address this problem. Most of our Cold War pollution was in rural, very isolated portions of the

TABLE 11.5.   The Timeline of EPA

U.S. President and EPA Administrator	
Richard Nixon *William D. Ruchelshaus*	**1970** • December 2, U.S. EPA forms • Clean Air Act enacted to reduce auto emissions, to require states to create plans to improve air quality, and to set national air quality standards **1971** • EPA works with the Department of Housing and Urban Development to treat the effects of and ban the use of lead-based paints • EPA defines air pollution levels **1972** • EPA bans the pesticide DDT due to its carcinogenic effects and its accumulation in the food chain • EPA works to build a network of sewage treatment facilities, with ultimate goal that by 1988, every U.S. city will have a safe sewage treatment program and plan • United States and Canada's International Great Lakes Water Quality Agreement • Federal Environmental Pesticide Control Act passed • United States and U.S.S.R. sign an environmental cooperation treaty • Federal Water Pollution Control Act Amendments passed • Ocean Dumping act passed
Russell E. Train	**1973** • Safe Drinking Water Act mandating standards to control ground waste injection **1975** • Automobile manufacturers install catalytic converters in new cars in order to satisfy EPA emission standards to reduce toxic compounds • The United Nations designates the EPA as an International Data Center
Jimmy Carter *Douglass M. Costle*	**1976** • EPA radiation monitoring van inspects cities • EPA provides jobs for recipients of welfare • The Toxic Substances Control Act is passed regulating PCBs and other harmful chemicals • Resource Conservations and Recovery Act (RCRA) passed and enacted, attacking midnight dumpers and introducing the cradle-to-grave policy **1977** • EPA sets the first national industrial water pollution standard • Clean Air Act Amendments passed • Clean Water Act passed

(Continued)

TABLE 11.5. The Timeline of EPA (continued)

U.S. President and EPA Administrator	
	1978 • EPA sets new standards for airborne lead • EPA bans the manufacturing of Heptachlor and Chlorodane • EPA bans aerosol fluorocarbons to protect the ozone layer • Great Lakes Water Quality Agreement passes **1979** • EPA bans the manufacturing of PCBs • EPA begins Hazardous Waste Enforcement/Emergency Response System • EPA bans two herbicides containing dioxins
Ronald Reagan (EPA budget sharply reduced)	**1980** • Comprehensive Environmental Response Compensations and Liability Act (CERCLA, also known as Superfund) passed, mandating $1.6 billion to set new standards and repair old dumping sites • EPA supervises Three Mile Island cleanup • EPA joins forces with New York State to relocate the residents of Love Canal
Ann M. Gorsuch (scandals ensue) *Ruckelshaus returns*	**1981** • Superfund finances the clean up of the Valley of Drums and names 114 other priority sites for Superfund cleanup **1982** • EPA announcement of National Contingency Plan for Superfund cleanups • Asbestos School Hazard Abatement Act passed • Superfund finances cleanup of Love Canal **1983** • EPA relocates Times Beach residents • EPA mandates cleanup of DDT in Triana, Alabama • EPA begins "Fishbowl Policy" • EPA bans EDB **1984** • EPA signs Chesapeake Agreement along with Department of Defense • Hazardous and Solid Waste Amendments of 1984 pass • Amendments made to Superfund policy that hazardous waste be treated before underground storage. Additionally, "hammer provisions" apply if EPA should fail to achieve fulfill requirements
Lee M. Thomas	**1985** • EPA responds to Bhopal toxic chemical spill in India • EPA sets new restrictions on lead in gasoline • EPA approves the first use of gene-altered bacteria

(Continued)

TABLE 11.5.   The Timeline of EPA (continued)

U.S. President and EPA Administrator	
	1986
	• EPA responds to the Chernobyl, U.S.S.R. nuclear power plant explosion
	• Asbestos Hazard Emergency Response Act passes
	• Superfund Amendments and Reauthorization, increasing Superfund budget, creating technology to expedite cleanups, and updating community emergency plans and corporate toxic release inventory reports.
	• Safe Drinking Water Act Amendments passes, granting EPA with more authority over water suppliers
	• EPA addresses concern about the explosion and leakage of toxic substances and works toward the passage of the first public right-to-know law, whereby those who manufacture and store toxic substances are required to keep detailed documents about the substances, and to make those documents available to the public, through the EPA
	1987
	• Clean Water Act passes, despite Reagan's veto
	• The United States becomes one of the 24 countries to sign the Montreal Protocol to phase out the manufacture of chlorofluorocarbons (CFCs), a component of aerosols and refrigerants, that depletes the ozone.
	• EPA calls for sanctions against states that fail to meet air standards
	• Hazardous chemical reporting rule passes
	• EPA authorizes thermal destruction of dioxin at Love Canal
George Bush, Sr.	**1988**
	• Federal Insecticide, Fungicide, and Rodent Act (FIFRA) amendments pass, which shifts much of the financial responsibility to the manufactures, rather than the EPA
	• -Ocean Dumping Ban Act passes banning sewage sludge and industrial waste
William K. Reilly	**1989**
	• EPA and the Department of Transportation respond to the Exxon Valdez accident
	• EPA tracks medical waste
	• EPA publishes the first Toxic Release Inventory
	1990
	• Clean Air Act Amendments of 1990 passes
	• Pollution Prevention Act passes
	• EPA restricts land disposal of hazardous waste
	• EPA Science Advisory Board recommends a strategy that EPA consider environmental risks more seriously when making decisions

(Continued)

TABLE 11.5. The Timeline of EPA (continued)

U.S. President and EPA Administrator	
	1991
	• Exxon Valdez and Shipping agree to pay the largest environmental criminal damage settlement in history
	• Establishment of voluntary toxics reduction program
	• EPA commits to environmental education
	• Signing of the federal recycling order mandating that federal agencies use recycled products where possible
Bill Clinton *Carol Browner*	**1992** • EPA issues final drinking water standards for 23 chemicals
	• U.N. Earth Summit promotes sustainable development
	1993
	• EPA identifies passive smoke as a human carcinogen
	• Sulfur dioxide trading rule passes
	• Federal facilities ordered to reduce toxic emissions
	• EPA requires complete phase-out of CFCs and other ozone depleters
	1994
	• EPA issues the first citizen right-to-know list of toxins
	• Federal environmental justice order signed, making federal departments and agencies consider environmental justice implications before action
	• Chemical industry air toxics reduction rule passes
	• EPA's Cabinet bill blocked on the House floor because of, among other factors, support for risk–cost–benefit analysis
	• Superfund cleanups remarkably accelerated
	• EPA launches gain to help revitalize 50 communities' inner-city brownfields—the abandoned, contaminated sites of former industries—resulting in both economic and environmental gain
	• Clinton Administration almost doubles the list of toxic chemicals that must be reported under the community right-to-know laws
	1995
	• U.S. commits to monitoring environment using remote-sensing data
	• EPA issues new requirements for municipal incinerators to reduce toxic emissions by 90%
	• EPA grows as an agency to include 18,000 employees, two-thirds of which are employed across 10 regional offices
	1996
	• Safe Drinking Water Act Amendments pass
	• EPA finalizes ban on leaded gasoline
	• EPA enacts lead-based paint right-to-know policy
	1997
	• EPA commits to children's health, regulatory reinvention, and right-to-know

(Continued)

TABLE 11.5.   The Timeline of EPA (continued)

U.S. President and EPA Administrator	
	• EPA implements Food Quality Protection Act
	• U.S. and Canada move to eliminate toxics in the Great Lakes
	• EPA website provides access to watershed data
	1998
	• EPA common sense initiative receives Hammer Award
	• Federal Clean Water Action Plan issued
	1999
	• EPA plans cleaner cars and fuel standards
	• EPA and Department of Energy present the first Energy Star building awards
	• Superfund reform accelerates the cleanup of hazardous waste
	• EPA demonstrates Clean Air benefits far outweigh its costs
George Bush, Jr	**2000**
	• EPA endorses cleaner diesel fuels plan
	• EPA bans most uses of Dursban
	• EPA proposes the cleanup of PCBs in the Hudson River
Christie Whitman	**2001**
	• EPA responds to the September 11 attacks, involving the deployment of vacuum cleaner trucks
	• EPA and the White House honor young environmentalists
	• EPA officially suspends stricter limits on arsenic in drinking water
	• Under public pressure, EPA adopts higher standards for arsenic in drinking water
	• United States signs Convention on Persistent Organic Pollutants
	2002
	• EPA awards Brownfields grants to assess the contamination of abandoned properties
	• Study shows a big drop in enforcement of environmental laws
	• Environmental education funding reduced
	• Bush administration delays enforcement of mercury and sulfur dioxide emissions for 10 years
	• Proposed Bush budget asks taxpayers to fund Superfund instead of polluters
	• Top EPA official resigns in protest to White House efforts to weaken environmental rules
	• EPA exempts large category of power plants from lawsuits for Clean Air Act violations
	• EPA issues Strategic Plan for Homeland Security
	• EPA and agriculture work together to improve America's water
	• Bush administration cuts funding for toxic cleanups to half of that requested by EPA
	• Another top EPA official resigns in protest
	• EPA deletes global-warming section of major pollution report
	• EPA water-quality report shows that U.S. water is getting dirtier

(Continued)

TABLE 11.5.   The Timeline of EPA (continued)

U.S. President and EPA Administrator	
	• Report shows Superfund cleanups drop to 42% per year from an average of 76% under the Clinton Administration
	• A report shows that polluters paid 64% less in fines under the Bush Administration than in the last two Clinton years
Michael O. Leavitt	**2003**
	• The Bush Administration and the EPA exempts thousands of older power plants, refineries, and factories from having to install costly clear air controls when they modernize
	• National Academy of Sciences panel criticizes U.S. global-warming plan
	• EPA allows sludge dumping in the Potomac River to continue to an additional seven years
	• EPA report notes that toxic cleanups still lag by 41%
	• EPA rules ignore mercury pollution from chlorine plants
	• Bush signs "Healthy Forests" bill, criticized by environmental as allowing more logging and less species protection on federal lands
	2004
	• EPA is no longer required to consult with Federal wildlife agencies on pesticide approval that could potentially harm endangered plants and animals
	• EPA Office of Regulatory Enforcement report shows a 76% decline in civil lawsuits filed by the federal government fro violations of environmental laws

Sources: Andrews, 1995; www.epa.gov/history/timeline.htm; *Sierra Magazine*, 2004.

western United States. Russia lived (and still lives) in the middle of their Cold War pollution.

Many environmental leaders and organizations have also made negative statements about the EPA. The EPA has done a relatively good job at addressing visible, higher pollutants, but has done a considerably poorer job at addressing "invisible toxics" that are persistent, are found at very low concentrations, and bioaccumulate (with the obvious exceptions of PCBs and DDT). A new and highly aggressive area of research is the ubiquitous presence of many endocrine disrupting compounds, including human-made estrogens, plasticizers, and pharmacecticals that have been found in many, if not all, of our waterways. Another major criticism is that the EPA has done well at regulating large Fortune 500-type industries while largely ignoring the hundreds of thousands of small businesses and municipalities that contribute significantly to stream and air pollution.

Many also criticize the handling of Superfund sites and the slow and costly cleanup of these sites. Much of the problem with cleaning up these sites is of course associated with assigning legal blame for who is responsible for the abandoned polluted site. However, considerable legal maneuvering is also involved in determining *how* the site should be cleaned and *to what level* the site should be cleaned. The Superfund problem is an issue that will take decades to settle. One last concern is the shift in environmental policy from equal "protection of the environment and human health " to a human risk-based approach to regulation and cleanup that has the potential to ignore or lessen the emphasis on protection of the environment where risk cannot be quantified. Still, we must give the EPA credit; the large issues have been addressed, and the public and the EPA are moving on to the smaller, but still significant, issues facing us today.

So, what does industry think of the EPA and its regulations? Most industry leaders truly want a strong, consistent, and effective regulatory agency such as the EPA (Baum, 1995), not only on the national level but also on the global level. A consistent set of rules and controls on pollution, across the world, would provide a level playing field on which to base industry. The export of jobs to developing countries is not only due to the low wages that employers can pay workers in these countries but also do to lax or nonexistent environmental regulations. Consistent regulations also help ensure that if an industry uses a chemical today and disposes of it legally, the industry will not be responsible for the chemical in the future.

The pollution that occurs in the United States today is mostly due to regulated sources, unlike the "midnight dumping" that occurred prior to the strong regulation by the EPA and other Federal agencies. The 12 major environmental laws discussed in the previous section are largely responsible for this. But we still have violations. A recent study by the EPA and the Chemical Manufacturers Association (CMA) found that *noncriminal violations* largely occur due to a failure to understand the regulations or human error, and not intentional releases of chemicals. The most common violations include (McCoy, 1999): (1) facilities being unaware of the applicability of specific regulations, (2) human error in judgment or responsibility, (3) failure to follow procedures, (4) faulty equipment design or installation, (5) problems with compliance by contractors, and (6) various communication difficulties. But, as the old saying goes, "ignorance of the law [or not operating your equipment (factory) correctly] is no excuse, and the violators are usually fined or closed down.

Still, clearly, pollution must exist in a regulated environment. Not all pollution can be eliminated, given our population density and the relatively limited carrying capacity of the land we live on and the air we breathe. In today's world, if you breathe air, drink water, walk on soil, or live in a modern dwelling, you will incorporate some form of pollution into your body. It's simply a fact of modern life; our goal in enacting legislation is to limit this intake to an acceptable, below-threshold level.

Before we move on to the major U.S. international environmental laws, one point should be made. Recent presidential candidates have expressed the need for a cabinet-level environmental post but have conveniently forgotten this interest when elected to office. One would think that the quality of the air we breathe, the water we drink, and the soil we live on would rank as important as the economy and defense. But, sadly, no direct voice to the President has taken shape. The motivation for change often comes from disasters; this is true at local intersections in need of traffic lights or stop signs, and it is also true at the national level, where change is slow even as thousands of Americans die each year from pollution-related illnesses (asthma, cancer, etc.). Unfortunately, additional and probably significant environmental degradation will have to occur before our elected officials will take the dramatic actions ultimately necessary to truly protect the environment in the long-term.

11.5 INTERNATIONAL AGREEMENTS/TREATIES INVOLVING THE UNITED STATES

Two acts that we have already discussed have international implications: the Endangered Species Act, which protects migratory species such as birds and ocean species, and the Marine Protection, Research, and Sanctuaries Act, which regulates ocean dumping (although focusing on U.S. waters). Many other explicitly international environmental agreements have been made during the past century.

11.5.1 U.S.–Canada Environmental Agreements

One of the first international agreements that the United States made was with Canada. The 1909 Boundary Waters Treaty states the rights and restrictions of each nation in using boundary waters (LeMarquand, 1993, p. 67). It declares that each country's government maintains the exclusive jurisdiction and control over the use and diversion of all waters on its side of the boundary line. The Treaty also addresses the issue of pollution, citing that "boundary waters and waters flowing across the boundary shall not be polluted on either to the injury of health or property on the other" (LeMarquand, 1993, p. 67). Most significantly, the Boundary Waters Treaty created the International Joint Commission (IJC), which is composed of half Canadian and half U.S. advisory boards that prevent and resolves disputes between the United States and Canada under the 1909 Boundary Waters Treaty.

In the 1960s and 1970s, as the Great Lakes shoreline became the site of more and more industrial and residential development, the IJC found itself unable to handle the large volume of complaints related to pollution and contamination of the Great Lakes, especially dangerously high levels of phosphorous. In response, the governments of Canada and the United States created the Great Lakes Water Quality Agreement of 1972, which was later amended in 1978 and 1987. The stated goal of these agreements was to restore and maintain the chemical, physical, and biological integrity of the waters of the Great Lakes Basin Ecosystem, though they were largely aimed at removing toxic substances from the waters. The agreements also created the Great Lakes Water Quality Board, which is responsible for advising the IJC on all environmental issues involving the Great Lakes.

In the past 15 years, a number of national and bi-national programs and follow-up agreements have emerged to enable Environment Canada (EC) and the U.S. EPA to work toward elimination of toxic substances from the lakes, as required by the 1987 amendment. These additions were necessary because researchers still found dangerously high levels of PCBs, mercury, toxaphene, and other pollutants in the lakes. Among the programs is the 1997 Great Lakes Binational Toxics Strategy (GLBTS), which established a four-step process for the EC and EPA to work toward elimination of toxic substances in the Great Lakes. The GLBTS also defined Level I Substances (of the highest concern), including benzo(a)pyrene, chlordane, DDT, hexachlorobenzene, mercury and mercury compounds, PCBs, PCDD, and toxaphene.

Though water is arguably the most complex aspect of Canada–U.S. environmental relations, air quality and acid rain are also crucial concerns for each country. Canada struggled to convince the United States to establish emission regulations until the 1990s. The U.S. Department of Energy in 1976 began a research program on Long-Range Transport of Airborne Pollutants and two years later joined the Canada–U.S. Research Consultation Group on transport of these pollutants. In 1980 both governments signed a Memorandum of Intent to cooperate in fighting acid rain (Doern, 1994, p. 150). Yet, this agreement collapsed two years later when the Canadian government confronted the Reagan administration with a study attributing the pollution of the eastern Canadian lakes to sulfur and nitrogen oxide emissions from Midwestern U.S. industry. The United States declared that more research was necessary before regulations could be established (Harris, 2001, pp. 9–10).

Throughout the 1980s, the United States rejected repeated Canadian requests to limit emissions harmful to Canadian environmental quality. Meanwhile, in 1984 Canada and nine European countries vowed to reduce their sulfur dioxide emissions by 30% in the next 10 years. The United States refused to join what became known as the "Thirty Percent Club." In 1986, the U.S. Court of Appeals overturned a lower court ruling that would have compelled seven states to reduce acid rain-causing emissions, while a year later an agreement was formed among six Canadian provinces to cut sulfate emissions in half. It is interesting to note that the United States did sign an agreement with Mexico in 1987 concerning air quality and transboundary pollution; however, this followed the publication of a report that stated that Mexico's pollution emissions were causing environmental damage in the United States. The Bush (senior) administration finally gave into pressure from environmentalists and the Canadian government and passed the new Clean Air Act in 1990, which included a provision to reduce sulfur dioxide emissions and acid rain-causing pollution (Doern and Esty, 2002, pp. 150–151).

Another U.S.–Canada agreement addressed ground-level ozone and the Antarctic ozone hole. The 2000 Protocol amended the 1991 Air Quality Agreement to include provisions for the reduction of ground-level ozone, following studies by Health Canada showing that ground-level ozone is a major contributor to premature pollution-related ozone and that emissions from the U.S. Midwest and Eastern seaboard were largely responsible for high ozone concentrations. The Protocol designates Pollution Emission Management Areas (PEMAs), sets vehicle emission standards, and requires that both countries cut nitrogen oxide emissions in their respective PEMAs.

11.5.2 Multinational Agreements

Obviously, pollutant transport is not limited to states, regions, countries, or even continents. Only by joining forces and working toward common environmental goals can countries achieve broad environmental goals. This is especially true for issues of atmospheric pollution, which affects a globally shared resource. More and more international treaties like the ones discussed are needed. Citizens of all countries must force government leaders to enter well-planned international agreements.

The Montreal Protocol to Limit Ozone-Depleting Chemicals. In 1985, a team of British researchers discovered unusually low levels of ozone above Halley's Bay in Antarctica, representing a "hole" in the stratosphere that consists of nine million square miles with little to no ozone protection (Makhijami, 1995, pp. 30–31). Continuing studies have shown that his hole appears every spring over Antarctica; there are reports of thinning ozone in other parts of the world as well. This information, along with the framework already established from the Vienna Convention, led to the Montreal Protocol on Substances that Deplete the Ozone Layer (http://sedac.ciesin.columbia.edu/entri/). This protocol established measures by which controlled substances, such as CFCs, would be phased out in developed nations. Those countries that signed the protocol were under obligation to follow the guidelines to help reduce destruction of the ozone layer. Twenty-five countries signed the agreement on September 16, 1987 (Newton, 1995, p. 102). By 1999, all countries that signed the Protocol had agreed to cut CFC production by 50% (Newton, 1995, p. 16).

The United States, the world's largest producer and consumer of ozone-depleting chemicals, acted very aggressively. The Amendments to the Clean Air Act of 1990 allowed government agencies to ban the use of many ozone-depleting substances. As a result, the EPA ruled that after February 16, 1993, certain products using CFCs would be banned. The EPA limited the sale of certain types of material only to essential products, such as medical sprays, after January 17, 1994, and it planned to completely ban the production and importation of methyl bromide. However, recently the EPA has delayed its phase-out of certain agricultural ozone-depleting chemicals (*Environmental Science and Technology*, 2004a).

The Kyoto Protocol for Limitation of Greenhouse Gas Emissions. With the success of the Montreal Protocol, environmentally conscious leaders soon sought a similar agreement for controlling global warming gases, especially carbon dioxide. Extensive scientific evidence has led to near-consensus on the fact that humans are rapidly contributing to global warming by their unrestricted use of fossil fuels. Important data sets include the CO_2 monitoring data from Mauna Loa and long-term (tens of thousands of years) data from the Vostok ice core monitoring project. Even the U.S. Department of Defense recognizes climate change as a threat to national security (*Environmental Science and Technology*, 2004b). Among the numerous voices calling for action, perhaps one of the most convincing groups is the usually conservative insurance companies, who realize the future potential for economic disaster. Predictions of not acting soon include increased ocean levels, shifting ocean currents, warmer atmospheric temperatures (especially at the poles), more dramatic and destructive violent weather, further loss of species diversity, changing habitats, and increased development and spread of disease vectors.

On December 1–11, 1997, leaders from 160 nations meet in Kyoto, Japan, to consider and develop an agreement with binding limitations on greenhouse gases.

This agreement came to be known as the Kyoto Protocol and calls for developed nations to limit their greenhouse gas emissions, relative to 1990 levels. From the period between 2008 and 2012, countries were to decrease their greenhouse emissions to below 1990 levels: 7% for the United States, 8% for the European Union, and 6% for Japan. Specific points of the Kyoto Protocol are as follows (EPA, 1999):

- A five-year period for the initial phase reductions (for an average of 9% below 1990 level)
- Reductions in six major greenhouse gases, including carbon dioxide, methane, nitrous oxide, and three synthetic ozone-depleting CFCs,
- Using activities that absorb carbon, such as planting trees, to offset emissions

But what are the implications for the U.S. Economy? The Energy Information Administration of the U.S. Department of Energy (www.doe.gov) estimates that U.S. compliance with the Kyoto Protocol would require (1) a reduction in CO_2 emissions requiring the use of 18–77% less coal and 2–13% less petroleum, (2) an increase of between 2% and 16% in natural gas use and an increase in renewable energy sources between 2% and 16%, and (3) a market-based means of reducing energy use by increasing the price of energy from 17% to 23%.

While the Clinton administration agreed to the terms of the Kyoto Protocol, the subsequent Bush administration rejected these terms, citing economic concerns due to a slowing of the economy. This ignores potential economic costs of not acting to avoid or minimize the impacts of climate change on the inundation of coastal areas, on a potential increase in violent weather, on water scarcity in some areas, and on agriculture. For ratification, the Protocol requires that industrialized countries producing 55% of the world's greenhouse gases sign the Treaty. Even without the cooperation of the United States, the Treaty is likely to come into force with the recent commitment of Russia to sign on (www.newscientist.com, October 30, 2004 issue). This may put pressure on the United States and Australia, which could be restrictive enough to really change the course of global warming.

More information can be obtained from the Intergovernmental Panel on Climate Change (IPCC), which was established by the World Meteorological Organization and the United Nations Environment Programme. Their website and publications contain the definitive source of information on global warming.

The Basel Convention on Hazardous Waste. The most publicized treaty regarding hazardous waste came out of the Basel Convention on the Control of Transboundary Movement of Hazardous Wastes and Their Disposal, held in Basel, Switzerland. The Convention is a response by the international community to the problems caused by the annual worldwide production of hundreds of millions of tons of wastes. The convention was adopted in 1989 and came into effect in May 1992. This global environmental treaty strictly regulates the transboundary movement of hazardous wastes and requires that its parties ensure that such wastes are managed and disposed of in an environmentally sound manner. So far, the total number of parties (countries) who have ratified the Treaty is up to 154: 35 from Africa, 40 from Asia and the Pacific, 27 from Western Europe and others, 22 from

Central and Eastern Europe, and 30 from Latin America and the Caribbean. Countries that have yet to ratify the treaty are Afghanistan, Haiti, and the United States (the largest generator of hazardous wastes). The global acceptance of this Treaty without U.S. ratification clearly calls into question the quality of U.S. environmental leadership.

Stockholm Convention on Persistent Organic Pollutants. The most recent treaty that the United States has entered into resulted from the 2004 Stockholm Convention on Persistent Organic Pollutants (POPs). Under this agreement, participating countries will stop the production of aldrin, chlordane, DDT, dieldrin, endrin, heptachlor, morex, toxaphere, and furans. Some exemptions are possible, including the use of DDT to control malarial mosquitoes in some countries. Still, the agreement prioritizes ways to reduce or eliminate these uses.

11.6 ENVIRONMENTAL POLICY IN THE EUROPEAN UNION

This chapter has primarily focused on environmental movements and legislation in the United States, but similar movements occurred elsewhere concurrently or shortly after the movements in this country. As a comparison, we will give a summary of environmental legislation in the European Union. The EU is one of the predominant global economic forces as well as a significant body of industrialized countries (and therefore a major source of industrial pollution). While reading this section, compare the laws and actions of the EU to the corresponding circumstance in the United States.

11.6.1 Brief Introduction to the European Union

Within the last few decades, the politics and economy of Europe have changed significantly in many respects, with the introduction and progression of the EU. Officially founded in 1957 after member states created and signed the Treaty of Rome, the EU has evolved into what is today: one of the most influential political and economical institutions in the world. The evolution of the EU has led to a complex agreement between the 15 original member states, which gave up their control over specific policy areas to create a common body of law (McCormick, 1999, pp. 121–122). Among those policy areas that the EU has recently begun to emphasize is the environment; the majority of policy focusing on environmental standards in the member states is now controlled by EU law. Since the creation of the Treaty of Rome, in which no mention of environmental policy was made, the union turned its focus toward the important role the environment, among other factors, plays in the economy (McCormick, 1999, p. 120).

11.6.2 History of Environmental Policy

No reference to environmental policy was made in the Treaty of Rome; therefore, early environmental actions were based on a creative interpretation of Articles 100 and 235 of the Treaty (Grant et al., 2000, p. 9). Article 100, related to the common market, allows the EU to engage in actions that "directly affect the establishment or functioning of the common market"; Article 235 is a more general statement that applies when "action by the community should prove necessary . . . and this treaty has not provided the necessary powers" (Lévêque, 1996, pp. 10–11). Despite the lack of a solid foundation for environmental action, the EU created an estimated 150 pieces of legislation between 1967 and 1987, when the Single European Act introduced a specific environmental section into the Treaty (Grant et al., 2000, p. 9). During this period, European countries' attempts to recover from war drove them to create environmental legislation that focused on making progress with the common market. However, it eventually became clear that broader environmental protection was an important part of economic growth, and the emphasis began to shift toward a genuine concern for the environment (McCormick, 1999, p. 133).

The new attitude toward environmental protection led to the 1972 Paris Summit, which took place following both encouragement from environmentally conscious member states and the EU's participation in the United Nations Conference on the Human Environment in Stockholm. The Paris Summit focused on the development of the first Environmental Action Program (EAP), with the goal of harmonizing national politics between member states (Grant et al., 2000, pp. 9–10). The publication of the first EAP in 1973 was an important step in environmental policy and led to a series of five more programs, the sixth of which is still in progress today. However, the first three programs, which lasted from 1973 to 1976, 1977 to 1981, and 1982 to 1986 respectively, still relied on Articles 100 and 235 of the Treaty of Rome as their legal basis.

The 1986 The Single European Act (SEA) gave EU environmental policy a new face by providing a legal basis for making environmental decisions. The basic environmental objectives addressed in Article 130(r-t) were as follows (Grant et al., 2000, p. 11):

- to preserve, protect, and improve the quality of the environment,
- to protect human health, and
- to ensure a prudent and rational utilization of natural resources (Clinch, 1999)

The Act also incorporated environmental protection within every other policy area, highlighted its main principles and goals, and made environmental policy subject to the cooperation procedure in the European Parliament (although not initially) (McCormick, 1999, p. 136). The Treaty focused on the idea of "polluter-pays." After the creation of the SEA, there was a flurry of policy produced, with more legislation created between 1989 and 1991 than has been created in the last 20 years of policy making (Grant et al., 2000, p. 11).

The Maastricht Treaty (formally the Treaty of the European Union) was created simultaneously with the Fifth Environmental Action Program, "Towards Sustainability." This Treaty brought about a more defined goal for environmental policy in the EC by creating a new environmental act, Act 130, which created a more ambitious environmental policy than the SEA (Lévêque, 1996, p. 14). Most notably, the Act extended qualified majority voting to most environmental issues and introduced subsidiarity, which proposes that collective solutions be made in the EU only when the member states cannot achieve them on their own (Clinch, 1999, p. 4). The main ambitions of the Treaty (Article 130R 1 and 2) included prioritizing environmental quality, human health, resource use, and international cooperation, as well as stating that

Community policy on the environment shall aim at a high level of protection, taking into account the diversity of situations in the various regions of the Community. It shall be based on the precautionary principle and on the principles that preventative action shall be taken, that environmental damage should as a priority be rectified at the source and that the polluter should pay.

During the years after Maastricht, another conference was held in Amsterdam in 1999, which was focused on a more thorough integration of the environment into other EU policy areas. Particular focus was put on a campaign to sustain economic and social development, which sought to fight poverty and foster integration of the developing countries in the world (Europa, 2003a). As the EU continues to develop its goals in relation to the environment, it is clear that it holds the environment as a main focus of discussion and law, rather than a policy area considered only in relation to other "more important" issues.

11.6.3   Economy and the Environment

Often, the criticism directed toward the environmental policy in Europe is aimed at the community's tendency to put economic development and trade before environmental considerations (Europa, 2003b). The history of this problem was prominent previous to the Maastricht Treaty and was seen in two general areas. First was the intertwinement of environmental policy with economic policy. Environmental policies were intended to improve the building of the single market and, as a side note, to contribute to the environment. Second was the specialization in technical standards, which concerned the uniformity of standards to ensure the absence of unfair competition advantage (Lévêque, 1996, p. 12). The fear of countries losing their competitive advantage due to environmental regulations made them hesitant to accept unilateral action and unwilling to give power to international environmental organizations.

However, as trade barriers came down and a new light was shed on the importance of the environment, these worries became less important and created a strong encouragement for the member states to replace their national environmental policy with uniform regulations (McCormick, 1999, p. 135). Although environmental policy in the EU today can still affect the competitiveness and locations of certain industries, it is not designed specifically for gaining a competitive advantage (Lévêque, 1996, pp. 22–23). The recently completed Fifth Environmental Action Program focused on integrating the economy and the environment and worked to create more instruments that would mobilize the power of the single market (Clinch, 1999, p. 10). Although the Fifth EAP has done its best to move the member states in the same direction, the weight each state places on the environment still varies and impacts their involvement in the creation of the policy at an EU level.

11.6.4   The Union Versus Member States

There is an intricate system of policymaking in the EU, which leaves room for member states to encourage or discourage the creation of environmental policy based on their own national needs and attitudes (Anderson and Liefferink, 1997, p. 9). Overall, the relationship between the Union and national interests when creating environmental policy can be described as reciprocal.

Although the EU is a voluntary arrangement, it is unlikely that any country would leave, because the economic ties within the Union would make it extremely costly (McCormick, 1999, p. 124). However, member states are protected by certain legal provisions, including unanimity voting. The introduction of subsidiarity after the Treaty of Amsterdam also allows the member states to guard their sovereignty in specific categories of environmental policy, including fiscal provisions, measures concerning land use, planning of towns and country and management of water sources, as well as measures regarding the member state's choice about energy sources (Lévêque, 1996, p. 14).

Over the years, some countries have earned reputations for being "greener," while others are considered laggards in their views and motivation on environmental policy. Among the "green" countries, Germany, Denmark, and the Netherlands have been recognized for their pioneering role in the Union. In the 1980s, Germany earned a reputation for being the "engine" of EU environmental policy and, with the rise of the Green Party in national policy, proposed a combination of extreme and relatively harmonized solutions to EU problems. Germany is committed to being a part of the EU and has a strong position economically and politically in the Union (Anderson and Liefferink, 1997, pp. 26–27).

Denmark's presence in the EU has been somewhat reluctant, but their concern for the environment is seen specifically in their desire to maintain extremely high standards. Denmark is responsible for the introduction of Article 100A(4), which allows member states to maintain high levels of environmental protection even when EU regulations are less stringent. Recently, as the EU standards have risen, Denmark has taken a more active role in policymaking and has allowed the European Environmental Agency to be located in Copenhagen.

The Netherlands, although small, has had a significant amount of influence through its development of ambitious domestic policies, known as National Environmental Policy Plans (NEPP). The Fifth European Action Plan was modeled after the preceding Dutch NEPP. The Netherlands has recognized that its own desires to maintain high standards must be balanced by a need to get Union agreement, a distinct contrast with the typical Danish view.

More recently, Finland, Sweden, and Austria have become the leaders in environmental policy and protection since they joined the Union in 1995. Finland has not always maintained an active environmental policy in the past, and it was considered more of a follower in environmental issues until it created a Ministry of the Environment in 1983. However, Finland was the first country to enforce a tax on CO_2 emissions, despite the Finns' often humble attitude toward taking initiative (Anderson and Liefferink, 1997, pp. 21–25).

Sweden has always played an important role in international environmental policy; it hosted the UN's Stockholm conference in 1972, which encouraged the EU to create their own environmental regulations. Sweden is also involved in other international environmental organizations and supports the Stockholm Environmental Institute—an international network of independent institutes that work to find environmental solutions. Since joining the EU, Sweden had been able to (a) promote its ambitious environmental goals in a more formal manner and (b) positively improve the status of environmental policy at an international level.

Austria, the third country to join in 1995, consolidated its environmental laws in the 1980s. Most notable has been Austria's concern about the ozone layer, which led to the 1985 Vienna Convention and has made Austria a leader in the area of EU transport laws and the goals to reduce NO_x emissions from trucks by 60%.

The other states within the EU find themselves in a position to either (a) follow in the footsteps of the mentioned leaders or (b) oppose their high environmental standards. The most resistance has come from countries such as Greece, Portugal, and Spain, none of whom had an environmental ministry when they joined the EU in the 1980s. These and other less environmentally oriented countries tend to have varying positions on environmental topics, opposing those policies that they feel will harm their domestic economies while only occasionally making the environment a priority (Anderson and Liefferink, 1997, pp. 21–24). However, since the introduction of majority voting on environmental policy and the addition of the environmentally conscious member states, the tendency for laggard and middle states to be brought along environmentally has significantly increased (Grant et al., 2000, p. 29).

11.6.5 The Making of Environmental Policy

The EU represents one of the best models for environmental policymaking, considering that international cooperation is the most logical and potentially effective way to address a problem, like environmental quality, that does not respect national borders (McCormick, 1999, p. 135). The Union places three parties—the Directorate of the European Commission, the Council of Ministers, and the European Parliament—in charge of the bulk of the policymaking, with a fourth for enforcement.

The Directorate of the European Commission is responsible for the early stages of legislation, with commissioners drafting legislation and proposing it to the Council of Ministers. However, the Commission is admittedly weakened by the Directorate's lack of control over the administrations, combined with the general absence of a single authority in the decision-making process. Such lack of definitive power in an environment of diverse opinions creates problems when the goal is to spread a single environmental policy across the EU (Grant et al., 2000, pp. 17–18).

The Council of Ministers is a body of elected ministers from each member state, which works with a permanent body of representatives to make decisions. Presidents of the Council hold office for six months and generally approach the office from a nationalist perspective, as a chance to move their national interests to the top of the EC agenda. Currently the Council presidency lies in the hands of Greece, one of the less environmentally oriented countries (www.europeangreens.org/info, accessed 2003). The Council is important to the European public because it helps them recognize their ability to influence decisions within the EU (McCormick, 1999, pp. 97–99). The bargaining that takes place within the Council tends to temper the ambitiousness of the Commission's policy proposals and weaken the restrictions they have drafted (Grant et al., 2000, p. 31).

The European Parliament became more significant after the SEA and the Maastricht Treaty came into play. Although the Parliament has the power to reject legislation, it now tends not to do so, for fear that such rejection would prevent any legislation from being adopted (Grant et al., 2000, pp. 34–35). Despite its tendency not to reject legislation, as a general rule the Parliament is considered to be a "greener" institution than the Council (Lévêque, 1996, 16). Made up of hundreds of Members of the European Parliament, it has fair representation from Green Groups, with 30 current members belonging to a Green Party (www.europeangreens.org/info, accessed 2003). Although the Green Party remains significant in the EP and promotes the environmental stance that the Parliament tends to take, its members regularly run into problems, including bureaucratic inertia. The Green Party also prefers "no action" over "bad action," which makes its members hard to rely on for definite support of all of the environmental policy that is proposed. Ultimately, the decisions that the Parliament and the Commission make do not come to anything unless the Council is in agreement, due to the policy of majority vote within the Council.

Considered the "judicial branch" of the EU, the European Court of Justice still plays an important role in policymaking. The court, made up of 15 judges, ensures that member states follow policy, gives judgments on cases involving the member states and the community, and gives opinions as to the compatibility of international agreements with the EU Treaty. Although environmental policy now has a stable place within the EU treaties, the Court's role in strengthening and expanding its status should not be underestimated (Grant et al., 2000, pp. 36–39).

Europe uses three main legal instruments to enforce policies. The first and most common instrument in the area of the environment is the *directive*. Directives are proposed by the Commission and approved by the Council and the Parliament. The member states must achieve the required result of each directive, and the directive must be transposed into national legislation within each member state, although the method of implementation is left up to the individual countries (Lévêque, 1996, p. 9). The second type of instrument is the *regulatory act*, which must be directly implemented by each member state. The third and weakest form of policy instrument is the *recommendation*, which is simply a proposal by the Commission or the Council and is nonbinding. The promotion of paper recycling, for example, was done through a recommendation (Lévêque, 1996, pp. 9–10).

Recently, there has been an increase in the use of "voluntary agreements" in the EU, in which a firm or group of firms commits to operate in a certain way or to achieve specific operating goals that improve environmental performance. The idea is that firms participate in these agreements to avoid the threat of regulation that they believe could be more costly to implement. Within the EU, the more environmentally advanced countries, such as Germany and the Netherlands, have created such agreements at a national and subnational level. However, these agreements are probably most effective as complements to more formal instruments; despite the rise of such agreements, the role of the political institutions within the EU is still very apparent (Clinch, 1999, p. 25).

11.6.6 Existing Environmental Policy

The first effective step toward an environmentally conscious European Union occurred with the creation of the first Environmental Action Program (1972), which was followed by a series of five more plans. The sixth remains in progress today. Titled "Towards Sustainability," the fifth EAP (created in accordance with the Maastricht Treaty) was considered much more ambitious than the previous four plans, and it concentrated on sustainable development. Rather than focusing on a command-and-control style of regulation, the fifth EAP hoped to involve all the major groups—government, enterprise, public, and industry—when considering new policy. It also encouraged the use of a wider range of instruments, including voluntary approaches and market incentives (Lévêque, 1996, pp. 14–15).

The goals proposed in the fifth EAP, along with the 1998 Treaty of Amsterdam, which encouraged the integration of the environment into all policy areas, has led to a "horizontal" approach to environmental regulation. This approach takes into account all the causes of pollution and environmental problems: industry, energy, tourism, transport, agriculture, and so on. The main areas of environmental concern, however, remain the same, and there is an abundance of policy in the areas of waste management, noise pollution, water pollution, air pollution, nature conservation, and natural and technological hazards (Europa, 2003a, p. 2).

Waste Management Legislation. In the area of waste management, a number of specifics have been targeted in recent legislation. Three complementary strategies have been created to deal with problems:

- Minimizing waste through product design
- Encouraging recycling and re-use of waste
- Reducing pollution caused by waste incineration

(European Council, 1999). Included is a directive concerning landfill of waste, which prevents and reduces negative environmental effects of landfills by introducing stringent and technical requirements. For example, before waste can be accepted in a landfill, it must follow the following criteria:

- Waste must be treated before being landfilled.
- Hazardous waste, within the meaning of the directive, must be assigned to a hazardous waste landfill.
- Landfills for nonhazardous waste must be used for municipal waste and for nonhazardous waste.
- Landfill sites for inert waste must be used only for inert waste.

A system for attaining operating permits is also established, and member states must ensure that existing landfills are not allowed to operate unless they comply with the directive's provisions as quickly as possible. In order to ensure compliance in a timely and constant manner, the member states are required to report on the status of their landfills every three years (European Council, 1999).

The Commission also established a system of coordinated management of waste within the community, in order to limit waste production. Cooperation between the member states enables the creation of an adequate network of disposal installations (European Commission, 2000).

A separate directive was created for the management of hazardous waste, which works to manage, recover, and correctly dispose of such waste. Member states identify hazardous waste using a list created by the directive, and all wastes within each state must be recorded and identified. Wastes from different categories cannot be mixed, and no hazardous waste can be mixed with nonhazardous waste except where special safety measures concerning the environment and human health have been taken. Permits must be obtained by establishments wishing to undertake disposal operations; once accepted, the sites are subject to occasional inspections. Transporters, producers, and establishments must keep a record of their activities for use by member states' authorities. The authorities, in turn, publish plans for the management of hazardous waste, which are evaluated by the Commission. Member states may stray from policy only when the hazardous waste poses no threat to the population or the environment, and must first consult the Commission on their decision.

Among the methods of disposal is incineration of waste. The EU has adopted policy that addresses incineration, including regulations for existing and future incineration plants. The goal of the directive on general waste incineration is to prevent or reduce, as much as possible, water and soil pollution caused by the incineration of waste, as well as reduce resulting risk to human health. New policy covers all wastes, including nonmunicipal nontoxic wastes such as sewage, sludge, tires, and hospital wastes, and previously unaddressed toxic wastes such as waste oil and solvents. The directive also covers co-incineration facilities, which produce energy or material products using waste as a regular or additional fuel. All incineration and co-incineration plants must be authorized, and permits listing the categories and quantity of waste permitted are issued by member state authorities. Before a plant can accept hazardous waste, its operator must have information on the composition, generation, and hazardous characteristics of the waste. Plants must reduce the harmful residue and do their best to recycle it. Applicants for new permits must make their information available to the public so the latter can voice their opinions on the proposal.

Incineration of hazardous waste falls under the control of another directive that works to reduce both the effects of hazardous waste incineration on the environment and the ensuing risks for public health. Technology is an important part of the directive, requiring plants to strive for the highest technology possible to reduce emissions and residue, as well as ensure compliance with threshold values and operating conditions. Measurements of emissions must be taken on a monthly basis, and if these show threshold values exceeded, the plant must shut down until it can comply with the directive (European Council, 1994).

In terms of radioactive waste, a 1992 EU directive created a system of prior authorization for all movement of radioactive waste, in order to increase protection against the dangers arising from ionizing radiation. The directive applies to all shipments of radioactive waste between member states and shipments entering or leaving the EU. The directive provides a common, mandatory system of notification and a uniform control document for the transfer that occurs. Additionally, EU countries are not allowed to export waste to countries that are not equipped to receive radioactive waste due to reasons of either location or economic resources. A later, 1996 communication also discussed the illegal traffic of nuclear materials and radioactive wastes. The EU has been working with other organizations to fight the unfortunate rise of this illegal transport (European Council, 1992).

Water Legislation. In the area of water policy, a number of directives have been introduced to address water issues relating to drinking, bathing and other fresh waters, as well as shellfish cultures. A community water policy and a measure that deals with urban wastewater have also been created. The EU has also participated in international conventions concerned with the protection of a safe environment for marine life (Europa, 2003a, pp. 3–4).

The community water policy is aimed at the development of an integrated water management policy. The directive is very general and addresses maintaining secure supplies of drinking and nondrinking water to meet human consumption and economic needs, protecting and preserving the aquatic environment, and restricting natural disasters (drought and floods). The directive introduces a number of principles, with the hope that member states will implement the highest standards in the various areas addressed in environmental needs (European Commission, 1996).

The drinking water and surface water directives are two of the oldest directives of environmental policy in the EU, established in 1980 and 1975 respectively. The drinking water directive lays down EU-wide minimum quality and control standards for drinking water in the member states. The water standards are defined by the properties of the water and any undesirable or toxic substances in the water. Member states fix their values at desired levels that satisfy the minimum requirements of EU policy. Monitoring of water quality is the responsibility of the member states themselves and must once again meet the minimum standards set out by the union (European Council, 1980).

The regulation of surface water follows similar criteria, as it hopes to prevent pollution of surface water that may be used as drinking water. Standardized techniques are used to measure the quality of the water, and member states are responsible for creating provisions and testing them. In this case, member states have the explicit right to set more stringent requirements that those set by the community (European Council, 1975).

The directive on urban wastewater treatment seeks to harmonize treatment measures throughout the EC by evaluating the collection, treatment, and discharge of urban waste and from industrial sectors. The treatment requirements depend on the sensitivity of the area that will be receiving the treated water. Once again, member states are responsible for monitoring water quality—of both discharges from treatment plants and the waters receiving them. National authorities are required to publish situation reports every two years (European Council, 1991). A second directive clarified the first, setting specific criteria for nitrogen content and effluent temperature, and incorporating an alternative method for testing temperature climatic zones (European Council, 1998a, 1998b).

Air Quality Legislation. Recognizing that the reduction of air pollution is a global priority, the EU has held several conferences and worked to create goals and policies. The EU has participated in the United Nations Framework Convention on Climate Change in 1992 and the Kyoto Protocol in 1997. In working for air quality improvement, the EU also addresses the global goals of reducing concentration of ozone in ambient air, fixing national ceilings for other atmospheric pollutants, and limiting polluting emissions from large combustion plants (Europa, 2003, p. 4).

Commitment to the Kyoto Protocol has been a recurring theme in EU-specific conferences, and an international plan was created for implementing the legislation proposed. The Kyoto Protocol requires the EC to cut its greenhouse gas emissions by 8% between 2008 and 2012 in relation to 1990 levels. EU policy has aimed to cutting emissions from industrial activities and road vehicles by

- Reducing pollutant emissions (using catalytic converters, roadworthiness tests, etc.)
- Reducing the fuel consumption of private cars (in collaboration with manufacturers)
- Promoting clean vehicles (through tax incentives)

A Commission communication also launched a public debate over implementing emissions trading before the Kyoto Protocol is applied in 2008. Rapid implementation of its own emissions trading scheme would give the EU experience before the international trading scheme is introduced in 2008.

Under a system of emissions trading, allowances for pollutant emissions (specifically, in the case of Kyoto, greenhouse gas emissions) are distributed to companies, allowing the government to control overall emissions. Trading then allows the companies some flexibility, because those who exceed their allotted pollution can buy "spare" allowance from a company that is not reaching its allowance. This is often considered a practical system, which does not cause any added environmental damage. It is stated that EU-wide participation in emissions trading could reduce the cost of EU implementation of the Kyoto Protocol and would provide a smoother functioning of the internal market by creating a single price for allowances traded by companies. Whether the EU ends up proving a supportive role or a regulative authority, it is important that they follow certain guidelines that will

- Ensure equal treatment for companies of comparable size under the emissions trading scheme
- Minimize the possibility of competition being distorted
- Ensure cooperation with existing legislation
- Ensure that the scheme is applied effectively
- Ensure compatibility with the scheme established by the Kyoto Protocol

The Commission believes that the entire community needs to go through the process step by step so that they are provided with practical experience and competition is not distorted within the single market. The Kyoto Protocol will have to be accompanied by policy in the areas of air, transport, and energy, thereby making all programs compatible.

Legislation Addressing Environmental Disasters. In the realm of natural and technology hazards, a number of issues could arise concerning the safety of humans and the environment. The EC has recognized this possibility and created an Action Program on Civil Protection, which extends through 2004. The EC has also signed the Transboundary Impacts on Industrial Accidents, which protects human beings and the environment against industrial accidents capable of causing transboundary effects. The program promotes active international cooperation between the contracting parties during and after an accident, if one occurs. The program was signed by 14 member countries and the EU institution itself during a convention in Helsinki on March 18, 1992. The convention outlines a set of measures that will protect human beings and the environment against accidents, specifically

- Nuclear accidents or radiological emergencies
- Accidents at military installations
- Dam failures
- Land-based transport accidents
- Accidental releases of genetically modified organisms
- Accidents caused by activities in the marine environment and the spillage of harmful substances at sea

If any of these disasters occur within a country, the affected parties must be alerted of the problem and discussions must be held on the identification of problems capable of causing transboundary effects. Countries must take appropriate action to prevent these accidents, and if there is potential for one to occur, the public that could be affected must be informed. If there is dispute between countries, the program proposed three methods for settling the dispute: (1) negotiation, (2) submission of the dispute to the International Court of Justice, and (3) arbitration. Countries must keep the others informed of their efforts at implementation of the program (European Council, 1998b).

11.6.7 Implementation of Environmental Policy

Although policymaking has improved over the years since the environmental movement began in Europe, implementation does not have as successful a record overall. The EU institutions have shown limited ability to ensure successful implementation in member states, and long-term EU policy has suffered from the lack of legal basis. Poor implementation has also been blamed on disorganization within the EU institutions, lack of financial and technical resources, and EU institutions' failure to recognize the difficulty in implementing the policy they create.

On a national level, there have been issues with the varying ability of or motivation for member states to implement EU policy. States with poorer environmental records, such as Spain and Greece, have had more infringement cases started against them by the ECJ than the other states, while countries such as Germany and the Netherlands are lazy about adopting legislation because their domestic environmental departments are already successful. Denmark, meanwhile, has a good record of implementation despite past resistance to EU policy, and it now has an active role in creating new EU policy (McCormick, 1999, p. 137). The most common system of implementation in member states is a cooperative, conciliatory approach by enforcement officers who seek to achieve compliance through negotiation, persuasion, and public education or awareness campaigns. Other techniques are legally based and involve referring cases for prosecution, issuing fines or closing commercial premises. This approach is not common, however, and it is used primarily when there is a high risk of a pollution accident, where pollution has already occurred, or where advice is not achieving the desired outcome (Grant et al., 2000).

Although the implementation of directives is left to member states, the EU is responsible for the implementation of the regulations it imposes on the states. At the EU level, a number of programs have been introduced for the implementation of EC measures through funding, including the LIFE program, the eco-label project, and the Community Eco-Management and Audit Scheme, in addition to measures that define the criteria for environmental inspections and the assessment of the effects of plans and programs on the environment.

The LIFE program was born in 1992, and its goal is to develop, implement, and update EC environmental policy and environmental legislation, particularly in relation to the integration of the environment into other policy areas. LIFE finances the EC member states as well as countries not in the union that border the Mediterranean or Baltic Seas or that, alternatively, are CEEC and have applied to become union members. The most recent, third phase of funding had a lifespan of 2000–2004 and a budget of €640 million. It followed the first phase, from 1992 to 1995, with a budget of €400 million and followed the second phase, from 1996 to 1999, with a €450 million budget. LIFE uses the budget to finance proects that meet the following three criteria:

- Projects are of community interest and contribute to life objectives.
- Projects are carried out by technically and financially sound participants.
- Projects are feasible in terms of technical proposals, timetable, budget, and value of money.

The persistence of programs such as LIFE helps the countries with less successful environmental programs and funding, to reach EC standards.

The eco-label project promotes products that have less environmental impact than other products in the same product group, providing consumers with accurate information and guidance on a number of products. Product in the system are judged by the European Union Eco-Labeling Board, according to a number of environmental concerns and criteria (European Parliament and European Council, *Eco-label*, 2000).

The Community Eco-Management and Audit Scheme promotes the continuous improvement of environmental performance within EC organizations, and it also provides the public and interested parties with the information it acquires. This regulation replaces one made in 1993, which allowed voluntary participation of industrial companies in an audit system. The member states are required to establish a system for accrediting independent environmental verifiers and for supervising their activities. The regulation requires member states to encourage the participation of small and medium-sized undertakings in the EMAS and requires them also to promote EMAS so that its presence maximizes public awareness (European Parliament and European Council, *Community Eco Management and Audit Scheme (EMAS)*, 2000).

The two other implementation programs that exist within the EC help to (1) integrate the environment into the preparation and adoption of plans and programs that may have significant environmental consequences (European Parliament and European Council, *Assessment of the Effects of Plans and Programmes on the Environment*, 2001) and (2) ensure greater compliance and a more uniform application and implementation of EC environmental policy by creating the minimum criteria for the execution of environmental inspections (European Parliament and Council, *Environmental Inspections: Minimum Criteria*).

In considering the availability of funds for implementation, the EU's lack of ability to introduce taxes must be noted. The unanimity rule for such matters makes it extremely difficult for an EU-wide tax to be introduced. A uniform level of tax would be difficult because the tax would be based on the level of damage being done in a specific area, and this could vary between countries. In addition, countries such as Sweden and Denmark that have domestic environmental taxes have reported a number of difficulties with small and vulnerable plants in isolated regions, who demonstrate that their alternative options are limited (Clinch, 1999, pp. 39–40).

As in other areas, policymakers are working to move away from the tendency to create too much ambitious environmental legislation that has little hope of implementation.

11.6.8 Public and the Environment

Public support for environmental protection within the EC has remained strong since the Community's formation, with eurobarometers indicating that the public places the environment above finance, defense, or employment as an EU issue. The public recognizes that environmental issues are not country-specific, and most agree that protection of the environment should be addressed at an international level (McCormick, 1999, p. 136).

However, the variety of concerns among member countries has had both a negative and a positive impact on the EC as an organization. Opinionated anti-EC groups within Denmark (preferring more stringent Danish legislation) have used environmental policy as an argument for withdrawing from the union, and it fought to create the previously mentioned Article 100A(4) (Anderson and Liefferink, 1997, p. 29). On the other hand, the public interest in various member states has led to the strong Green Party presence in the Parliament, as well as a variety of euro-interest groups.

Although grassroots Green parties play only a small role in the devising phase of policies, and few organizations are prepared to work internationally, they have provided enough influence at the local level (Lévêque, 1996, p. 20) to get their representatives elected to nine different countries' Parliament positions. On a domestic level, seven EC member countries now have Green cabinet ministers as a result of public support of the Green Party (European Greens, 2003, p. 2). Interest groups have also been active in the implementation process of environmental policy; their concern is focused on the member states' levels of successful implementation, and their work involves rallying toward the inspection of weak environmental areas (McCormick, 1999, p. 137).

Created in 1994, the European Environmental Agency is an organization that works with the EC to provide information to the public and to policymakers. The public can use the EEA to access accurate and current information on policies or developments in the policymaking process. The EEA sees itself as a networking organization for everyone, and it has provided an important link between the public and policymakers.

11.6.9   The Future of Environmental Policy

The European Union is currently involved in the sixth European Action Program, in place from 2001 to 2010, and titled "Our Future, Our Choice." The goals of the sixth EAP are very similar to those of the recently completed fifth EAP, but the sixth EAP takes a more strategic approach to finding solutions to environmental problems. The new EAP has a focused plan that includes four priority areas as well as five key approaches to their solutions. The priority areas are climate change, nature and biodiversity, environment and health, and natural resources and waste. But the new, innovative aspects of the sixth EAP appear in its method, which includes

- Ensuring the implementation of existing environmental legislation
- Integrating environmental concerns into all relevant policy areas
- Working closely with business and consumers to identify solutions
- Developing a more environmentally conscious attitude toward land use (Europa, 2003b)

A primary concern in the future of EU environmental policy is the looming accession of the Central and Eastern European Countries (CEEC) to the union. These countries (Bulgaria, the Czech Republic, Cyprus, Estonia, Latvia, Lithuania, Hungary, Malta, Poland, Romania, Slovenia, and Slovakia) have recently become subject to a number of EU efforts to improve their general standings in all policy areas in preparation. For the past decade, the EU has been making efforts in these countries to clean up contamination and to develop environmental strategies, training, and investment, yet the environment still poses a major problem to the CEEC accession.

CEEC involvement in environmental issues began with the Rio de Janeiro Earth Summit in 1992 and continued in an effort to break away from the legacy of environmental destruction facing all former Communist bloc countries. After CEEC countries moved away from centralized, state-controlled economies, the public began to support cleanup of the environment, recognizing a connection between environmental health and public health. In terms of legislation, two areas were in desperate need of support, after years of considering natural resources in the CEEC as "free goods": pollution research and monitoring (Mannin, 1999, pp. 158–163).

The EC has created legislation that outlines strategies for each country to improve its environmental policy as it incorporates EU policy into its own legislation. The 10 countries constitute more of an environmental issue than any other member state that has vied for accession in the past, including Greece, Spain and Portugal, which also had poor environmental standards at the time of their accession. As presented by the EU Commission, the CEEC countries face particular challenges in the areas of air pollution, water pollution, and waste management. Furthermore, they must address problems in the areas of legislation, institutions, and finance. Another goal for the candidate countries is the creation of national programs to address the specific problems of concern. There is no deadline for policy implementation within the CEEC (European Commission, 1998).

Although some improvement has occurred during the years since the CEEC adopted a free market approach, the ability of member states to decrease, halt, or reverse environmental damage is largely hindered by a lack of funds. From the beginning, it was clear that investment would be needed in environmental hardware as well as training and education. The CEEC would have to clean up the errors of the past, slow the current rate of deterioration, and ensure that future economic development would be carried out through environmentally sustainable methods; the CEEC did not have the resources to do all of this (Mannin, 1999, p. 163).

As a result of insufficient funding in the CEEC, the EU has made a considerable effort to provide financial support through the Phare program, the main instrument of financial and technical cooperation between the EU and the CEEC. Set up in 1989 as a support system for Poland and Hungary in their economic and political transitions, Phare now covers the 10 associate CEEC countries as well as four others. Phare's main goal is to help the candidate countries acquire the capacity to implement EU policy by helping them bring their industries and infrastructure up to standards and by mobilizing the required investment. The money put into Phare by the EU is generally provided through grants, which are either funded by the community or co-financed by member states, the European Investment Bank, and third countries. Phare's budget has been growing since its formation in 1989, and since 2000 it has reached its maximum of €1560 million a year (European Council, *The Phare Community*, 1989).

A second funding program implemented by the EC is the Instrument for Structural Policies and for Pre-accession. The regulation provides assistance to the Phare program between 2000 and 2006 with a budget of €1 billion a year. The objective of this regulation is to provide financial assistance with a view to contributing to the preparation for accession to the European Union of the applicant countries of CEE. The plan focuses on the environment and grants assistance to projects that allow the countries to comply with the standards in EC environmental law.

A third form of assistance comes from the European Bank, which works to fund and promote environmental activities and sustainable development. The Bank produces strategies for countries and sectors and carries out environmental appraisals. Two fairly recent examples of EBRD's support for environmental improvement are the 1995 "Municipal and Environmental Infrastructure" initiative and a 1997 agreement with some EC countries to provide financial help. A final example of assistance is Debt-for-Environment-Swaps, a program initiated by the

United States and Bolivia in 1987. Allowing debtor countries to reduce their external debt obligations, the program requires that the debtor make a commitment to mobilize domestic resources for environmental protection.

The CEEC reaction to external funding has not been entirely positive, but their reliance on outside assistance both in management of environmental problems and in financial investment requires that the CEEC cooperate. Without the help of donor countries and organizations, the preparation of the CEEC for EC accession would be significantly hindered.

ACKNOWLEDGMENTS

Many students contributed to this chapter, including members of my spring 2004 section of Chemistry of the Natural World and my spring 2004 section of Environmental Chemistry. The noted authors of this chapter wrote, in part, the following sections: Section 11.1 was taken from work of Lance DeMuth, Loren Sackett, and Rose Kormanyos; Section 11.2 was written by Rose Kormanyos; Section 11.3.3 was prepared by Savanna Ferguson; Section 11.5 was taken from the work of Jill Schulte; the Section 11.6 was written by Sarah McConnell. I thank you all for your contributions.

REFERENCES

Anderson, M. S. and D. Liefferink. *European Environmental Policy: The Pioneers*, Manchester University Press, New York, 1997.

Andrews, R. L. Reform or reaction: EPA at a Crossroads. *Environ. Sci. Technol.* **29**(11), pp. 505A–510A (1995).

Barton, G., ed. *American Environmentalism*, Greenhaven Press, San Diego, 2002, p. 70.

Baum, R. M. EPA's silver anniversary, Editor's Page, *Chem. Eng. News* **October 30**, 5 (1995).

Baun, M. J. *A Wider ECrope; The Process and Politics of the European Union Enlargement*, Rowman & Littlefield, Oxford, England, 2000.

Bryner, G. C. *Blue Skies, Green Politics*, A Division of Congressional Quarterly Inc., Washington, DC, 1995.

Caldwell, L. K. Foreword in *The National Environmental Policy Act: Judicial Misconstruction, Legislative Indifference, and Executive Neglect*, by M. J. Lindstrom and A. A. Smith, Number 17, the Environmental History Series. Dan Flores, ed., Texas A&M University Press, College Station, TX, 2001.

Clinch, J. P. *Environmental Policy Reform in the EC*. Environmental Studies Research Series, University College Dublin, Dublin, Ireland, 1999.

Doern, B. G. and T. Conway. *The Greening of Canada*, University of Toronto Press, Toronto, p. 150, 1994.

Doern, B. G. and D. C. Esty. *Greening the Americas: NAFTA's Lessons for Hemispheric Trade*, MIT Press, Cambridge, MA, 2002.

Ember, L. R. EPA at 25. *Chem. Eng. News* **October 30**, 16–17 (1995a).

Ember, L. R. EPA administrators deem agencies first 25 years bumpy but successful. *Chem. Eng. News* **October 30**, pp. 18–23 (1995b).

Environ. Sci. Technol. **38**(12), 217–218A (2004a).

Environ. Sci. Technol. **38**(10), 179A (2004b).

EPA. Fact Sheet on the Kyoto Protocol, October 1999 www.epa.gov (1999).

EPA. The History of Drinking Water Treatment, Office of Water, EPA-816-F-00-006, February 2000.

EPA. Plain English Guide to Clean Air Act. Accessed February 6, 2003. www.epa.gov/oar/oaqps/peg_caa/pegcaa02.html.

EPA. A Look at EPA Accomplishments: 25 years of protecting public health and the environment. EPA Press Release—December 1, 1995. http://epa.history/topics/epa/25b.htm, accessed February 5, 2003.

Europa. Activities of the European Union; Summaries of Legislature. *Environment: Introduction.* http://ECropa.EC.int/scadplus/leg/en/lvb/128066.htm, January 22, 2003a.

Europa. Sixth Environmental Action Programme. *Environment 2010: Our Future, Our Choice.* http://ECropa.EC.int/comm/environment/newprg/, February 10, 2003b.

European Commission. Community Water Policy. Communication, Brussels, Belgium, 1996.

European Commission. *Accession Strategies for the Environment, 1998,* May 20, 1998.

European Commission. Decision 2000/738/EC. *Official Journal* L 298 (2000).

European Council. *Surface Freshwater: Quality and Control Requirements,* Directive No. 75/440/EEC, Brussels, Belgium, 1975.

European Council. *General Provisions on the Quality of Drinking Water.* Directive No. 80/778/EEC, Brussels, Belgium, 1980.

European Council. *Urban Waste Water Treatment,* Directive No. 91/271/EEC, Brussels, Belgium, 1991.

European Council. *Transfer of Radioactive Waste: Supervision and Control,* Directive No. 92/3/Euratom, Brussels, Belgium, 1992.

European Council. Packaging and Packaging Waste (Brussels, Belgium: Directive No. 941621 EC, 1994.

European Council. Directive 98/15/EC, 1998a.

European Council. Directive 98/83/EC, 1998b.

European Council. *Instrument for structural policies for Pre-accession,* Regulation No. 1267/1999, Brussels, Belgium, 1999.

European Council. *The Phare Community,* Regulation No. 3906/89, Brussels, Belgium, 1989.

European Greens. Information, Greening Europe-More Urgent Than Ever, http://www.europeangreens.orglinfo/policy/manifesto99_1.html. Feb. 2003.

European Parliament and European Council. *Community Eco Management and Audit Scheme (EMAS),* Regulation No. 761/2000, Brussels, Belgium, 2000.

European Parliament and European Council. *Assessment of the Effects of Plans and Programmes on the Environment,* Directive No. 2001/42/EC, Brussels, Belgium, 2001.

European Parliament and Council. *Environmental Inspections: Minimum Criteria,* Reccomendation, Brussels, Belgium, 2001.

European Parliament and European Council. *LIFE: A Financial Instrument for the Environment, 2000,* Regulation No. 1655/2000, Brussels, Belgium, 2000.

European Parliament and European Council, *Eco-label, 2000,* Regulation No.1980/2000, Brussels, Belgium, 2000.

Grant, W., D. Matthews, and P. Newell. *The Effectiveness of Ecropean Union Environmental Policy,* St. Martin's Press, New York, 2000.

Harris, P. G. *The Environment, International Relations, and United States Foreign Policy,* Georgetown University Press, Washington, DC, 2001.

Hileman, B. Pesticides registration streamlined: EPA will no longer consult with wildlife agencies on pesticides. *Chem. Eng. News* **August 9**, 9 (2004).

Hileman, B. Environmental leaders give EPA mixed reviews on its performance. *Chem. Eng. News* **October 30**, 30–37 (1995).

Hix, S. *The Political System of the European Union,* St. Martin's Press, New York, 1999.

Kirschner, E. M. Industry See Maturation, Contradiction in EPA's Quarter-Century history. *Chem. Eng. News* **October 30**, 24–29 (1995).

LaGrega, M. D., P. L. Buckingham, J. C. Evans, and the Environmental Resources Management Group. *Hazardous Waste Management,* McGraw-Hill, New York, 1994.

Landy, M. K., et al. *The Environmental Protection Agency: Asking the Wrong Questions from Nixon to Clinton.* Oxford University Press, New York, 1994.

Lave, L. B. and G. S. Omnenn. *Clearing the Air: Reforming the Clean Air Act.* The Brookings Institution, Washington, DC, 1981.

LeMarquand, D. The International Joint Commission and changing boundary relations. *Nat. Resour. J.* **Winter**, 59–91 (1993).

Lepkowski, W. Government seeks new balance in environmental protection. *Chem. Eng. News* **October 30**, 44–49 (1995).

Lévêque, F. *Environmental Policy in Europe*, Edward Elgar Publishing Company, Brookfield, VT, 1996.

Lewis, J. The birth of the EPA. *EPA J.* **November**, (1985). http://www.epa.gov/history/topics/eap/15c.htm accessed on February 5, 2003.

Lindstrom, M. J. and Z. A. Smith. *The National Environmental Policy Act: Judicial Misconstruction, Legislative Indifference, and Executive Neglect*, by Lindstrom, M. J. and A. A. Smith, the Environmental History Series, Number 17, D. Flores, ed., College Station, TX, Texas A&M University Press, 2001.

Makhijani, A. and K. R. Gurney. *Mending the Ozone Hole: Science, Technology, and Policy*, MIT Press, Cambridge, MA, 1995.

Mannin, M., ed., *Pushing Back the Boundaries; The European Union and Central and Eastern Europe*, Manchester University Press, New York, 1999.

McCormick, J. *Understanding the European Union*, St. Martin's Press, New York, 1999.

McCoy, M. 1999. Why companies go astray. *Chem. Eng. News* **July 19**, 9 (1999).

National Environmental Policy Act of 1969, Public Law 91–190, 42 U.S.C. 4321–4327, January 1, 1970, as amended by Public Law 94–52, July 3, 1975, Public Law 94–83, August 9, 1975, and Public Law 97–258, 4(b), September 13, 1982 SEC. 2 [42 U.S.C. 4321].

Newton, D. *The Ozone Dilemma*, Instructional Horizons, Inc. Santa Barbara, CA, 1995.

Quarles, J. R. *Cleaning Up America: An Insider's View of the Environmental Protection Agency*, Boston, Houghton Mifflin, 1976.

Rawls, R. L. Environmental scientists fault EPA for its shifting, short-term research focus. *Chem. Eng. News* **October 30**, 38–43 (1995).

Seignecr, C., K. Vijayaraghavan, K. Lohman, P. Karamchandani, and C. Scott. Global source attribution for mercury deposition in the United States. *Environ. Sci. Technol.* **38**(2), 555–569 (2004).

Sierra Magazine, The Bush record, pp. 3–88. September–October (2004).

Switzer, J. V. *Environmental Politics: Domestic and Global Dimensions*, St. Martin's, New York, 1994.

Taylor, B., ed. *Ecological Resistance Movements: The Global Emergence of Radical and Popular Environmentalist*. State University of New York, Albany, 1995.

Wallace, H. and W. Wallace. *Policy Making in the European Union*, Oxford University Press, New York, 2000.

PART VI

POLLUTANT CASE STUDIES

"If we are going to live so intimately with these chemicals—eating and drinking them, taking them into the very marrow of our bones—we better know something about their nature and their power."

—from *Silent Spring*, Rachel Carson

CHAPTER 12

CASE STUDIES OF SELECTED POLLUTANTS

Frank Dunnivant and Emily Welborn

This chapter focuses on several "world class pollutants"—that is, pollutants that are spread around with world and that can be found in every environmental compartment (air, surface water, groundwater, oceans, soil, etc.). While there are several chemicals that could be presented here, we will focus on four classic pollutants: mercury, lead, PCBs, and DDT. We will close this chapter with a different type of ubiquitous pollutant, endocrine disruptors, which may turn out to be one of the most serious anthropogenic threats to human and animal health.

12.1 MERCURY

12.1.1 Sources

Mineral. Cinnabar (HgS) is by far the most important source of mercury, typically consisting of 0.6–0.7% of mined mercury ore (Gribble, 1988). The most important mine in the world is in Almaden, Spain, and has been in operation since Roman times (Berry and Mason, 1959). In 1985, the major producers of mercury were the former USSR (2415 Mg), Spain (1725 Mg), China (1190 Mg), Algeria (862 Mg), the United States (100 Mg), Turkey (>690 Mg), Finland (100 Mg), and Yugoslavia (90 Mg).

Other Anthropogenic Sources. While mining ore is a major source of mercury for industrial use, other sources of mercury to the environment include (1) the combustion of coal, (2) oil production combustion, (3) cement production, (4) lead production, (5) zinc production, (6) pig iron and steel production, (7) caustic soda production (chloralkaline plants), (8) gold production, and (9) waste disposal (such as batteries in land fills) (Pacyna and Pacyna, 2002). The burning of coal is one of the major sources of mercury to the environment. Coal produced in the United States contains mercury at concentrations between 0.07 ppm (Uinta coal) and 0.24 ppm (northern Appalachian coal) (USGA Fact Sheet FS-095-01, 2001). With this mercury concentration, any other material would be classified as hazardous waste.

A Basic Introduction to Pollutant Fate and Transport, By Dunnivant and Anders
Copyright © 2006 by John Wiley & Sons, Inc.

Yet, in burning coal for energy, we emit all or most of its mercury to the atmosphere, from which it is deposited into our air, soil, and water. Mercury has even been found in atmospheric deposition falling on the Arctic and Antarctic (Rouhi, 2002). Major annual anthropogenic sources of Hg emissions to North America (southern Canada, United States, and northern Mexico) are (1) electrical utilities (52.7 Mg/yr; U.S. contribution 41.5 Mg/yr), (2) waste incineration (32.2 Mg/yr; U.S. contribution 28.8 Mg/yr), (3) residential, commercial, and industrial coal burning (12.8 Mg/yr; U.S. contribution 12.8 Mg/yr), (4) mining (6.7 Mg/yr; U.S. contribution 6.4 Mg/yr), (5) chlor-alkali facilities (6.75 Mg/yr; U.S. contribution 6.7 Mg/yr), (6) mobile sources (24.8 Mg/yr; U.S. contribution 24.8 Mg/yr), and (7) miscellaneous other sources (64.1 Mg/yr; U.S. contribution 30.9 Mg/yr). These add up to 200.1 Mg of mercury per year released to the environment (Seigneur et al., 2004). The United States contributes 76% of this.

Biosynthetic.   Mercury from all of these sources is generally deposited in its relatively less toxic, inorganic form. Yet, after mercury pollution has been deposited in soils and sediments, microorganisms in anaerobic environments can transform the inorganic mercury to organic forms, specifically methyl and dimethyl mercury. These forms are very toxic and are a major concern especially when they bioaccumulate, as they do in fish; this is discussed later in more detail.

12.1.2   Production/Use

Cinnabar (HgS) is processed by roasting the ore in the presence of oxygen where the sulfide is oxidized to sulfurous acid and the freed metal (Hg) is volatized and recovered through condensation. The major uses of processed mercury ore are in the manufacture of drugs and chemicals and as a cathode in chlorine and soda ash production. It is also used in (a) the electrical industry for batteries, rectifiers, automatic switches, and mercury vapor lamps and (b) the instrument industry in the construction of thermometers and barometers. Some insecticides and fungicides for agriculture contain mercury, and mercury is still used in dental fillings (dental amalgam contains 40–50% mercury, 25% silver, and 25–35% copper, zinc, and tin; www.epa.gov). At one time, mercury was used in the United States for extraction and recovery of gold and silver from mine tailings, as it still is today in some parts of the world (Gribble, 1988).

12.1.3   Fate and Environmental Distribution

Mercury enters the environment via two primary pathways: (a) into the atmosphere from the smelting of ores, fossil combustion sources (primarily coal), and chlor-alkaline bleach production facilities and (b) into waterways via atmospheric deposition and industrial pollution. Regardless of the original source, mercury is initially deposited into the water in its inorganic, ionic form (Hg^{2+}). Inorganic mercury is toxic and can be taken in through the intake of water or food. A more toxic form of mercury is the methylated form (CH_3Hg^-), which is produced by bacteria in anaerobic environments, such as those found in some groundwaters and especially in river and lake sediments. Both forms of mercury, organic and inorganic, bioaccumulate in the food chain and can easily reach health-threatening levels in fish.

Recently, governmental agencies and the news media have focused on mercury contamination in our foods due to its bioaccumulation tendency. Several fish consumption warnings for pregnant women and children have been issued (Renner, 2004; Crenson, 2002; Wright, 2005). Fish, especially ocean fish, are a major source of mercury exposure in the human diet. The higher the fish is on the food chain, the more mercury it will contain, as illustrated in a *New York Times* article on seafood consumption. The article reports mercury levels for animals at increasing trophic levels: The concentration increased from 0.023 ppm in clams, to 0.035 in salmon, to 0.042 in scallops, to 0.047 in shrimp, to 0.092 in flounder, to 0.117 in crab, to 0.121 in cod, to 0.150 in Pollock, to 0.206 in tuna (Saar, 1999). Hence, environmentalists have two reasons to avoid tuna: not only to protect dolphins, as has long been advertised, but also to avoid mercury intake. Still, not all tuna is a problem, since mercury levels vary significantly depending on the type of tuna processed into food. "White tuna" is one of the safest forms of tuna with respect to mercury levels.

Efforts are underway, although slowly, to reduce mercury emissions. Nine chlor-alkaline plants that manufacture sodium hydroxide, chlorine, and chlorine bleach in the United States are gradually replacing mercury in their systems. This changeover will account for approximately 65 tons of mercury that is "lost" in the system annually (*Chemical and Engineering News*, 2004a, 2004b). One of the major impediments to reducing mercury from coal burning is, not surprisingly, the coal and energy lobbyists, as well as the current Bush administration, who fear the economic costs to industries required to update their equipment and processes (*Chemical and Engineering News*, 2005a, 2005b). These impediments have slowed installation of mercury recovery equipment, even though existing technologies have the potential to virtually eliminate mercury from emissions passing through smokestacks (Betts, 2003). While the U.S. federal government has slowed the reduction of mercury emissions, other countries are calling for enforceable reductions in emissions (*Chemical and Engineering News*, 2005a, 2005b). Also, individual states in the United States are starting to require emission control efforts for coal burning facilities under their jurisdiction (*Chemical and Engineering News*, 2004a, 2004b; Christen, 2004).

12.1.4 Health Effects

Although it will come as a surprise to many college-age students, liquid mercury was a common medicine in the distant past, when people would even drink small quantities of the liquid metal to treat a variety of illness. Fortunately, this elemental form of mercury is not significantly absorbed across the intestinal lining. This is why doctors are not alarmed when a child breaks a mercury-filled thermometer. Still, there is a movement to replace mercury thermometers for medical use, in order to decrease the amount of mercury pollution created. Mercury vapor, on the other hand, is very toxic, since it can pass the lung–blood vessel barrier and proceed directly to the brain. Ionic mercury in water or food is easily absorbed into the blood and is toxic. Methylmercury, the most toxic from of mercury, can be directly absorbed through the skin, but this is mostly a concern for chemists working in the laboratory.

Sufficiently high exposure to any form of mercury can damage the gastrointestinal tract, the nervous system, and the kidneys. Symptoms of exposure to inorganic mercury include skin rashes and dermatitis, mood swings, memory loss, mental disturbances, and muscle weakness. The most common source of methylmercury to the average citizen is from eating fish that have accumulated methylmercury from the water. Infants can also be exposed via their mothers, who bioaccumulate mercury especially in their breast milk. Fetuses, infants, and children exposed to methylmercury have impaired neurological development (www.epa.gov). Methylmercury exposure also impacts cognitive thinking, memory, attention, language, and fine motor and visual spatial skills (www.epa.gov).

12.2 LEAD

12.2.1 Sources

Minerals. There are a number of minerals that contain lead. These include galena (PbS; the most economically important lead mineral in the Earth's crust), boulangerite ($Pb_5Sb_4S_{11}$; found in vein deposits associated with galena, stibnite, sphalerite, pyrite, quartz, siderite, and other lead sulphosalts), cerussite ($PbCO_3$; a common mineral in oxidized zones of ore deposits containing galena), anglesite ($PbSO_4$; a secondary mineral found in the oxidized zone of ore deposits containing galena), and pyromorphite ($Pb_5(PO_4)_3Cl$), minetite ($Pb_5(AsO_4)_3Cl$), and vanadinite ($Pb_5(VO_4)_3Cl$) (three minor secondary minerals found in the oxidized zone of galena deposits) (Berry and Mason, 1959, pp. 307, 345, 417, 430, and 455).

Synthetic Sources. The four most common sources of synthetic lead are lead–lead oxide batteries, lead in paints (typically present in the past, at a level of 0.5% or 5000 ppm), the pesticide lead arsenate ($Pb_3(AsO_4)_2$), and, most importantly, tetraethyl lead (as a gasoline additive).

12.2.2 Production/Use

To reduce lead from its principal ores (galena, anglesite, and cerussite) to its base metal (Pb), the ores are roasted (in a simple process also referred to as calcining or smelting). Lead has a variety of uses in our industrialized world and is common in (a) maintenance-free batteries, metal sheeting, piping, cable covers, ammunition, and foil and (b) alloys such as pewter, solder, babbitt-metal, bronzes, and anti-friction metals (Gribble, 1988). Lead has been used as pigments, especially in paints (as $Pb(OH)_2(CO_3)_2$, replaced in modern paints with TiO_2), glass-making (as PbO_2), and the rubber industry. One of the most common uses of lead was in an anti-knock compound (tetraethyl lead) in leaded gasoline. Production figures for 1985 show that 4.11 megatons (Mt) of lead were smelted, with the dominant producers being the United States (1.0 Mt), the former USSR (0.50 Mt), West Germany (0.36 Mt), the

United Kingdom (0.31 Mt), Canada (0.24 Mt), Australia (0.22 Mt), and France (0.22 Mt) (Gribble, 1988, pp. 180–183).

12.2.3   Fate

Lead is one of the most common forms of heavy metal pollution throughout the world, due primarily to its use as a gasoline additive. In the environment, lead is predominately present, and absorbed in organisms, as a free cation (Pb^{2+}) and is classified as a toxic heavy metal. Lead is only slightly soluble in water and is mostly found adsorbed to minerals in soil and sediments. Lead from smelting operations and leaded gasoline was dispersed into the atmosphere, so respiration is another possible intake pathway. Lead has not been found to be readily adsorbed by plants but does enter the food chain through deposits on food crops. Lead is not known to bioaccumulate in fish, nor is it methylated like mercury.

12.2.4   Environmental Distribution

As noted above, lead is one of the most commonly found heavy metals although it is at part per million and part per billion levels. The U.S. EPA requires action when the level of lead in drinking water reaches 15 ppb, but prefers a concentration of "zero," recognizing that almost any level of lead can be harmful to infants. Inhalation and ingestion of lead-contaminated paint chips are a common source of lead to infants, and they have received much public attention in recent years. In the past, however, widespread combustion of leaded gasoline was the major source of lead to all living organisms.

Lead has been known to be toxic for thousands of years, and the addition of lead to gasoline in the early 1900s came as a shock to most scientists, who certainly realized that if you put a toxic compound like tetraethyl lead in your gasoline tank, it would of course be spewed out into the world as you drive your car. However, the gasoline industry chose the cheaper anti-knocking agent regardless, and we faced decades of lead exposure even though other anti-knocking agents for our combustion engines were readily available (see Kitman, 2000, a must-read for environmentalists, listed as the suggested reading for Chapter 9, available for viewing at http://www.globalleadnet.org/advocacy/initiatives/nation.cfm or for download at http://www.globalleadnet.org/pdf/Toxicity_of_Lead.pdf).

Human intake and blood levels of Pb have been closely correlated with twentieth-century leaded gas use, and an excellent plot that demonstrates this trend has been published by Lippmann (1990). Fortunately, this means that biological lead levels dropped after leaded gasoline was replaced by unleaded, during the 1970s. Lead blood levels fell from 16 µg/dL in 1976, at the height of our leaded gasoline consumption, to less than 10 µg/dL in 1980, and they have continued to fall in countries using unleaded gasoline. (Lippmann, 1990) Today, our estimated intake of Pb from all sources is approximately 50 µg/day in the United States. In the United States, unleaded gasoline was introduced in the 1970s and slowly took over the market as all cars were equipped with catalytic converters. The U.S. EPA banned the use of leaded gasoline in highway vehicles in December 1995; however, it is still used in airplane fuel. These efforts resulted in a decrease of lead in the air over the United States by 94% between 1980 and 1999 and a 78% decrease in the levels of lead in human blood between 1976 and 1991. These improvements are clearly indicated by a survey of lead concentrations by EPA shown in Figure 12.1 and is a direct result of the reduction in lead shown in Figure 12.2*a,b*.

Many developing countries still use leaded gasoline. By the early 1990s, however, Brazil and Canada had phased out lead in gasoline, and many other countries (Argentina, Iran, Israel, Mexico, Taiwan, Thailand, and most EU countries) significantly reduced the lead concentration in gasoline. Unfortunately, lead exposure and emissions on the global scale are on the increase, due primarily to the continued use of leaded gasoline in Africa, Asia, and South America.

12.2.5 Health Effects

Humans. Lead has been known to be toxic to humans as far back as 3800 B.C., when the Greeks noticed that drinking acidic beverages from lead containers could result in illness. Once lead is absorbed into the body, it enters the bloodstream and is transported to all parts of the body. Some is deposited into our bones, where it replaces calcium because Ca^{2+} and Pb^{2+} have similar ionic radii. This can result in brittle bones after years of accumulation. Lead also inhibits the function of enzymes

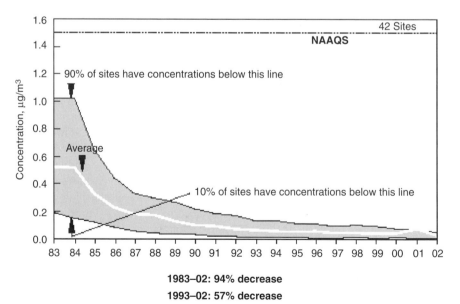

Figure 12.1. Lead air quality for a variety of sites in the United States from 1983 to 2002. (Source of data: U.S. EAP at www.epa.gov.)

involved in the synthesis of heme, the porphyrin binding complex in hemoglobin that serves as the binding site of O_2. Lead affects nerve cells by decreasing the nerve conduction velocity, even at relatively low blood levels. Lead also causes damage to the kidneys, liver, brain, and nerves and can result in seizures, mental retardation, behavioral disorders, memory problems, and mood changes. It also causes high blood pressure and increases heart disease and anemia. Lead toxicity primarily occurs due to lead's ability to bind to critical proteins that are also nitrogen and sulfur ligands, and thus to interfere with their function. Lead can be removed from the body by intravenous injection of metal chelators that compete for the binding of Pb with these proteins. Chelated Pb is then excreted from the body by the kidneys.

A common source of lead toxicity to wildlife is the deposition of fishing weights and lead shot from munitions in river and lake sediments. Waterfowl, sorting through these sediments for food, ingest the lead particles, which are then dissolved

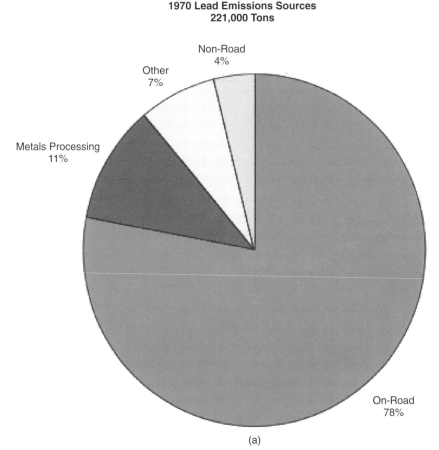

(a)

Figure 12.2. Lead emissions in the United States for years (*a*) 1970 and (*b*) 1997. (Source of data: U.S. EPA at www.epa.gov.)

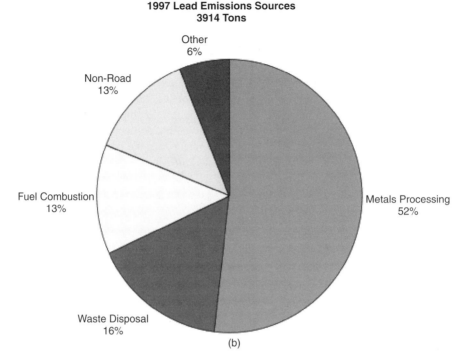

Figure 12.2. *Continued*

in their acidic stomachs and spread throughout their bodies in high doses. The United States has replaced lead shot with other metals.

12.3 PCBs

12.3.1 Sources

Synthetic. PCBs are produced by the chlorination of biphenyl with chlorine gas. There are 10 possible sites for chlorination on the biphenyl ring, and more than one site can be chlorinated. This leads to 209 possible PCB compounds, referred to as congeners. One such structure is shown in Figure 12.3, for 2,2′,4,4′,6,6′-hexachlorobiphenyl. For simplicity purposes, the governing body of chemical nomenclature, IUPAC, has developed a system for numbering the PCB congeners from 1 to 209.

12.3.2 Production/Use

The industrial synthesis of PCBs is a bit crude and does not result in pure congeners but in complex mixtures, referred to in the United States as Aroclors, the trade name patented by U.S. Swann Chemical Company in 1929. U.S. Swann Chemical

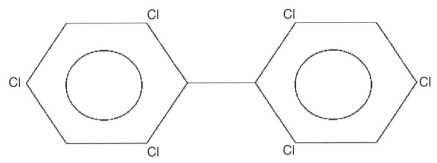

Figure 12.3. Structure of 2,2′,4,4′,6,6′-hexachlorobiphenyl, a PCB.

Company was later bought in 1935 by Monsanto Chemical Company, which became the main manufacture of PCBs in the United States. Trade names for industrial-grade PCBs include Clophen (Bayer, Germany), Phenoclor (Caffaro, Italy), Pyralene (Kanegafuchi Chemical Company, Japan), Kanechlor (Prodelec, France), Fenchlor (Chemko, former Czechoslovokia), and Delor (from the former USSR), but there are many others.

The common Aroclor mixtures include Arochlors 1221, 1232, 1242, 1248, 1254, 1260, 1262, 1268, and 1270. The basic rules of Aroclor nomenclature are that the first two numbers in the name refer to the 12-carbon basic structure of the biphenyl structure. The last two numbers refer to the average degree of chlorination of the biphenyl rings. Note that any given Aroclor mixture will contain tens of different PCB molecular structures. The degree of chlorination is a result of the reaction time with chlorine, which is determined by the desired chemical and physical properties. As a side note, it is interesting to note the shift in nomenclature when PCBs started receiving a "bad name." At this time, a "new" Aroclor was released that was supposed to have less detrimental environmental properties. The new Aroclor was numbered 1016. The selection of "10" was unclear, but in keeping with the established numbering system, the "16" was clearly meant to imply a lower degree of chlorination. In fact, Aroclor 1016 was essentially the same as Aroclor 1242.

During the early years of their production, PCBs were thought of as the perfect chemical—they were highly unreactive and did not degrade. However, this means that PCBs are also very persistent in the environment, as we will discuss in the next section. The unreactive nature of PCBs lends them to many functions, including use as dielectric fluids in capacitors and transformers, as heat transfer fluids, as hydraulic fluids, and in lubricating and cutting oils. PCBs were also employed as additives in pesticides, paints, copying paper, carbonless copy paper, adhesives, and sealants, as well as in plastic formulations.

Little was known about the dangers of PCBs prior to the 1980s. For example, at the height of the environmental movement in the mid-1970s, PCBs were presented in undergraduate microbiology classes as an ideal fluid for preparing microscope slides, due to the fact that they do not cause refraction of light. Students were neither told the name of the fluid nor advised to take precautions in handling it, and therefore they came into extensive contact with the fluid. Of course, in the 1970s, there was no such things as material safety data sheet (MSDS) or the Citizen's Right-to-Know Act. Still, this example demonstrates the importance of always having and reading the label of chemicals that you work with.

One of the most common and unfortunate uses of PCBs was as dielectric fluid, used as a fluid insulator in electrical systems. In order for a chemical to be a good dielectric fluid, it should be an excellent electrical insulator (a nonconductor) and be heat resistant and not thermally degrade, since most electrical equipment becomes very hot. Aroclor mixtures were ideal for this, and for a time they were legally placed in all transformers and capacitors. But transformers wear out, on occasion explode, and sometimes leak, spreading PCBs wherever transformers were used or disposed of.

The Toxic Substances Control Act of 1976 gave EPA the authority to regulate all aspects of the manufacture, distribution, use, and disposal of chemicals, including PCBs. The U.S. Congress banned the manufacture, processing, distribution in commerce, or use of PCBs after January 1, 1978. Congress further specified that after January 1, 1979, no person could process or distribute in commerce any PCBs unless granted permission by the EPA. But the United States Congress only regulates the United States, and some chemical companies simply shifted their operations to countries where PCBs were, and in some cases still are, legal.

12.3.3 Fate and Environmental Distribution

PCBs are hydrophobic compounds, and therefore they prefer to sorb to virtually any surface, especially soil and sediment media, rather than remain free in water. Thus, dissolved water concentrations will be very low in contaminated environments, while soils and sediments, especially those containing organic matter will contain relatively high concentrations of PCBs. This distribution of PCBs between phases occurs by a partitioning process, as discussed in Chapters 2 and 3. Even higher concentrations of PCBs will be found in the lipid deposits of biota, enabling PCBs to readily bioaccumulate in animals.

In general, PCBs are considered to be nonbiodegradable, although some laboratory studies do show considerable degradation in enriched, highly specialized bacterial cultures. Some in the chemical industry like to use these studies to claim that the PCBs are not a persistent problem, since they have been shown to degrade. But if this were true, and the laboratory rates and conditions were applicable to the natural world, PCBs would have been degraded decades in the past. The clearly documented presence of PCBs in soils, sediments, and biota around the world show that PCBs do in fact persist in natural environments.

PCBs are truly one of our ubiquitous pollutants, in that they can be detected in every environmental sample around the world, given sufficiently low detection limits. In general, PCB levels around the world are decreasing—but slowly, given the refractory nature of PCBs. A few examples of PCB concentrations in animals are marine organisms (0.003 to 212 µg/g), fish (0.1 to 190 µg/g), birds (0.1 to 14,000 µg/g), and humans (0.001 to 75.5 µg/g). Concentrations for environmental compartments range from 0.004 to 36 ng/m³ for ambient air, from 5 to 10,800 36 ng/m³ for indoor air, from 9.8 to 58,000 ng/m³ for stack gas, from 0.01 to 10 × 10⁶ ng/m³ for occupational sites, from 0.004 to 4200 ng/L for water, from 0.1 to 250 ng/L for rain, from 0.02 to 17,800 μg/g for soil, and from 0.00008 to 61,000 μg/g for sediment.

12.3.4 Health Effects

PCBs are listed by EPA as a probable human carcinogen and studies indicate that toxicity is directly related to the chlorine substitution pattern. The lack of ortho-chlorine (chlorines in the 2, 2′, 6, or 6′ position) give the PCB a similar conformation as chlorinated dioxins (2,3,7,8-tetrachlorodioxin is one of the most toxic chemicals known to humans). Toxicology for PCBs is also very species-specific, with some mammals showing high susceptibility to birth defects and cancer related to exposure.

12.4   DDT

12.4.1 Sources

Synthetic. DDT was first synthesized in 1873 by Othmar Zeidler, who was working at the University of Strasbourg. The structure for DDT and its degradation produces, DDD and DDE, are shown in Figure 12.4.

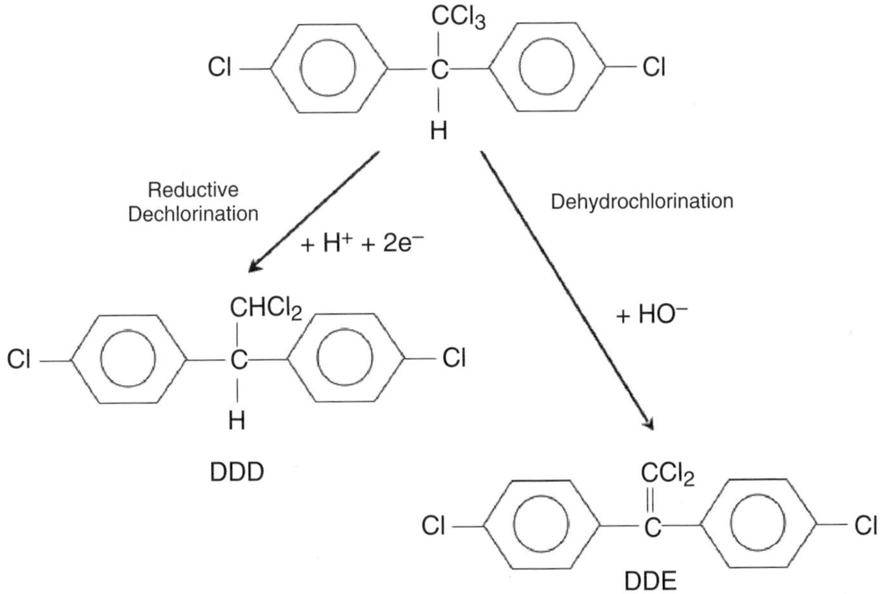

Figure 12.4.   Structure of DDT and its major degradation products.

12.4.2 Production/Use

DDT is one of many chlorinated pesticides and Persistent Organic Pollutants (POPs) that have been in the news for decades. Recently, efforts have focused on an international treaty to ban their production and use (*Chemical and Engineering News*, 2004; Burke, 2004; Hogue, 2004). POPs of immediate interest include aldrin, chlordane, DDT, dieldrin, dioxins, eldrin, furans, heptachlor, hexachlorobenzene, mirex, PCBs, and toxaphene. In 2001, the Stockholm Convention on Persistent Organic Pollutants entered into effect, stating that treaty partners will begin evaluating additional chemicals for possible control starting in May 2005. Table 12.1 shows the cumulative world production and use of each chemical.

DDT was (and is) primarily used as an insecticide for mosquito control. While it is not used in most developed countries, it is commonly used in areas affected by malaria, although its effectiveness as an insecticide has diminished over time due to adaptation and evolution of the mosquito. Many people make the statement that "DDT has saved more lives than it has hurt." This is definitely true if you are only concerned with human life. DDT has been a very effective tool during wartime and in mosquito-infested countries in fighting human disease. But it has also been a disaster for other species, especially raptors, or birds of prey, such as the American bald eagle.

12.4.3 Fate

All of the POPs, including DDT, are hydrophobic chlorinated hydrocarbons and exhibit very low solubility in water (sub-ppm levels). Their behavior in the environment is similar to that of PCBs. Chlorinated hydrocarbons have high soil/water partition coefficients and highly partition to organic matter in soil and sediment, increasing the pollutant concentration on the soil and decreasing the concentration in the water. POPs are well known to bioaccumulate and are found in lipid regions in plants and organisms.

12.4.4 Environmental Distribution

DDT, and many other POPs, have been found in every environmental compartment in the world, including remote areas such as high alpine lakes (Catalan et al., 2004; Vives et al., 2004) and in the Antarctic food web (Chiuchiolo et al., 2004).

12.4.5 Health Effects

DDT and the other POPs are considered to be endocrine disruptors, interfering with the hormonal systems of the human body (and other animals). An increase in reproductive abnormalities in humans and wildlife has occurred over the last 20–40 years, and many scientists are looking closely at chlorinated hydrocarbons as the cause of these observations. Studies indicate a decrease in fertility in men and a significant increase in still births, neonatal deaths, and congenital defects.

TABLE 12.1. Production and Use of 12 POPs (Christen, 1999)

POP	Production Start Date	Cumulative World Production (tons)	Designed Use
DDT	1942	2.8–3 million	Domestic and agricultural insecticide against mosquitoes
Aldrin	1950	240,000	Insecticide to control soil pests (termites) on corn and potatoes; fumigant
Dieldrin	1950	240,000	Insecticide used on fruit, soil, and seeds
Endrin	1947	70,000	Rodenticide and insecticide used on cotton, rice, and corn
Chlordane	1952	No data	Insecticide used in fire ant control and on a variety of crops
Heptachlor	1945	1–2 million	Insecticide in seed grain and crops; termiticide; used for fire ant control
Hexachlorobenzene	1959	No data	Used as a fungicide; also a byproduct in pesticide manufacture.
Mirex	1948	1.33 million	Insecticide and fire retardant
Toxaphene	1929	1–2 million	Insecticide, especially against ticks, mites, maggots; used on cotton
PCBs			Used as weatherproofers, hydraulic fluids in transformers, liquid insulators, dielectrics, and to prolong residual activity of plastics
Dioxins			By-product of combustion-especially plastics- and of chlorine products manufacture and paper production
Furans			By-product, often bonded to dioxins

DDT adversely affects animals by causing reproductive and developmental failure, possible immune system effects, and widespread deaths of birds. Long-term exposure to DDT has been tied to neurological, hepatic, renal, and immunologic effects in animals. DDT prevents androgen from binding to its receptor, thereby blocking androgen from guiding normal sexual development in male rats. The best-known DDT effect is the interference of the estrogen system in birds of prey that results in the thinning of eggshells and premature breaking of the eggs during nesting. DDT almost single-handedly decimated the populations of bald eagles, hawks, and owls in the U.S. countryside.

12.5 ENDOCRINE DISRUPTORS

Compounds that disrupt endocrine systems in animals have been widespread for decades, but were generally thought to be active only in animals, not humans, at the low concentrations commonly found in the environment. Today, however, scientists have discovered that many widely used chemicals can act as endocrine disruptors in the human body, in much the same way that DDT is an endocrine disruptor for bald eagles and other raptors. The story of endocrine disruptors is long and involved, and is still unclear, but scientists are making daily headway into the cases and effects of endocrine disruptors. The following section is a summary of what we currently know or suspect.

12.5.1 Sources

Synthetic. Endocrine disruptors (EDCs) include a variety of persistent organic pollutants, including DDT and many compounds used in the formulation of plastics.

12.5.2 Uses and Points of Contact

The group of commonly used endocrine disruptors is highly varied, and they are used in a wide variety of modern consumer products. Basically anything resembling a type of plastic contains potential endocrine disruptors.

Perhaps the most direct source of endocrine disruptors to people is from foods, which are exposed to EDCs either by direct application or as a result of leaching from packaging materials (Casajuana and Lacorte, 2003). Many food products are contained in material, including plastic bottles and lined aluminum cans, that contain or consist of endocrine disrupting compounds that can leach into the food (Casajuana and Lacorte, 2003).

However, there are many sources of endocrine-disrupting compounds in our environment other than food, including medical supplies, plastic children's toys, and the food and waste produced by livestock treated with hormones. Estrogen hormones are excreted by both sexes and all species of farm animals in both urine and feces, creating the potential for environmental contamination (Hanselman et al., 2003). Bovine waste in the United States is a much larger source of hormones, even, than human waste, due to the widespread use of hormones and large scale of cattle production (Raman et al., 2004). However, the use of livestock waste in agricultural fields for fertilizer does dilute the estrogen chemicals, whereas human waste from wastewater treatment plants are dispersed directly into the environment, acting as point sources of EDCs (Thacker, 2004). The persistence of EDCs also makes them difficult to eliminate in the wastewater treatment process.

One specific type of endocrine disruptors, the polybrominated diphenyl ethers (PBDEs), are used as flame retardants and are used on indoor objects such as furniture, carpeting, mattresses, televisions, and certain plastic products (Stapleton et al., 2005). They are also used in (a) certain manufactured products such as polymers, resins, electronic devices, building materials, textiles, and polyurethane foam padding (Oros et al., 2005), (b) many consumer products such as electrical and office equipment, including computers, televisions, copiers, and printers, and (c) house products such as upholstery, carpeting, wall coverings, and ceiling materials. Because of their effectiveness in suppressing fires, they have been applied to many items in large amounts, to prevent property damage. They may also be spread into the environment by combustion, either in routine garbage incineration or in major events such as the destruction of the World Trade Center in 2001 (Litten et al., 2003). The health effects of PDBEs include reproductive and developmental toxicity and cancer, as discussed below (Schecter et al., 2004).

Much public attention has been given to the possible dangers of polycarbonate water bottles. Polycarbonate contains the EDC bisphenol-A, which interferes with estrogen uptake by binding to estrogen receptors (Whittelsey, 2003). There have been many warnings against washing polycarbonate bottles with harsh detergents and against the use of polycarbonate bottles that have been harshly used or are old, because they are more likely to release bisphenol-A into their contents. Many water bottles are made of polycarbonate and are thought to release bisphenol-A under these conditions (*Non-Toxic Times*, 2004). Unfortunately, the specific effects of small bisphenol-A to the human body are not well enough understood to know whether the levels leached from water bottles could be harmful.

12.5.3 Fate and Environmental Distribution

Endocrine disruptors, although they include a wide variety of compounds, are generally hydrophobic, with behavior in the environment like that of PCBs and DDT. EDCs are also known to bioaccumulate through food chains.

A study of the bioaccumulation of two perfluorinated compounds and known endocrine disruptors—PFOS (perfluoroocotanesulfonate) and PFOA (perfluorooctanoate)—in an Eastern Arctic marine food web showed that there exists a positive linear relationship between PFOS concentrations in organisms based on their wet weight and trophic level, indicating that a higher trophic level corresponds to more PFOS (Tomy et al., 2004).

Endocrine disruptors are found in nearly every natural and human environment. PBDEs in particular have been measured in harbor seals, fish, and a local municipal wastewater treatment plant effluent in San Francisco. Additionally, a study in 2002 measured PBDEs in water, surface sediments, and bivalves in the San Francisco Estuary. Bivalves populations are known to reflect water conditions in an ecosystem and to demonstrate the level of contaminant bioavailability in the water column. They are also an important food source for various fish. When compared to levels of PBDEs in European water sources, the concentrations in the San Francisco Estuary were much higher but similar to other water sources in the United States, even though California has some of the strictest state regulations regarding the use of PBDEs.

A study comparing PBDE concentrations in supermarket salmon from eight different farming areas in North America and Europe found that the highest concentration of PBDEs in salmon came from farmed salmon in Europe, followed by farmed salmon in North America, and then wild salmon from the Pacific Ocean. However, all the different fish samples showed the presence of PBDE. The highest PBDE levels, in Chinook salmon, have been attributed to the Chinook perhaps being at a higher trophic level than other salmon, resulting in more bioaccumulation of these compounds (Hites et al., 2004).

Some studies have revealed possible solutions to the presence of endocrine disrupting compounds in various environmental compartments. As previously mentioned, there are high amounts of estrogen compounds in sewage from sources such as human and animal waste. A few previously suggested solutions to this problem have been ozonation, UV-radiation, membrane filtration, and activated carbon adsorption. However, all of these suggestions would add significant costs to the sewage treatment process. An article by Andersen et al. (2003) states that an "activated sludge system for nitrification and denitrtification including sludge recirculation can appreciably eliminate natural and synthetic estrogens." For example, some data reveal that E1 and EE2 (two forms of estrogen) were reduced by 50% and 70%, respectively, after the first denitrification tank. Natural estrogens can then be reduced further, biologically, in the second denitrification tank, achieving an overall removal of more than 98%. A better understanding of the fate of estrogens is necessary in order to develop more efficient and less expensive methods to eliminate estrogens from wastewater (Andersen et al., 2003).

Another suggested method of removing endocrine disrupting compounds from the environment is by using plants to "mobilize and translocate DDT" and other persistent organic pollutants (POPs) from soil environments. With more weathering of a soil, it is more difficult to remove POPs, which tend to bioaccumulate due to their persistence in the environment. Plants may ingest these contaminants through soil water, where they are translocated to the upper part of the plant and may be metabolized or mineralized. This process, called *phytoremediation*, uses vegetation to treat polluted soils, sediments, groundwater, or surface. One study tested five plants for their ability to phytoremediate: zucchini, tall fescue, alfalfa, rye grass, and pumpkin. Pumpkin had the highest value of pollutants removed from the soil, suggesting that it is the most valuable plant for phytoremediation (Lunney et al., 2004).

12.5.4 Health Effects

Endocrine disrupting compounds interfere with the many functions of the endocrine system. These functions include the regulation of cellular proliferation and differentiation, growth, development, reproduction, senescence, behavior, regulating mood, tissue function, and metabolism (Berne and Levy, 1990, p. 779). Endocrine cells or glands send messages through the body in the form of hormones or other molecular signals, which are carried through the bloodstream and act on other targeted cells that contain receptors for specific hormones. The hormones come in contact with many cells but only interact with those containing the correct receptor (*The New Encyclopaedia Britannica*, 2002, p. 288). The response is to either stimulate or repress the transcription of proteins that are necessary to perform certain bodily functions (Campbell and Reece, 2002, p. 960). This pathway controls the performance of many bodily functions.

Endocrine disrupting compounds (EDCs) cause harm by either mimicking hormones or blocking their receptors, and they have been a source of concern for humans and wildlife for many years because of the serious effects they have on the endocrine system. Some possible effects of EDCs on the body are increased likelihood of testicular, prostate, and breast cancer, decreased sperm count and quality, and other reproductive disorders (Lee et al., 2003). In recent years, the concern about endocrine disrupting compounds present in the environment has increased as more people have become aware of their widespread use and potentially harmful effects. In response, scientists around the world have performed many studies to determine the levels of EDCs in various consumer products, identify their specific effects, and evaluate the potential dangers associated with their ubiquity in the environment.

In addition to DDT, other endocrine disrupting compounds present in natural waters mimic steroidal estrogen hormones and can have strong effects on the reproductive abilities of aquatic wildlife such as fish, turtles, and frogs (Hanselman et al., 2003). In 2003, a group of scientists from Europe performed a study examining the extent of the effects of endocrine disrupting compounds on fish from aquatic ecosystems in Europe. Estrogen in fish stimulates the synthesis of the protein vitellogenin, used in the development of oocytes. Therefore, the presence of estrogen-active endocrine disrupters in the fish's environment also stimulates the production of vitellogenin, which results in feminization of males. Although not every fish responds to contact with contaminated water, an overall increase in vitellogenin levels in a population can indicate the presence of endocrine disrupters in their environment (Pickering and Sumpter, 2003).

In the Potomac River near Sharpsburg, Maryland, feminization of male fish has been observed to produce eggs inside their sex organs. The discovery has been linked to endocrine disrupters inhibiting normal sexual development due to inhibition of hormonal signals. Of course, the Potomac also supplies drinking water for the Washington, D.C. area (Associated Press, 2004). Studies such as this have led to wider concern about the effects of EDCs and their widespread presence in the environment.

PBDEs, a group of endocrine disruptors discussed above, have been detected in human serum, adipose tissue, and breast milk. Their levels are 17 times higher in people living in the United States than in those living in Europe (Stapleton et al., 2005). The presence of PBDEs in mice have resulted in neurotoxicity (Oros et al., 2005), and the metabolism of PBDEs is thought to be similar to that of PCBs because of their similar structures. Therefore, PBDEs accumulate in organisms and may affect them similarly to PCBs (Tuerck et al., 2005). PBDEs alter the thyroid hormone homeostasis and weakly prevent the function of estrogen receptors. More specifically, in neonatal mice, they "disrupt spontaneous behavior, impair learning and memory, and induce other neurotoxic effects. . . ." (Hites et al., 2004)

So what do we do with all this information about endocrine disrupting compounds in our bodies and the environment? The various studies mentioned previously reveal the many different sources of EDCs: wastewater effluent, food, plastic, furniture, medical supplies, livestock waste, etc. The levels (concentrations) of EDCs in the human body seem to be on the same magnitude as the concentrations of our own hormones. Additionally, many of these chemicals bioaccumulate through the food chain, so animals that eat higher on the food chain will contain even higher levels of EDCs. Ongoing animal testing seeks to discover the effects of EDCs on humans, although, as always, interspecies extrapolation will no doubt prove very difficult.

Endocrine-disrupting compounds may seem completely unavoidable in modern society, due their widespread presence, persistence, and tendency to bioaccumulate. Clearly, continued research into their effects, their alternatives, and their environmental fates is necessary to provide the knowledge necessary to evaluate the costs and benefits of their omnipresence in our lives.

REFERENCES

For the Mercury Section

Baird, C. *Environmental Chemistry*, W. H. Freeman, New York, 1995.

Berry, L. G. and B. Mason, *Mineralogy: Concepts, Descriptions, and Determinations*, W. H. Freeman, San Francisco, 1959.

Betts, K. Dramatically improved mercury removal. *Environ. Sci. Technol.* **August 1**, 283–284A (2003).

Chemical and Engineering News, Senators criticize mercury controls at chlorine plants, **May 31**, p. 18 (2004a).

Chemical and Engineering News, Massachusetts cuts mercury at power plants, **May 31**, p. 18 (2004b).

Chemical and Engineering News, EPA hit for mercury air standard, **February 14**, p. 24 (2005a).

Chemical and Engineering News, EPA's mercury analysis biases, GAO says, **March 14**, p. 29 (2005b).

Christen, K. States going it alone on mercury controls. *Environ. Sci. Technol.* **38**(9), 157A (2004).

Crenson, S. L. Study records elevated mercury. *Associated Press*, Sunday, **October 20** (2002).

Gribble, C. D. *Rutley's Elements of Mineralogy*, 27th edition, Unwin Hyman, London, 1988.

Pacyna, E. G. and J. M. Pacyna. Global emission of mercury from anthropogenic sources in 1995. *Water, Air, and Soil Pollution* **137**, 149–165 (2002).

Renner, R. Mercury woes appear to grow. *Environ. Sci. Technol.* **38**(8), 144A (2004).

Rouhi, A. M. Mercury showers. *Chem. Eng. News*, **April 15**, p. 40 (2002).

Saar, R. A. *New York Times*, New Efforts to Uncover the Dangers of Mercury, Health and Fitness section, p. D7, Tuesday, **November 2**, 1999.

Seigneur, C., K. Vijayaraghavan, K. Lohman, P. Karamchandani, and C. Scott. Global source attribution for mercury deposition in the United States. *Environ. Sci. Technol.* **38**(2), 555–569 (2004).

Spiro, T. G. and W. M. Stigliani. *Chemistry of the Environment*, 2nd edition, Prentice-Hall, Upper Saddle River, NJ, 2003.

USGA Fact Sheet FS-095-01, 2001.

Wright, K. Our Preferred Poison, *Discover*, **March** (2005).

www.epa.gov

For the Lead Section

Baird, C. *Environmental Chemistry*, W. H. Freeman, New York, 1995.

Berry, L. G. and B. Mason. *Mineralogy: Concepts, Descriptions, and Determinations*, W. H. Freeman, San Francisco, 1959.

Gribble, C. D. *Rutley's Elements of Mineralogy*, 27th edition, Unwin Hyman, London, 1988.

Kitman J. L. *The Secret History of Lead The Nation*, March 20, 2000.

Lippmann, M. Lead and human health: Background and recent findings. *Environ. Res.* **51**, 1–24 (1990).

Spiro, T. G. and W. M. Stigliani. *Chemistry of the Environment*, 2nd edition, Prentice-Hall, Upper Saddle River, NJ, 2003.

USGS Fact Sheet FS-095-01, Mercury in Coal—Abundance, distribution, and modes of occurrence, U. S. Department of the Interior, USGS, **September**, 2001.

www.epa.gov

www.usgs.gov

For the PCB Section

Erickson, M. D. *Analytical Chemistry of PCBs*, 2nd edition, Lewis Publishers, New York, pp. 1–98, 1997.

For the DDT Section

Baird, C. *Environmental Chemistry*, W. H. Freeman, New York, 1995.

Burke, M. POPs treaty takes flight. *Environ. Sci. Technol.* **38**(9), 157A (2004).

Catalan, J., M. Ventura, I. Vives, and J. O. Grimalt. The roles of food and water in the bioaccumulation of high mountain lake fish. *Environ. Sci. Technol.* **38**(16), 4269–4275 (2004).

Chemical and Engineering News, Persistent pollutant pact enters into force, **May 24**, p. 24 (2004).

Chiuchiolo, A. L., R. M. Dickhut, M. A. Cochran, and H. W. Ducklow. Persistent organic pollutants at the base of the Antarctic marine food web, *Environ. Sci. Technol.* **38**(13), 3551–3557 (2004).

Christen, K. U.N. negotiations on POPs snag on malaria. *Environ. Sci. Technol.* **33**(21), 444–445A (1999).

Hogue, C. U.S. vote at treaty meetings threatened. *Chem. Eng. News* **March 29**, 22–23 (2004).

Spiro, T. G. and W. M. Stigliani. *Chemistry of the Environment*, 2nd edition, Prentice-Hall, Upper Saddle River, NJ, 2003.

Vives, I., J. O. Grimalt, J. Catalan, B. O. Rosseland, and R. W. Battarbee. Influence of altitude and age in the accumulation of organochlorine compounds in fish and high mountain lakes. *Environ. Sci. Technol.* **38**(3), 690–698 (2004).

For the Endocrine Disruptor Section

Andersen, H., H. Siegrist, B. Halling-Sorensen, T. A. Ternes. Fate of estrogens in a municipal sewage treatment plant. *Environ. Sci. Technol.* **37**(18), 4021–4026 (2003).

Associated Press. Male Fish Growing Eggs Found in Potomac. **December 21**, 2004, 8:40 A.M. ET.

Berne, R. M. and M. N. Levy. *Principles of Physiology*, C. V. Mosby, St. Louis, 1990.

Brossa, L., E. Pocurull, F. Borrull, R. M. Marcé. A rapid method for determining phenolic endocrine disrupters in water samples. *Chromatographia* **56**(9/10), 573–576 (2002).

Campbell, N. A. and J. B. Reece. *Biology*. Benjamin Cummings, San Francisco, 2002.

Carson, R. *Silent Spring*, The Riverside Press Cambridge, Boston, 1962.

Casajuana, N. and S. Lacorte. Presence and release of phthalic esters and other endocrine disrupting compounds in drinking water. *Chromatographia* **57**(9/10), 649–655 (2003).

Dunlap, T. R. *DDT: Scientists, Citizens, and Public Policy*, Princeton University Press, Princeton, 1981.

Guenther, K., V. Heinke, B. Thiele, E. Kleist, H. Prast, and T. Raecker. Endocrine disrupting nonylphenols are ubiquitous in food. *Environ. Sci. Technol.* **36**(8), 1676–1680 (2002).

Hanselman, T. A., D. A. Graetz, and A. C. Wilkie. Manure-borne estrogens as potential environmental contaminants: A review. *Environ. Sci. Technol.* **37**(24), 5471–5478 (2003).

Hileman, B. Clash of views on bisphenol A: Finding gives rise to controversy over interpretation of scientific literature. *Chem. Eng. News* **May 5**, pp. 40–41 (2003).

Hileman, B. FDA suggests replacing DEHP in plastics. *Chem. Eng. News* **September 16**, ••–•• (2002).

Hileman, B. Government regulation, phthalates in toys: EU bans some phthalates in toys and child care products. *Chem. Eng. News* **October 4**, p. 6 (2004).

Hites, R. A., J. A. Foran, S. J. Schwager, B. A. Knuth, M. C. Hamilton, and D. O. Carpenter. Global assessment of polybrominated diphenyl ethers in farmed and wild salmon. *Environ. Sci. Technol.* **38**(19), 4945–4949 (2004).

Inoue, K., K. Kato, Y. Yoshimura, T. Makino, and H. Nakazawa. Determination of bisphenol A in human serum by high-performance liquid chromatography with multi-electrode electrochemical detection. *J. Chromatogr. B* **749**, 17–23 (2000).

Lee, L. S., T. J. Strock, A. K. Sarmah, P. Suresh, and C. Rao. Sorption and dissipation of testosterone, estrogens, and their primary transformation products in soils and sediment. *Environ. Sci. Technol.* **37**(18), 4098–4105 (2003).

Litten, S., D. J. McChesney, M. C. Hamilton, and B. Fowler. Destruction of the World Trade Center and PCBs, PBDEs, PCDD/Fs, PBDD/Fs, and chlorinated biphenylenes in water, sediment, and sewage sludge." *Environ. Sci. Technol.* **37**(24), 5502–5510 (2003).

Lunney, A. I., B. A. Zeeb, and K. J. Reimer. Uptake of weathered DDT in vascular plants: Potential for phytoremediation. *Environ. Sci. Technol.* **38**(22), 6147–6154 (2004).

The New Encyclopaedia Brittanica, Vol. 18, 15th edition, Encyclopaedia Britannica, Inc., Chicago, 2002.

North, K. D. Tracking polybrominated diphenyl ether releases in a wastewater treatment plant effluent, Palo Alto, California. *Environ. Sci. Technol.* **38**(17), 4484–4488 (2004).

On the trail of water bottle toxins . . . Are hikers and others quenching outdoor thirst with H_2-UH-OH? *The Non-Toxic Times* **April**, (2004).

Oros, D. R., D. Hoover, Francois Rodigari, David Crane, Jose Sericano. Levels and distribution of polybrominated diphenyl ethers in water, surface sediments, and bivalves from the San Francisco Estuary. *Environ. Sci. Technol.* **39**(1), 33–41 (2005).

Personal communication with Paul H. Yancey, Biology Department, Whitman College, 2004.

Pickering, A. D. and J. P. Sumpter. Comprehending endocrine disrupters in aquatic environments. *Environ. Sci. Technol.* **37**(17), 331A–336A (2003).

Raman, A. R., E. L. Williams, A. C. Layton, R. T. Burns, J. P. Eater, A. S. Daugherty, M. D. Mullen, and G. S. Sayler. Estrogen count of dairy and swine wastes. *Environ. Sci. Technol.* **38**(13), 3567–3573 (2004).

Schecter, A., O. Papke, K.-C. Tung, D. Staskal, and L. Birnbaum. Polybrominated diphenyl ethers contamination of United States food. *Environ. Sci. Technol.* **38**(20), 5306–5311 (2004).

Sherwood, L., H. Klandorf, and P. H. Yancey. *Animal Physiology: From Genes to Organisms*, Brooks/Cole, Australia, 2005.

Stapleton, H. M., N. G. Dodder, J. H. Offenberg, M. M. Schantz, and S. A. Wise. Polybrominated diphenyl ethers in house dust and clothes dryer lint. *Environ. Sci. Technol.* **39**(4), 925–931 (2005).

Suzuki, T., Y. Nakagawa, I. Takano, K. Yguchi, and K. Yasuda. Environmental fate of bisphenol A and its biological metabolites in river water and their xeno-estrogenic activity. *Environ. Sci. Technol.* **38**(8), 2389–2396 (2004).

Thacker, P. D. Livestock flood the environment with estrogen. *Environ. Sci. Technol.* **38**(13), 241A–242A (2004).

Tomy, G. T., W. Budakowski, T. Halldorson, P. A. Helm, G. A. Stern, K. Friesen, K. Pepper, S. A. Tettlemier, and A. T. Fisk. Fluorinated organic compounds in an eastern Arctic marine food web. *Environ. Sci. Technol.* **38**(24), 6475–6481 (2004).

Tuerck, K. S. J., J. R. Kucklick, P. R. Becker, H. M. Stapleton, and J. E. Baker. Persistent organic pollutants in two dolphin species with focus on toxaphene and polybrominated diphenyl ethers. *Environ. Sci. Technol.* **39**(3), 692–698 (2005).

United States Department of Health and Human Services, 2002. Public Health Service. Toxicological Profile for DDT, DDE, and DDD. Agency for Toxic Substances and Disease Registry. Prepared by Syracuse Research Corporation Under Contract No. 205-1999-00024. September 2002.

Whittelsey, F. C. Hazards of hydration: Choose your plastic water bottles carefully. *Sierra Magazine* **November/December**, 16–18 (2003).

Yamamoto, T. and A. Yasuhara. Quantities of bisphenol A leached from plastic waste samples. *Chemosphere* **38**(11), 2569–2576 (1999).

PART VII

SUPPORTING LABORATORY EXPERIMENTS

"Tell me and I will listen. Show me and I will watch. Let me experience, and I will learn."

—Lao Tzu, 500 B.C.

CHAPTER 13

EXPERIMENTS

13.1   THE DETERMINATION OF ALKALINITY IN WATER SAMPLES

Purpose.   To determine the alkalinity of a natural water sample by titration.

Background.   Alkalinity is an expression of a water's ability to neutralize acids. Therefore, alkalinity measurements are important for considering the fate of acidic industrial pollution emitted directly into our waterways and as acid precipitation. Alkalinity is also a measure of a water's buffering capacity, or its ability to resist changes in pH upon the addition of acids or bases. Alkalinity in natural waters is due primarily to the presence of weak acid salts, although strong bases (e.g., OH^-) may also contribute in industrial waters. Bicarbonates represent the major form of alkalinity in natural waters and are derived from the dissolution of CO_2 from the atmosphere and the weathering of carbonate minerals in rocks and soil. Other salts of weak acids, such as borate, silicates, ammonia, phosphates, and organic bases from natural organic matter, may be present in small amounts. Alkalinity, by convention, is reported as mg/L $CaCO_3$, since most alkalinity is derived from the weathering of carbonate minerals rather than from CO_2 partitioning with the atmosphere. Alkalinity for natural water (in molar units) is typically defined as the sum of the carbonate, bicarbonate, hydroxide, and hydronium concentrations, such that

$$[\text{alkalinity}] = 2[CO_3^{2-}] + [HCO_3^-] + [OH^-] - [H_3O^+] \qquad (13.1)$$

Alkalinity values can range from zero in acid rain impacted areas, where the pH of the water will be below 5.4, to less than 20 mg/L (as $CaCO_3$) for waters in contact with non-carbonate-bearing soils, to 2000–4000 mg/L (as $CaCO_3$) for waters from the anaerobic digesters of domestic wastewater treatment plants.

Neither alkalinity nor its converse, acidity, has any known adverse health effects, although highly acidic or alkaline waters are frequently considered unpalatable. However, alkalinity can be affected by or affect other parameters. Below are some of the most important effects of alkalinity.

1. The alkalinity of a body of water determines how sensitive that water body is to acidic inputs such as acid rain. A water with high alkalinity better resists changes in pH upon the addition of acid (from acid rain or from an industrial input).

---

2. Turbidity is frequently removed from drinking water by the addition of alum, $Al_2(SO_4)_3$, to the incoming water followed by coagulation, flocculation, and settling in a clarifier. This process releases H^+ into the water through the reaction

$$Al^{3+} + 3H_2O \rightarrow Al(OH)_3 + 3H^+ \qquad \text{Reaction 1}$$

In order for effective and complete coagulation to occur, alkalinity must be present in excess of the H^+ released. Usually, additional alkalinity, in the form of $Ca(HCO_3)_2$, $Ca(OH)_2$, or Na_2CO_3 (soda ash), is added to ensure optimum treatment conditions.

3. Hard waters (with excessive metal ion concentrations) are frequently softened by precipitation using CaO (lime), Na_2CO_3 (soda ash), or NaOH. The alkalinity of the water must be known in order to calculate the lime, soda ash, or sodium hydroxide requirements for precipitation.

4. Maintaining alkalinity is important to corrosion control in piping systems. Corrosion is of little concern in modern domestic systems, but many main water distribution lines and industrial pipes are made of iron. Low pH waters cause corrosion in metal pipe systems, which are costly to replace.

5. Bicarbonate (HCO_3^-) and carbonate (CO_3^{2-}) can complex other elements and compounds, altering their toxicity, transport, and fate in the environment. In general, the most toxic form of a metal is its uncomplexed hydrated metal ion. Complexation of this free ion by carbonate species can reduce toxicity.

Safety and Hazards

- As in all laboratory exercises, safety glasses must be worn at all times.
- Avoid skin and eye contact with NaOH and HCl solutions. If contact occurs, rinse your hands and/or flush your eyes for several minutes. Seek immediate medical advice for eye contact.
- Use concentrated HCl in the fume hood and avoid breathing its vapor.

Student Procedure

Samples.   You will have several water samples to titrate. These will include one or more of the following: distilled water, deionized water, tap water, river water, and groundwater. You will determine the alkalinity of one or all of these samples (in triplicate), as per your instructor's instructions.

Procedures.   Characterizing the alkalinity of a water sample requires titrating to two different endpoints, each with its own significance. Titrating to pH 8.3 HCl gives you the total amount of hydroxide and carbonate alkalinity, while titrating farther to pH 4.5 will give you the total alkalinity.

1. First, an adequate sample volume for titration must be determined. This is accomplished by performing a test titration. Select a volume of your sample, such as 100 mL, and titrate it to pH 8.3 with standardized 0.02 MHCl solution to estimate

the carbonate alkalinity of your sample. This will tell you how much of your acid titrant is required to neutralize all the base (alkalinity) in the volume of sample that you chose. For best accuracy in adding the titrant, you should use at least 10 mL but not more than 50 mL from a 50-mL buret. Because this is the case for each of the titration endpoints, the volume required for titration to 4.5 should be at the high end of this range.

To titrate to pH 4.5, add bromcresol green or the mixed bromcresol green-methyl red indicator solution. Slowly add 0.02 M HCl, and note the color change indicating a pH value of 4.5. Alternatively, a pH meter can be used to determine the inflection point (pH 4.5).

Adjust your sample size so that it requires ~40–50 mL to titrate.

2. *Titration to pH 8.3*: Titrate the determined volume of sample with standardized 0.02 M HCl solution. Add phenolphthalein or metacresol purple indicator solution, and then add HCl slowly from a buret. The equivalence point is marked by a color change around pH 8.3. Alternatively, a pH meter can be used to determine the inflection point. This measurement will be a combination of the hydroxide and carbonate alkalinity.

3. Continue the titration to the ~4.5 endpoint (as described in 13.1 above) with the same HCl solution, either with the same sample aliquot or with a new one. Better results will be obtained by titrating a new sample to the ~4.5 endpoint. This will avoid potential color interferences between the 8.3 and 4.5 pH indicators.

4. Repeat the titrations of the sample, Steps 2 and 3, at least three times.

Calculate the hydroxide and carbonate (phenolphthalein endpoint described below) alkalinity and total alkalinities (alk.) for your samples. Report your values in mg $CaCO_3$/L. Show all calculations in your notebook.

Calculation:

$$\text{Phenolphthalein alk.} = \frac{\left(\text{L of acid to pH 8.3}\right)\left(\text{M of acid}\right)\left(100\,\text{g}/\text{mol } CaCO_3\right)\left(1\,\text{mol } CaCO_3/2\,\text{mol alk.}\right)\left(1000\,\text{mg}/\text{g}\right)}{\text{L of sample}}$$

$$\text{Total alk.} = \frac{\left(\text{L of acid to pH 4.5}\right)\left(\text{M of acid}\right)\left(100\,\text{g}/\text{mol } CaCO_3\right)\left(1\,\text{mol } CaCO_3/2\,\text{mol alk.}\right)\left(1000\,\text{mg}/\text{g}\right)}{\text{L of sample}}$$

Waste Disposal. After neutralization, all solutions can be disposed of down the drain with water.

Assignment. Report the alkalinity (as $CaCO_3$) of each sample that you titrate. Summarize your data and turn in a hard copy to your instructor.

13.2   TOTAL SUSPENDED AND DISSOLVED SOLIDS IN WATER SAMPLES

Purpose.   To determine the amount of suspended and dissolved solids in water samples.

Background.   Environmental waters may contain a variety of solid or dissolved impurities. In quantifying levels of these impurities, suspended solids is the term used to describe particles in the water column. Practically, they are defined as particles large enough to not pass through the filter used to separate them from the water. Smaller particles, along with ionic species, are referred to as dissolved solids. In considering waters for human consumption or other uses, it is important to know the concentrations of both suspended and dissolved solids.

First, let's consider some implications of total suspended solids (TSS).

- High concentrations of suspended solids may settle out onto a streambed or lake bottom and cover aquatic organisms, eggs, or macro-invertebrate larva. This coating can prevent sufficient oxygen transfer and result in the death of buried organisms.

- High concentrations of suspended solids decrease the effectiveness of drinking water disinfection agents by allowing microorganisms to "hide" from disinfectants within solid aggregates. This is one of the reasons the TSS, or turbidity, is removed in drinking water treatment facilities.

- Many organic and inorganic pollutants sorb to soils, so that the pollutant concentrations on the solids are high. Thus, sorbed pollutants (and solids) can be transported elsewhere in river and lake systems, resulting in the exposure of organisms to pollutants away from the point source.

Second, consider the importance of monitoring total dissolved solids (TDS).

- The total dissolved solids (TDS) of potable waters ranges from 20 to 1000 mg/L. In general, waters with a TDS less than 500 mg/L are most desirable for domestic use.

- Waters with TDS greater than 500 mg/L may cause diarrhea or constipation in some people.

- Water with a high TDS is frequently hard (i.e., has a high Ca^{2+} and/or Mg^{2+} concentration) and requires softening (the removal of hardness cations) by precipitation. The TDS of a water sample can be used to determine the most appropriate method of water softening, since precipitation reduces TDS while some ion exchange processes may increase TDS.

- Waters with high TDS may result in clogged pipes and industrial equipment through the formation of scale (Ca and Mg solids precipitated in the pipes).

Procedure

Note: Each of these procedures will require you to come in at unusual times during the next week. You must observe safety procedures set up by your school pertaining to working in the laboratory alone or in pairs.

Total Suspended Solids (TSS) and Total Dissolved Solids (TDS) Measurements

Overview. In this procedure, you will first use 100 mL of your sample to perform the most commonly used solids measurement, the total suspended solids (TSS). This requires you to filter a known volume of sample through a pre-heated and pre-tared glass–fiber filter, to remove the suspended solids. The difference between the initial (just filter) and final (filter with solids) weights, divided by the volume of sample, will yield the TSS. The TSS measurement accounts for all solids that do not pass through the filter (typically 0.2–0.45 μm in size). The filter is weighed after drying at 104°C, both before and after the filtering procedure. The filtrate (the remaining sample liquid) is then used in the dissolved solids determination (TDS, the next procedure). The TDS is a simple experiment where the filtrate is placed in a tared (pre-weighed) beaker and evaporated to dryness.

Student Procedures

PREPARING YOUR FILTERS

1. Rinse three filters with 20–30 mL DI to remove any solids that may remain from the manufacturing process. Place the filters in separate, labeled aluminum weight pans, dry them in a 104°C oven for 30 minutes, place them (filter and pan) in a desiccator, and obtain a constant weight by repeating the oven and desiccation steps.

OBTAINING THE TSS MEASUREMENT

2. Filter 100 mL of sample through each tared filter. Save the filtrate.

3. Place each paper in its aluminum weight pan in the 104°C oven for 1 h. Cool the filter and pan in a desiccator and obtain a constant weight by repeating the drying and desiccation steps. (This step will be completed after your normal lab meeting time.)

Calculation:

TSS mg/L

$$= \frac{(\text{average weight from step 3 in g} - \text{average inital weight from step 1 in g})(1000 \ mg/L)}{\text{sample volume in L}}$$

TOTAL DISSOLVED SOLIDS (TDS) MEASUREMENT

4. Obtain a constant weight for three 150-mL beakers, using the same procedure used for the filters in step 1.

5. Add 100 mL of your filtered sample to each beaker and evaporate it in the 98°C oven overnight.

6. The next day, place the beakers in the 104°C oven and heat them for 1 h.

7. Place the beakers in a desiccator until cool and obtain a constant weight.

8. Repeat steps 5 and 6 until you obtain a constant weight (within 0.5 mg of each other). The differences between this weight and the weight of the beakers originally will be the total masses of dissolved solids in the 100-mL samples.

Calculation:

TDS mg/L

$$= \frac{(\text{average crucible weight from step 7 in g} - \text{average crucible weight empty in g})(1000 \; mg/g)}{\text{sample volume in L}}$$

Hints for Success

- Always completely mix your sample before removing any solution/suspension. The soil/sediment particles will settle and bias your results if you do not completely mix the sample every time you remove an aliquot.

- Perform all measurements in triplicate.

- Carefully clean all containers and pre-wash all filters with DI water prior to use. As noted in the procedures, you must heat filters to the maximum temperature that you will use experimentally, before filtering. Also as noted in the procedures, you must obtain a constant weight (generally within 0.5 mg) before you end each experiment. (Fingerprints and dust weigh enough to significantly affect your results.)

- Your balances have been calibrated, but for best results you should still use the same balance for every measurement. Even if the calibration on a balance is slightly off, the change in weight will probably be accurate.

Assignment. Turn in a table showing your measurements for each sample, your calculations TSS and TDS, and your final averages.

13.3 THE DETERMINATION OF HARDNESS IN A WATER SAMPLE

In the past, water hardness was defined as a measure of the capacity of water to precipitate soap. However, current laboratory practices define total hardness as the sum of divalent ion concentrations, especially those of calcium and magnesium, expressed in terms of mg $CaCO_3$/L. There are no known adverse health effects of hard or soft water, but the presence of hard waters results in two economic considerations: (1) Hard waters require considerably larger amounts of soap to foam and clean materials, and (2) hard waters readily precipitate carbonates (known as scale) in piping systems at high temperatures. Calcium and magnesium carbonates are two of the few common salts whose solubility decreases with increasing temperature. This is due to the removal of dissolved CO_2 as temperature increases due to the changing Henry's law constant. The advent of synthetic detergents has significantly reduced the problems associated with hard water and the "lack of foaming."

However, scale formation continues to be a problem in domestic hot water heaters and for industry.

The source of a water sample usually determines its hardness. For example, surface waters usually contain less hardness than groundwaters, which spend longer in contact with geologic materials that are sources of divalent cations. Carbonates in surface soils and sediments also increase the hardness of surface waters and subsurface limestone formations also increase the hardness of ground waters. Hardness values can range from a few to hundreds of mg $CaCO_3$/L. Therefore, depending on your water's source, some modifications to the procedure below may be necessary.

The divalent metal cations responsible for hardness can react with soap to form precipitates in hot water pipes (the version known as scale occurs where particular anions are present). The major hardness-causing cations are calcium and magnesium, while strontium, ferrous iron, and manganese can also contribute. It is common to compare the alkalinity values of a water sample to the hardness values, with both expressed in mg $CaCO_3$/L. Hardness up to the value of the alkalinity is referred to as the "carbonate hardness." Thus, when the hardness is equal to or less than the total alkalinity, all hardness is carbonate hardness. When the hardness value is greater than the total alkalinity, the amount in excess is referred to as the "noncarbonate hardness."

The method described below relies on the competitive complexation of divalent metal ions by ethylenediaminetetraacetic acid (EDTA) or an indicator. The chemical structure for the disodium salt of EDTA is in Figure 13.1. Note the lone pairs of electrons on the two nitrogens. These, combined with the dissociated carboxyl groups, enable formation of a one-to-one hexadentate complex (one cation bonded to six sites on one EDTA molecule) with each divalent ion in solution. Although the complexation constant (describing the location of the complexation equilibrium) is a function of pH, virtually all common divalent ions will be complexed at pH values greater than 10, the pH used in this titration experiment and in most hardness tests. Thus, the value for hardness calculated from the experimental results includes all divalent ions in a water sample.

Three indicators are commonly used in the EDTA titration: Eriochrome Black T (Erio T), Calcon, and Calmagite. The use of Eriochrome Black T requires that a small amount of Mg^{2+} ion be present at the beginning of the titration. Calmagite is used in this experiment because its endpoint is sharper that that of Eriochrome Blank T.

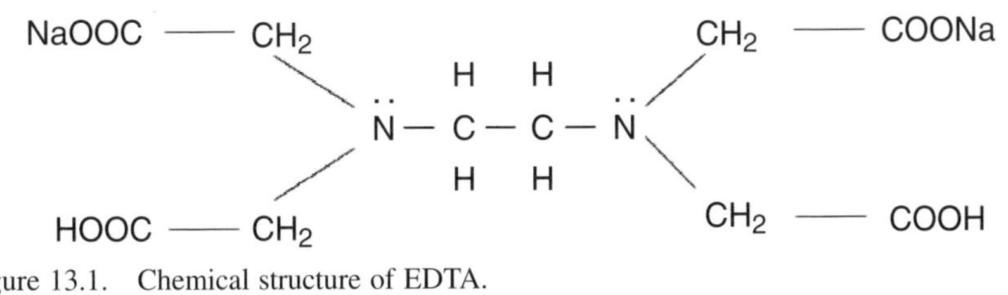

Figure 13.1. Chemical structure of EDTA.

Safety and Hazards

- As in all laboratory exercises, safety glasses must be worn at all times.
- Avoid skin and eye contact with pH 10 buffer. In case of skin contact, rinse the area for several minutes. For eye contact, flush eyes with water and seek immediate medical advice.

In the Laboratory. Two methods are available for determining the hardness of a water sample. The one described and used here is based on a titration method using a chelating agent. The basis for this technique is that at specific pH values, ethylenediaminetetraacetic acid (EDTA) binds with divalent cations to form a strong complex. Thus, by titrating a sample of known volume with a standardized (known molarity) solution of EDTA and a complexing indicator that competes with EDTA, you can measure the amount of divalent metals in solution. The endpoint of the titration is observed using a colorimetric indicator, in our case Calmagite. When a small amount of indicator is added to a solution containing hardness (at pH = 10.0), it combines with a few of the hardness ions and forms a weak wine red complex. During the titration, EDTA complexes more and more of the hardness ions until it has complexed all of the free ions and "out-competes" the weaker indicator complex for hardness ions. At this point, the indicator returns to its uncomplexed color (blue for Calmagite), indicating the endpoint of the titration, where only EDTA-complexed hardness ions are present.

Student Procedures

- Pipet an aliquot of your sample into a 250-mL Erlenmeyer flask. The initial titration will only be a trial, and you will probably need to adjust your sample volume to obtain the maximum precision from your pipeting technique (obtained by adding more than 10 mL but less than 50 mL of EDTA titrant). Increase or decrease your sample size as needed depending on the results of your initital titration.
- Add 3 mL of the pH 10 buffer solution and ~1 mL of the Calmagite indicator. Check to ensure that the pH of your sample is at or above a pH value of 10. Add additional buffer solution if needed.
- Titrate with EDTA solution and note the color change as you reach the endpoint. Continue adding EDTA until you obtain a stable blue color with no reddish tinge. (*Note*: incandescent light can produce a reddish tinge at and past the endpoint.)
- Repeat until you have at least three titrations that are in close agreement.
- Titrate at least two types of water.
- Calculate the hardness for each of your samples. Express your results in mg $CaCO_3$/L. If the stockroom gave you a 0.0100 M solution of EDTA, 1.00 mL of EDTA solution is equivalent to 1.00 mg $CaCO_3$.

Calculation:

Hardness is usually expressed in terms of $CaCO_3$ (mg $CaCO_3$/L). We calculate hardness from your titration results by

$$mg\ CaCO_3/L = \frac{0.0100\ mol\ EDTA}{L}(L\ of\ EDTA\ used)\frac{1\ mol\ divalent\ metal}{1\ mol\ EDTA}*$$

$$\frac{1\ mol\ CaCO_3}{2\ mol\ divalent\ metal}\ \frac{100\ g}{mol\ CaCO_3}\ \frac{1000\ mg}{g}\ \frac{1}{liters\ of\ sample}$$

Waste Disposal.   After neutralization all solutions can be disposed of down the drain with rinsing.

Assignment.   Prepare a summary table showing all of your results with the averages for each water tested.

13.4   THE DETERMINATION OF DISSOLVED OXYGEN IN WATER USING THE WINKLER METHOD (IODIOMETRIC TITRATION METHOD)

Purpose.   To determine the dissolved oxygen (DO) concentration in a water sample. To learn the chemical reactions involved in the Winkler DO method.

Background.   It is a common perception that all life is dependent upon the presence of oxygen, either in the atmosphere or in the water. However, this is anything but true. The first lifeforms to evolve on Earth are thought to have been anaerobic, requiring an oxygen-free environment to grow. In fact, free oxygen is toxic to anaerobic organisms' biochemical machinery. Oxygen was actually a waste product from these organisms, and their emission of oxygen over hundreds of millions of years enabled the evolution of aerobic organisms. Even today there are many types of respiration (and organisms) that do not require the presence of oxygen as their terminal electron acceptor (TEA). Every lifeform needs a terminal electron acceptor, to accept the excess electrons from their reduced food sources. For example, look at how we oxidize glucose with atmospheric oxygen to yield energy. Electrons on glucose are removed and added to diatomic oxygen and in this process oxygen is reduced from an oxidation state of zero to -2 while carbon is oxidized to $+4$. The net result is a generation of 2863 kJ of energy per mole of glucose oxidized, a higher energy yield than that achieved with more primitive TEAs (i.e. NO_3^-, SO_4^{2-}, etc.). Therefore, the first lifeforms only yielded small amounts of energy from their oxidation of food substrates. This is one reason why organisms that use oxygen as their TEA out-compete other lifeforms.

Oxygen is considered poorly soluble in water. It is interesting to note that air-breathing organisms have available around 19% oxygen (19,000 ppm on a volume per volume basis or 262 mg/L from $PV = nRT$) in air for consumption, while organisms respiring in water have only a maximum of ~0.15% oxygen (14.6 mg/L). As temperature increases or as salt content increases, the dissolved oxygen concentration decreases. The range of dissolved oxygen concentrations in water under normal conditions is shown in Table 13.1. Note that the range of dissolved oxygen (DO) in pure water (no salt content) is from 7.6 mg/L at 30°C to 14.6 mg/L at 0°C. Although this may seem like a narrow range, many organisms have become specialized so that they can live only in a small portion of this range. Important examples are mountain trout and several species of invertebrate insect larva that require very cold waters with the highest concentrations of dissolved oxygen.

Theory. Two methods are commonly used to determine the concentration of dissolved oxygen in water samples: the Winkler or iodometric method and the mem-

TABLE 13.1. Solubility of Dissolved Oxygen for Water in Contact with the a Dry Atmosphere (at 1.0 atmosphere containing 20.9% oxygen)[a]

Temperature (°C)	Chloride Concentration (mg/L)				
	0	5000	10,000	15,000	20,000
0	14.6	13.8	13.0	12.1	11.3
1	14.2	13.4	12.6	11.8	11.0
2	13.8	13.1	12.3	11.5	10.8
3	13.5	12.7	12.0	11.2	10.5
4	13.1	12.4	11.7	11.0	10.3
5	12.8	12.1	11.4	10.7	10.0
6	12.5	11.8	11.1	10.5	9.8
7	12.2	11.5	10.9	10.2	9.6
8	11.9	11.2	10.6	10.0	9.4
9	11.6	11.0	10.4	9.8	9.2
10	11.3	10.7	10.1	9.6	9.0
11	11.1	10.5	9.9	9.4	8.8
12	10.8	10.3	9.7	9.2	8.6
13	10.6	10.1	9.5	9.0	8.5
14	10.4	9.9	9.3	8.8	8.3
15	10.2	9.7	9.1	8.6	8.1
16	10.0	9.5	9.0	8.5	8.0
17	9.7	9.3	8.8	8.3	7.8
18	9.5	9.1	8.6	8.2	7.7
19	9.4	8.9	8.5	8.0	7.6
20	9.2	8.7	8.3	7.9	7.4
21	9.0	8.6	8.1	7.7	7.3
22	8.8	8.4	8.0	7.6	7.1
23	8.7	8.3	7.9	7.4	7.0
24	8.5	8.1	7.7	7.3	6.9
25	8.4	8.0	7.6	7.2	6.7
26	8.2	7.8	7.4	7.0	6.6
27	8.1	7.7	7.3	6.9	6.5
28	7.9	7.5	7.1	6.8	6.4
29	7.8	7.4	7.0	6.6	6.3
30	7.6	7.3	6.9	6.5	6.1

Increasing the salt content of water decreases the solubility of any dissolved gas.

Source: Wipple G. C. and M. C. Wipple Solubility of oxygen in sea water *J. Amer. Chem. Soc.*, **33**, 362, (1911).

brane electrode technique. Details on each of these methods can be found in Standard Methods (1998) and in Sawyer and McCarty (1978). The iodometric method will be discussed first and is the focus of this laboratory procedure. A more recent development is the use of Ru (bipy)$_3$ optical sensor for O$_2$, described in the environmental chemistry marketing literature and on the Internet.

The iodometric method, the more accurate of the two standard methods, determines the dissolved oxygen concentration through a series of oxidation-reduction reactions and is the reaction we will use here. First, Mn^{2+} (as MnSO$_4$) is added to a 250 mL or 300 mL sample. Next, the alkali-iodide reagent (KI in NaOH) is added. Under these caustic conditions, if oxygen is present in the water sample, the Mn^{2+} will be oxidized to Mn^{4+}, which precipitates as a brown hydrated oxide. This reaction is relatively slow and the solution must be shaken several times to complete the reaction. This reaction can be represented by the following expressions:

$$2Mn^{2+} + 4OH^- + O_2 \Rightarrow 2MnO_{2(S)} + 2H_2O$$

or

$$2Mn(OH)_2 + O_2 \Rightarrow 2MnO_{2(S)} + 2H_2O$$

After the MnO$_2$ precipitate settles to the bottom of the flask, sulfuric acid is added to make the solution acidic. Under these low pH conditions, the manganese is dissolved as MnO$_2$ oxidizes the iodide (I$^-$) to free iodine (I$_2$) through the following reaction

$$MnO_2 + 2I^- + 4H^+ \Rightarrow Mn^{2+} + I_2 + 2H_2O$$

Now, the sample is ready for titration with standardized sodium thiosulfate (Na$_2$S$_2$O$_3$·5H$_2$O). In this step, thiosulfate ion is quantitatively added in order to convert the I$_2$ back to I$^-$. The reaction can be represented by

$$2S_2O_3^{2-} + I_2 \Rightarrow S_4O_6^{2-} + 2I^-$$

The amount of I$_2$ present at this stage in the procedure is directly related to the amount of O$_2$ present in the original sample. The titration is complete when all of the I$_2$ has been converted to I$^-$. The endpoint of this titration can be determined through potentiometry or by using calorimetric indicators. The most common indicator is starch, which turns from deep blue to clear.

The dissolved oxygen concentration can be determined using the following equation, which also reflects the series of redox reactions in the equations given above:

$$mg\ O_2/L$$

$$= \frac{[(L\ S_2O_3^{2-})(MS_2O_3^{2-})]\left(\frac{I_2}{2S_2O_3^{2-}}\right)\left(\frac{MnO_2}{I_2}\right)\left(\frac{O_2}{2MnO_2}\right)\left(\frac{32\ g}{mol\ O_2}\right)\left(\frac{1000\ mg}{g\ O_2}\right)}{L\ sample}$$

Several modifications of the Winkler method have been developed to overcome interferences. The azide modification, the most common, effectively removes interference from nitrite, which is commonly present in water samples from biologically treated wastewater effluents and incubated biochemical oxygen demand samples.

Nitrite interferes by converting I^- to I_2, resulting in overestimation of the dissolved oxygen in the sample. The nitrite is then regenerated by oxidation. This is illustrated in the following equations:

$$2NO_2^- + 2I^- + 4H^+ \Rightarrow I_2 + N_2O_2 + 2H_2O$$

and

$$N_2O_2 + \frac{1}{2}O_2 + H_2O \Rightarrow 2NO_2^- + 2H^+$$

Note that N_2O_2 is oxidized by oxygen, which enters the sample during the titration procedure, and is converted to NO_2^- again, establishing a cyclic reaction that can lead to erroneously high results. This interference therefore yields apparent oxygen concentrations that are far in excess of the amounts that would be normally expected.

Nitrite interference can be easily overcome through the addition of sodium azide (NaN_3). Azide is usually added with the alkali-KI reagent, and when the sulfuric acid is added, the following reactions result in the removal of NO_2^-:

$$NaN_3 + H^+ \Rightarrow HN_3 + Na^+$$
$$HN_3 + NO_2^- + H^+ \Rightarrow N_2 + N_2O + H_2O$$

Other methods can also be used to remove ferrous iron (the permanganate modification), ferric iron (the potassium fluoride modification), and suspended solids (the alum flocculation modification). We will only be using the azide modification in this laboratory experiment.

The electrode method offers several advantages over the titration method, including speed, elimination or minimization of interferences, field compatibility, continuous monitoring, and *in situ* measurement. However, it is also associated with some loss in accuracy. Modern electrodes rely on a selectively permeable membrane that allows only dissolved oxygen to enter the measurement cell, thus eliminating most interferences. A detailed description of the operation of this electrode can be found in Sawyer and McCarty (1978). The calibration and measurement is relatively simple, and a direct readout of the oxygen concentration (in mg/L) is given.

In the Laboratory. You will be given one or more samples by your instructor for titration using the Winkler method. We suggest using cold water that has been equilibrated for 24 hours open to the atmosphere. For this laboratory exercise, you do not have to be concerned with preservation of the sample or sample handling practices, but in the real world there are many precautions that need to be taken. Most important is the preservation of field samples that need to be analyzed in the laboratory. The easiest way to avoid this is to use a field meter to determine the concentration of DO. This method is quick and relatively reliable. However, DO meters are expensive and some monitoring programs may specify the use of the Winkler titration method because of its greater accuracy.

Two approaches are used to preserve samples for later DO determination. First, you can "fix" your samples using the procedures describe below and then perform the titration when the samples are brought to the laboratory. Samples should be stored in the dark and on ice until titration. This preservation technique will allow you to delay the titration for up to 6 hours. However, this procedure may give low results for samples with a high iodine demand. In this case, it is advisable to use the second option which is to add 0.7 mL of concentrated sulfuric acid and 0.02 g of sodium azide. When this approach is used, it is necessary to add 3 mL of alkali-iodide reagent (below) rather than the usual 2 mL. In addition, avoid of any sample treatment or handling that will alter the concentration of DO, including increases in temperature and the presence of atmospheric headspace in your sample container.

You will titrate your samples using the procedures described below. As in all titration experiments, you should do a quick titration to determine the approximate volume of titrant needed. Follow this first titration with at least three careful titrations. Average your values for each sample. Students should work independently or in pairs since each student will be using the Winkler titration in the next experiment (the BOD determination) for which lab technique must be comparable for all students in the class.

Safety and Hazards

- As in all laboratory exercises, safety glasses must be worn at all times.
- Avoid skin and eye contact with caustic and acidic solutions. If contact occurs, rinse your hands and/or flush your eyes for several minutes. Seek immediate medical advice for eye contact.
- Use concentrated acids in the fume hood and avoid breathing their vapors.
- Sodium azide is a toxin and should be treated as such.

Student Procedures

1. To 250-mL or 300-mL sample bottle full of sample, add 1 mL $MnSO_4$ solution, followed by 1 mL alkali-iodide-azide reagent. If your pipets are dipped into the sample (as they should be), rinse them before returning them to the reagent bottles. If the solution turns white, no dissolved oxygen is present.

2. Place stoppers on the sample bottles in a manner to exclude air bubbles and mix by rapidly inverting the bottle a few times. When the precipitate has settled to half the bottle volume, repeat the mixing and allow the precipitate to re-settle.

3. Add 1.0 mL concentrated H_2SO_4.

4. Replace the stoppers and mix by rapidly inverting the bottles to dissolve the precipitate. You may open the bottle and pour the sample at this point since the DO and reagents have been "fixed" and will not react any further.

5. Titrate 200 mL of the fixed sample (delivered by a 200-mL pipet or with a graduated cylinder) with your standardized thiosulfate solution. First titrate to a pale straw color, add starch indicator, and titrate to a clear endpoint.

6. Repeat the titration for two more samples and average your results.

Assignment

1. Create a flow chart showing all of the oxidation–reduction reactions involved in the Winkler titration method. Explain each reaction.

2. Calculate an average and standard deviation for each type of water sample.

Waste Disposal. After neutralization, all solutions can be disposed of down the drain with water.

References

Sawyer, C. N. and P. L. McCarty. *Chemistry for Environmental Engineering*, 3rd edition, McGraw-Hill, New York, 1978.

Standard Methods for the Examination of Water and Wastewater, 20th edition, American Water Works Association, Washington. D. C., 1998.

13.5 THE DETERMINATION OF THE BIOCHEMICAL OXYGEN DEMAND (BOD) OF SEWAGE INFLUENT: BOD$_5$ AND/OR BOD$_{20}$

Purpose. To determine the biochemical oxygen demand (BOD) in a domestic wastewater sample.

Background. The focus of this laboratory exercise will be to determine the amount of oxidizable organic matter (sewage) in a wastewater sample. As we discussed in the dissolved oxygen chapter, the term DO refers to the chemical measurement of how much dissolved oxygen is present in a water sample, expressed in mg/L. The biochemical oxygen demand (BOD) is an estimate of how much total DO is required to oxidize the organic matter in a water sample. Thus, we will actually be measuring the change in DO in our experiments to estimate the BOD originally present in the water. But before we discuss the details of this experiment, it is important to gain an appreciation for the extent of the global sewage problem and environmental issues surrounding wastewater.

Our standard of living in the United States is a direct result of having adequate water and wastewater treatment, which are distinguishing features of developed countries. As early as 1700 B.C.E., people began to obtain the luxury of running water and then to deal with the disposal of associated wastes. Though there is evidence of plumbing and sewage systems at many age-old sites, including the *cloaca maxiumn*, or great sewer, of the ancient Roman empire, the common use of sewer and plumbing systems did not become widespread until modern times. Along with providing drinking water and disposing of sewage comes the challenge of preventing the rapid spread of disease within populations that utilize a common water source and treatment facilities.

The focus of modern sewage treatment is to remove turbidity, readily oxidizable organic matter, and pathogenic organisms. These three goals can easily be achieved at a minimal cost. Turbidity is removed in primary and secondary clarifiers and in sand bed filters. Organic matter is removed in biological contact units such as trickling filters and activated sludge lagoons. Most pathogens are naturally removed in the various treatment process, but removal is ensured with the use of sand bed filtration, chlorination, and ozonation. One of the major design criteria for a wastewater treatment plant, and in fact a daily monitoring parameter, is the biochemical oxygen demand (BOD) of the incoming and outgoing waste, a measure of the effectiveness of organic matter removal. In this laboratory exercise, we will be measuring the five-day BOD (BOD_5) or the ultimate BOD (BOD_L).

Theory. In general, the utilization of oxygen by microorganisms is considered to be a pseudo-first-order process which for a closed system (no re-aeration) is commonly described by

$$L = L_0 e^{-kt}$$

where L is the concentration of oxygen at time t, L_0 is the original concentration of oxygen in a sample, k is the rate constant, generally around 0.17/day for sewage waste, and t is time. A similar expression can be used to describe the amount of remaining BOD in the sample, since it is the inverse of the oxygen consumption,

$$L = L_0 - L_0 e^{-kt}$$

where L is the concentration of biodegradable organic matter at time, t, L_0 is the original or ultimate concentration of biodegradable organic matter, k is the rate constant, generally around 0.17/day for sewage waste, and t is time.

Traditionally, we are concerned with the amount of oxygen required to oxidize the BOD over a 5-day period. This time period was established years ago in England and results from the fact that it requires 5 days for the water in most English streams to reach the ocean. The BOD continues to exert an oxygen demand on the stream after this time and the ultimate BOD, determined over a 20-day period, is becoming commonly used in the United States.

In the Laboratory. A BOD determination is made by taking a sample and incubating it over a 5- or 20-day period (we will be determining the 5-day BOD) and then monitoring the dissolved oxygen concentration at intervals of 12 or more hours. For high concentrations of BOD, the sample must be diluted in order to avoid depleting all of the original oxygen present in the water sample. There are several requirements for the dilution water. For example, pure distilled water should not be used since microorganisms require certain salts for proper metabolism. Thus, potassium, sodium, calcium, magnesium, iron, and ammonium salts are added to the dilution water. Also, the water's pH should be buffered between 6.5 and 8.5 with phosphate buffers. Some water samples require a "seed" of viable microorganisms to complete the degradation process.

A general rule-of-thumb has been developed to provide sufficient accuracy in determining BOD values. This states that at least 2 mg/L of oxygen must be used over the course of the experiment (5 or 20 days), but at least 0.5 mg/L must remain in the final sample. The oxygen concentration can be measured by one of two methods described in the previous laboratory experiment, "The Determination of Dissolved Oxygen in Water Using the Winkler Method."

Safety and Hazards

- As in all laboratory exercises, safety glasses must be worn at all times.
- Avoid skin and eye contact with caustic and acidic solutions. If contact occurs, rinse your hands and/or flush your eyes for several minutes. Seek immediate medical advice for eye contact.
- Use concentrated acids in the fume hood and avoid breathing their vapors.

Student Procedures

Dilution of Your Waste Sample. A rule-of-thumb for estimating the dilution factor of your wastewater can be determined from Table 13.2 (from *Standard Methods for the Examination of Water and Wastewater*, 1998). The best way to determine the appropriate dilution factor is to consult the operator of the sewage treatment plant where you obtain your wastewater sample. Your lab instructor will help you with this calculation.

Seeding. A "seed" is needed when your sample does not have sufficient microbial community to immediately support exponential microbial growth. A seed usually consists of a small amount of sewage added to your samples. If you are using domestic wastewater, you will probably not have to seed your water, since viable microbial communities are already present. For the purposes of this experiment, we will assume that you do not need to seed your samples, but keep in mind that river, lake, and groundwater samples often need to have a seed added for their BOD determination. When you do use a seed, you must also run a blank for your BOD determination, since the seed will consume a small amount of the DO.

TABLE 13.2. BOD Measurable from Various Dilutions of Sample (Standard Methods for the Examination of Water and Wastewater, 1988)

Using Percent Mixtures		By Direct Pipeting into 300-mL Bottles	
% Mixture	Range of BOD	mL Sample	Range of BOD
0.01	20,000–70,000	0.02	30,000–105,000
0.02	10,000–35,000	0.05	12,000–42,000
0.05	4,000–14,000	0.10	6,000–21,000
0.1	2,000–7,000	0.20	3,000–10,500
0.2	1,000–3,500	0.50	1,200–4,200
0.5	400–1,400	1.0	600–2,100
1.0	200–700	2.0	300–1,050
2.0	100–350	5.0	120–420
5.0	40–140	10.0	60–210
10.0	20–70	20.0	30–105
20.0	10–35	50.0	12–42
50.0	4–14	100	6–21
100	0–7	300	0–7

pH Adjustment.   Some domestic wastewater samples have industrial inputs to the sewer system and as a result may have extreme pH values (very high or low). In these cases, it will be necessary to adjust the pH of your original wastewater sample prior to making dilutions according to Table 13.2. Use 1 M HCl or 1 M NaOH for these adjustments.

Chlorine Removal.   Some samples may contain residual chlorine compounds that will inhibit the growth of microorganisms and will interfere with the BOD determination. If your sample contains residual chlorine compounds, these can be removed with sodium sulfite. Domestic sewage samples rarely have residual chlorine compounds and we will not use sulfite in this procedure, but be aware that this is not always the case.

Setup and Titration of BOD Samples

1. Determine the appropriate dilution of your wastewater based on Table 13.2 and/or data from the wastewater treatment plant operator. It is best to have three dilutions, one 20–30% less dilute than suggested, one as suggested, and one 20–30% more concentrated than suggested. This approach should allow the determination of the BOD_5 or BOD_{20}.

2. Before you make your dilutions, homogenize your wastewater sample by blending it in a blender or food processor at high speed for 5 minutes. Also adjust the temperature to 20°C.

3. Add the desired volume of sample wastewater to each BOD bottle and fill with equilibrated, nutrient-added, 20°C dilution water. (Alternatively, you may mix a larger volume of diluted wastewater and fill your BOD bottles.)

4. Make sure the bottles are filled to the top with dilution water. Insert the tapered cap in a manner to exclude any air bubbles from the BOD bottle.

5. Incubate the dilutions at 20°C for 5 days, taking bottles from each dilution at each 12-hr increment and titrating them using the Winkler method to obtain a plot of BOD versus time. The necessary sampling times are dependent on the microbial oxidation rate, k, but we will use 12-hr sample intervals. After you have all of your samples set up, collect the members of the class together and assign sampling and titration times to cover the next 5 days.

6. Refer to the Winkler method (in the previous chapter) for the fixing and DO titration procedures.

7. Analyze your data and determine the BOD_5 or BOD_{20}.

Waste Disposal.   After neutralization, all solutions can be disposed of down the drain with water.

References

Standard Methods for the Examination of Water and Wastewater, 20th edition, American Water Works Association, Washington. D. C., 1998.

Assignment. Make a table summarizing the class data. What is the BOD$_5$ of your sample?

13.6 DETERMINATION OF A CLAY–WATER DISTRIBUTION COEFFICIENT FOR COPPER

Purpose. To determine the distribution coefficient of a metal on a characterized soil.

Background. Perhaps the most important fate and transport parameter is the distribution coefficient, K_d. The distribution coefficient is a measure of adsorption phenomena between the aqueous and solid phases and is fundamental to understanding processes responsible for the distribution of pollutants in aquatic systems. For applications of the distribution coefficient to fate and transport modeling of groundwater, lakes, and riverine systems, refer to the modeling Chapters 5, 6, and 8. Mathematically, it can be represented as the ratio of the equilibrium pollutant concentration in the solid (sediment or soil) phase to the equilibrium pollutant concentration in the dissolved (aqueous) phase:

$$K_d = \frac{C_{\text{Solid}}(\text{mg}/\text{kg})}{C_{\text{Aqueous}}(\text{mg}/\text{L})}$$

The purpose of the distribution coefficient is to quantify the pollutant's relative preferences for the two phases (solid and aqueous) and thus to determine the mass of pollutant present in each phase. The distribution coefficient is used in virtually every fate and transport model for the estimation of pollutant concentrations in aqueous systems. The aqueous phase concentration is important because the free aqueous phase concentration is usually the most toxic form of pollutants, especially in the case of dissolved metal cations. Inorganic and organic colloids and suspended solids in natural waters will increase the apparent water phase concentration, but pollutants adsorbed to these particles are usually not available for biological uptake. These particles can eventually settle out in quiescent regions of the natural water body or in estuaries and remove sufficient amounts of pollutant from the aquatic system.

Distribution coefficients are relatively easy to determine by allowing a pollutant–soil–water mixture of known composition to equilibrate, separating the mixture into solid and aqueous phases, and determining the pollutant concentration in each phase. This technique can be simplified by measuring (or knowing) the total mass of pollutant added to each sample (as determined from a blank sample), measuring the pollutant in the dissolved phase after equilibration, and estimating the mass of pollutant on the solid phase by difference (total mass of pollutant in blank minus aqueous phase mass). The distribution coefficient is then calculated using the equation given above.

The major problem with designing K_d experiments for the laboratory is the variability (and unpredictability) of results that are obtained given the variety of solid phases available, the nature of the pollutant used (ionic metals or hydrophobic organic compounds), and the experimental aqueous conditions used (pH values, ionic strengths, solids concentrations, and pollutant concentrations). Aqueous conditions are especially important when measuring the K_d for ionic pollutants. Without conducting the experiment under the exact experimental conditions to be used in lab, it is difficult to tell whether aqueous solutions will contain sufficient pollutant in the aqueous phase to be measured, or whether all of the pollutant will be present in the aqueous phase. Given these experimental design problems, it is not surprising that this vital experimental parameter (K_d) is not typically taught in environmental chemistry lab courses, but is usually covered in lecture material. This chapter contains a procedure, using standardized materials and conditions, for the determination of a distribution coefficient for copper. The procedure is also environmentally friendly since no (or limited) hazardous waste is generated.

In the Laboratory

Safety and Hazards

- As in all laboratory exercises, safety glasses must be worn at all times.
- Avoid skin and eye contact with NaOH, HCl, and HNO_3 solutions. If contact occurs, rinse your hands and/or flush your eyes for several minutes. Seek immediate medical advice for eye contact.

Student Procedures

Week 1. Pre-rinse all plastic vials and caps with DI water. "Sling" the vials several times to remove excess water.

TEAM 1. K_d as a Function of Cu Concentration (Kaolinite)

1. The mineral phase to be used as your adsorbent is kaolinite (KGa-1b).
2. Your overall goal is to measure the K_d as a function of Cu concentration for a kaolinite clay. In order to compare our various K_d results, we need to prepare solutions in which the mass of solid phase, the ionic strength, and the pH are as close to identical as possible. There are probably several ways that we can do this, but we will use the following approach.

   Prepare two vials for each Cu concentration. You will use a TSS concentration of 5000 mg/L. You will be using a total volume in each sample vial of 40.0 mL. Weigh 0.200 g of kaolinite (for the 5000 mg/L TSS) into each vial (except your blank vials). Be as close as you can to this mass, and record your significant figures to four decimal places. You will also need to have two blanks for each Cu concentration. These blank vials will contain ionic strength adjustor, Cu, and water (see step 4), but no solid phase. Label each with masking tape and a number (for example, "T1-1" means "Team one, vial 1"; "T1-B1" means "Team one, blank 1"; "T1-C1" means "Team one, Cu concentration 1"). Although the non-blank vials are identical at this point, labeling them prepares them for the addition of solutions of different copper concentrations.
3. Next, prepare the solutions to add to the clay-containing vials and blanks. Each vial will have a different amount of copper solution added, but the same ionic

TABLE 13.3. Cu Solutions Table for Teams 1 and 2

Desired Cu Solution Concentration in a Vial (ppm)	Addition Volume (mL) of the Cu Solution to the Right to Yield the Desired Cu Concentration to the Left (to the left)	Standard Cu Solution[a] (mg/L)
50.0	2.00	1000.0
25.0	1.00	1000.0
10.0	4.00	100.0
5.00	2.00	100.0
1.00	4.00	10.0
0.500	2.00	10.0

[a] Instructions for these solutions are given in the instructor procedures.

strength. Make the following solutions in 100-mL (or better yet 50-mL) graduated cylinders, using Cu^{2+} solutions of 1000 ppm, 100 ppm, and 10 ppm, provided by your lab instructor: Combine 2.00 mL of $Ca(NO_3)_2$–$4H_2O$ stock solution (ionic strength adjustor) with the appropriate amount of Cu solution for each concentration (Table 13.3) and fill to 40.0 mL with DI water.

4. Mix the solutions from step 3 immediately prior to adding them to the kaolinite-containing and blank vials, and cap. Again, be sure to prepare two blanks for each Cu concentration (containing everything, including Cu standard, but no solid phase). These will be necessary to determine if any Cu adsorbs to the container walls.

5. Place the vials on the mixer for at least three days. (You may leave these vials on the mixer for 7 days, until your next lab period, but it is best to remove them one day prior to performing the filtration step that occurs during week 2. This will allow the filtration to proceed faster.)

TEAM 2. K_d as a Function of Cu concentration (montmorillinite (STx-1 or SAz-1)

1. The mineral phase to be used as your adsorbent is montmorillinite (STx-1).

2. Your overall goal is to measure the K_d as a function of Cu concentration for a montmorillinite clay. In order to compare our various K_d results, we need to prepare solutions in which the mass of solid phase, the ionic strength, and the pH are as close to identical as possible. There are probably several ways that we can do this, but we will use the following approach: Prepare two vials for each Cu concentration. You will use a TSS concentration of 5000 mg/L. You will be using a total volume in each sample vial of 40.0 mL. Weigh 0.200 g of montmorillinite (for the 5000 mg/L TSS) into each vial (except your blank vials). Be as close as you can to this mass, and record your significant figures to four decimal places. You will also need to have two blanks for each Cu concentration. These blank vials will contain ionic strength adjustor, Cu, and water (see step 4), but no solid phase. Label each with masking tape and a number

(for example, "T2-1" means "Team two, vial 1"; "T2-B1" means "Team two, blank 1"; "T2-C1" means "Team two, Cu concentration 1"). Although the non-blank vials are identical at this point, labeling them prepares them for the addition of solutions of different copper concentrations.

3. Next, prepare the solutions to add to the clay-containing vials and blanks. Each vial will have a different amount of copper solution added, but the same ionic strength. Make the following solutions in 100-mL (or better yet 50-mL) graduated cylinders, using Cu^{2+} solutions of 1000 ppm, 100 ppm, and 10 ppm, provided by your lab instructor: Combine 2.00 mL of $Ca(NO_3)_2–4H_2O$ stock solution (ionic strength adjustor) with the appropriate amount of Cu solution for each concentration (Table 13.3) and fill to 40.0 mL with DI water.

4. Mix the solutions from step 3 immediately prior to adding them to the montmorillinite-containing and blank vials, and cap. Again, be sure to prepare two blanks for each Cu concentration (containing everything, including Cu standard, but no solid phase). These will be necessary to determine if any Cu adsorbs to the container walls.

5. Place the vials on the mixer for at least three days. (You may leave these vials on the mixer for 7 days, until your next lab period, but it is best to remove them one day prior to performing the filtration step that occurs during week 2. This will allow the filtration to proceed faster.)

Week 2. There will be several demonstrations at the beginning of lab to illustrate the use of the filter apparatus and mixing system.

1. Turn on the AAS to warm up the lamp.

2. Prepare calibration standards at concentrations of 0.100, 0.500, 1.00, 5.00, 10.0, 25.0, and 50.0 ppm Cu^{2+}. Prepare these in 1% HCl.

3. Filter the solutions that you prepared last week as illustrated by your lab instructor. First, filter them through the Gelman type A/E glass fiber filter, then through a 0.2-μm HPLC nylon filter with a syringe. Filter both the blanks and the actual samples. *Note*: You do not need to filter the entire 40.0-mL suspension, nor should you mix the vial, since it is already at equilibrium after the 7 days of mixing. Mixing the vial again will only slow the filtration step and delay your leaving lab. You only need a sufficient volume of the filtrate to run on the atomic absorption unit. Typically 10–20 mL will suffice, but check with your lab instructor to see how much you will need for your particular instrument.

4. Analyze the samples using AAS as demonstrated.

5. Calculate the K_d as illustrated in Chapter 3.

Waste Disposal. Check with your laboratory instructor for disposal instructions for the Cu waste.

Assignment. Prepare plots of Cu concentration in the dissolved phase (measured mg/L; placed on the *x*-axis) versus the calculated Cu concentration on the solid phase (mg/kg; placed on the y-axis). Calculate the slope of the line, which will be the K_d.

13.7   THE MEASUREMENT OF DISPERSION IN A SIMULATED LAKE SYSTEM

In Chapter 5, we presented two basic models for the transport of pollutants in lake systems. We made a simple and bold assumption for both models—that the lake system being modeled was completely mixed upon the addition of our pollutant. As you can imagine, this completely mixed system is rarely the case in the real world, but it is an assumption that must be made unless you have the knowledge to model the system using numerical methods of analysis. But for simple systems our assumption can be appropriate and sufficient. In this laboratory exercise, we will look at three different "model" lake systems, each containing different degrees of mixing. Our physical system of a lake system is simple, but this exercise will illustrate how complicated mixing can be, even in a system as simple as a bucket.

In the Laboratory

Safety and Hazards

- As in all laboratory exercises, safety glasses must be worn at all times.
- All of the chemicals used in this laboratory are relatively safe, but standard laboratory cautions and safety should be practiced.
- You will use dilute flourescein solution in this lab. Dilute concentrations are safe when spilled on your skin, but may turn your skin yellow for a brief time. To avoid this, we recommend wearing latex gloves during this laboratory exercise.

Student Procedures.   Your laboratory instructor will divide the class into three groups. Each group will have a specially prepared bucket (our simulated lake;), a dye for addition to your bucket, a set of test tubes for sampling the effluent from the bucket, and access to a spectrophotometer for measuring the absorbance of the dye. Each group will have a bucket set up to produce distinct mixing patterns, so that the results from the three groups can be compared.

1. Turn on the spectrophotometer and allow it to warm up. Zero the spectrophotometer with DI water. Complete step two as the spectrophotometer warms up.

2. Turn on the water source to your bucket. The water should enter through a top hole in the bucket and exit through a bottom hole connected to a tube, with the end of the tube adjusted to the height of the input port. This will allow the bucket to fill to a steady-state volume. After the bucket is full, if you are using a three-gallon bucket, adjust the water flow exiting the bucket to 1.0 L/min (±5%). If you are using a different volume, adjust the inlet flow with the faucet so that the hydraulic retention time (volume of bucket divided by the flow rate) is approximately 11.4 minutes. The bucket should be placed on a magnetic

(a)

Figure 13.2.   (*a–c*) Design of plastic buckets used to simulate a lake system.

stirrer, as illustrated by your instructor. All buckets should be placed on a stirrer of identical make (brand and model number) and adjusted to the same setting in order to allow comparison of the results from each bucket.

3. After the flow has stabilized to your desired value, rapidly add 100 mL of the dye in the middle of the mixing vortex. Wait 30 seconds for the contents to mix, and take your first sample with a test tube. This data point will be the initial concentration of dye.

(b)

Figure 13.2.   *Continued*

4. At 5- to 10-minute intervals, take a new sample and note the exact time of the sample. Sample the bucket until the absorbance falls to zero. Read the absorbance of the samples as you collect them.

5. Plot all of the class data on one plot (time on the *x*-axis and "reduced absorbance values" (each absorbance divided by the initial absorbance) on the *y*-axis) and answer the following questions:

(c)

Figure 13.2.   *Continued*

(a) Was the tracer (dye) input a step or pulse input?

(b) Calculate the detention time of water in your model lake system if it is different from 11.4 minutes.

(c) Which system showed the least dispersion?

(d) Which system showed the highest dispersion?

13.8 THE MEASUREMENT OF DISPERSION IN A SIMULATED RIVER SYSTEM

In Chapter 6 we presented two basic models for the transport of pollutants in river systems. We made a simple assumption for both models—that the river system was completely mixed upon the addition of our pollutant and that dispersion only occurred in the longitudinal direction (the direction of water flow downhill). This is essentially true in most cases, but the degree of longitudinal mixing changes as you move from calm regions of the river to riffle areas or eddies. In this laboratory exercise, we will look at several different "physical model" river systems, each containing different degrees of mixing. Our system is simple, but it will clearly illustrate how increasing the number of riffle areas and pools can be affect dispersion.

In the Laboratory

Safety and Hazards

- As in all laboratory exercises, safety glasses must be worn at all times.
- All of the chemicals used in this laboratory are relatively safe, but standard laboratory cautions and safety should be practiced.
- You will spill and spread dilute flourescein solution everywhere in this lab. Dilute concentrations are safe when spilled on your skin, but may turn your skin yellow for a brief time. We recommend wearing latex gloves during this laboratory exercise.

Student Procedures.   Your laboratory instructor will divide the class into several groups, corresponding to the number of different river systems have been prepared. Each group will have a specially prepared model river channel (refer to Figure 13.3*a*, *b*), a flourescein dye for addition to your river, a set of test tubes or beakers for sampling the effluent from river, and access to a spectrophotometer. Each group will have a river system with a different number of weirs, producing distinct mixing patterns, so that the results from the three groups can be compared.

Note: This procedure does not result in exact science. Your sampling times and volume of sample may be considerably off from ideal values, but this will be fine. We are not conducting a detailed chemistry experiment in this laboratory, but only illustrating the concept of dispersion.

1. Turn on the spectrophotometer and allow it to warm up. Zero the spectrophotometer with DI. Complete step two as the spectrophotometer is warming up.
2. Adjust the slope of your river system as illustrated by your instructor and allow it to directly drain into a sink.
3. Turn on the water source to your river system and adjust the flow to 1.0 L/min (adjust the flow as close to 1.0 L/min as possible; ±5%).
4. As one student adds 5.0 mL of the dye to the top of the river system, other students should be ready to rapidly sample the outlet of the river. For the river systems with no weirs, the sampling must be very fast. We suggest having test

(a)

Figure 13.3.   (*a*, *b*) Design of the PVC river system.

tubes lined up in a holder and taking a sample at the specified times below. For river systems with weirs, you will sample at slightly longer intervals. The sampling times are (1) for no weirs, sample every two seconds, (2) for one-weir systems, sample every 5 seconds, (3) for two-weir systems, sample every 10 seconds, and (4) for four-weir systems, sample every 10 seconds.

5. Sample the effluent from your system until you do not see any dye color in the test tubes. The more dams present in the river systems, the longer it will take to flush all of the dye from your model.

6. Analyze the samples with respect to absorbance using the spectrophotometer.

7. Repeat your experiment at least twice.

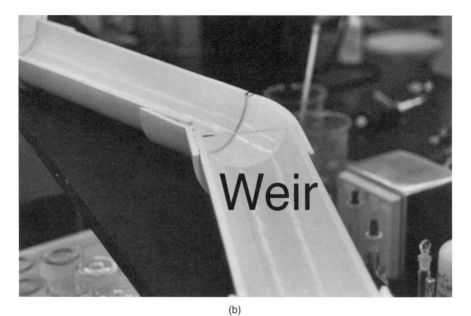

(b)

Figure 13.3. *Continued*

8. Plot all of the class data on one plot (*x*-axis: sample time in seconds; *y*-axis: absorbance at time *t* divided by absorbance of original dye solution) and answer the following questions:

   a. Was the tracer (dye) input a step or pulse input?

   b. Estimate the dispersion (in time units) of your river systems by taking the width of the dispersion peak at one-half the maximum absorbance value for each dye concentration. This will allow you to compare each river system.

   c. Which system showed the least dispersion?

   d. Which system showed the highest dispersion?

   e. Why are the dye concentrations (as measured by absorbance) highest for the system with no weirs?

   f. Why does the width of the dye plot increase with increasing number of weirs?

13.9 THE MEASUREMENT OF DISPERSION AND SORPTION IN A SIMULATED GROUNDWATER SYSTEM

In Chapter 8, we studied the fate and transport of pollutants in groundwater systems. A few scientists have had the opportunity to experiment and study with contaminated real-world groundwater systems, but this process can be very prohibitively expensive. To reduce the expense of studying these sites, many scientists use microscale systems. This involves taking soil from the contaminated site, packing it into a small column, studying the movement of pollutants through the column, and then relating the results to the large-scale system. In this experiment, we will study the movement of cadmium ion through a sand column.

Before we conduct the experiment, it is useful to review the concept of distribution coefficient, K_d, and the retardation factor, R. The distribution coefficient can be mathematically defined as

$$K_d = \frac{\text{Pollutant conc. in the solid phase}}{\text{Pollutant conc. in the aqueous phase}} = \frac{\text{mg/kg}}{\text{mg/L}}$$

and represents the relative affinity of the pollutant for the solid or liquid phases. K_d values greater than 1.0 show that the pollutant prefers the solid phase to the water phase. In order to incorporate this into the general transport equation for groundwater system, we must relate K_d to the degree to which the velocity of the pollutant in the system will be slowed down do to its preference for the solid phase. This is accomplished by

$$R = 1 + \frac{\rho_b K_d}{n}$$

where ρ_b is the bulk density (of the sand) and n is the volumetric water content of the soil column. R is directly used in step and pulse pollutant transport equations for groundwater systems. R can also be defined by a ratio of the rate of pollutant movement over the rate of water movement. Our "measure" of water and pollutant movement will be the volume of water needed to push the water or pollutant through the column. The volume of water needed for these measurements will be defined in units of column pore volumes. A pore volume is the volume of water contained in the soil column at any given time. This can easily be measured, as described in the experimental procedures section, by weighing the soil column with and without water in it. Measuring the water velocity is not easy, so we place a chemical tracer in the water that is not retained by the soil and moves at a velocity at or very near that of the water. In our experiment, we will use fluorescein dye as this conservative chemical tracer. We will measure the velocity of the dye (volume of water needed to push out the dye) and the velocity of the cadmium ions (volume of water needed to push out the cadmium).

In the Laboratory

Safety and Hazards

- As in all laboratory exercises, safety glasses must be worn at all times.
- Dilute concentrations of fluorescein dye are safe when spilled on your skin, but may turn your skin yellow for a brief time. In order to avoid this, we recommend wearing latex gloves during this laboratory exercise.
- Cadmium is a toxic heavy metal and should be handled with care. You will be using dilute concentrations. Wash any exposed skin with soap and water.

- Nitric acid will be used and should be handled with caution since it is a strong acid.

Chemicals and Equipment to Be Supplied by Your Instructor

- The experimental setup shown in Figure 13.4
- A 150-mg/L solution of Cd^{2+}
- Fluorescein dye solution
- A 0.50- to 1.0-mL syringe
- Up to 80 test tubes for collection of samples (your instructor will estimate how many you will need for your experiment)
- A 10-mL graduated cylinder
- A Spectronic-20 spectrophotometer
- A flame atomic absorption spectrophotometer (FAAS unit)

Student Procedures.   This experiment will last for two weeks and will involve a pulse injection of cadmium solution during the first week and a pulse injection of

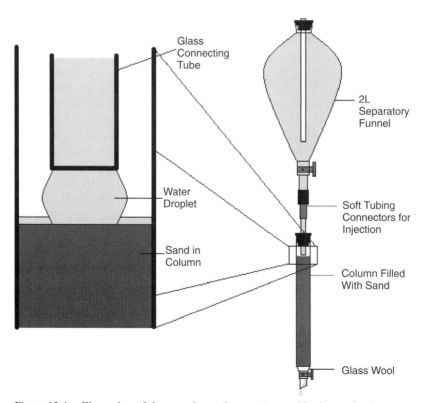

Figure 13.4.   Illustration of the experimental apparatus used in the sand column experiment.

dye during the second week. This will allow you to determine the relative velocity of each tracer. We have found that it is important to do the dye tracer experiment after the cadmium experiment.

Week 1

1. Observe your experimental setup and identify all of the parts and how they are put together as you read over the procedure.

2. Disconnect the glass column from the setup and place a very small ball of glass wool in the narrow base of the column. Weigh the column. Fill the column between 70% and 80% full with sand. Weigh the column again to determine the weight of sand in your column. Close the stopcock and completely fill the column with water. Place the top of the column in the palm of your hand to seal the column and invert it to displace all air pockets in the column. Add more water to remove the air and repeat the process until all air has been removed from the soil column. Hold the column upright and tap the column to settle the sand.

3. Wash the sand column by opening the stopcock and adding water to the top of the column. *Note*: Never let the water level fall below the top of the sand packing or you will introduce new air pockets to the sand. If you accidentally do allow the water level to fall below the top of the sand, repeat step 2 to remove the air pockets. After you washed the column with 100 mL of DI, drain the excess water from the column, stopping just above the top of the sand (to within ~2 mm). Weigh the sand column again to determine the volume of water in the column. This mass minus the weight of the sand and glass column, when converted to volume (1.00 mL = 1.00 g) is your column pore volume.

4. Reassemble the experimental setup, keeping the stopcock closed and completely sealing the stopper in the top of the column.

5. Fill the 2-L separatory funnel with DI.

6. Slowly open the stopcock on the separatory funnel to allow water to flow into the column. The water level will only slightly rise if you have a good seal between the rubber plug and the top of the column.

7. Slowly open the stopcock at the effluent of the sand column to obtain a flow rate of 0.25 mL/sec. Measure the flow rate of the column by placing a 10-mL graduated cylinder under the column, start a stopwatch, and time the collection of 5.00 mL of water. Adjust the top and bottom stopcocks to obtain the desired flow rate.

8. After you have adjusted the flow rate, allow the column to flow for a few minutes while you prepare the remainder of the experiment.

9. Obtain the cadmium solution and syringe from your instructor. Arrange your test tubes for collection. Weigh each test tube to obtain an initial test tube weight. Decide who will (a) inject the cadmium to the column, (b) collect the samples in test tubes, (c) weigh the test tubes containing water samples to determine the volume of water collected, (d) analyze the samples on the

FAAS unit to determine the Cd^{2+} concentration, and (5) record all data from the experiment.

10. Before you start your experiment, recheck the column flow and adjust it as necessary.

11. Start your experiment by injecting 0.50 mL of cadmium solution into soft rubber tubing (refer to diagram). Inject it in a manner so that all of the cadmium solution enters the narrow glass tube that leads directly to the sand column. This will minimize the artificial dispersion of the cadmium pulse caused by the injection. When the cadmium solution is injected, immediately start sampling. Collect approximately 10–15 mL in each test tube.

12. As you finish collecting each sample, weigh the test tube to determine the mass of water in each sample.

13. Measure the Cd^{2+} in your sample and your original Cd^{2+} stock solution using the FAAS (your instructor will illustrate how to do this).

14. Collect samples until you the cadmium pulse has moved through the column and the cadmium concentration returns to zero. Check with your instructor to make sure you can stop the experiment. Again, your instructor will have a good idea when this should occur.

15. To stop the experiment, close the two stopcock simultaneously.

16. Measure the absorbance of cadmium in your original solution (stock solution) and in the samples.

17. Convert all of your data to a data set consisting of two data columns: (1) cumulative number of pore volumes in each sample and (2) Cd^{2+} concentration.

Week 2

1. Slowly open the stopcock on the separatory funnel to allow water to flow into the column. The water level will only slightly rise if you have a good seal between the rubber plug and the top of the column

2. Slowly open the stopcock at the effluent of the sand column to obtain a flow rate of 0.25 mL/sec. Measure the flow rate of the column by placing a 10-mL graduated cylinder under the column, start a stopwatch, and time the collection of 5.00 mL of water. Adjust the top stopcock to obtain the desired flow rate. The flow should closely match the flow used in last week's experiment.

3. After you have adjusted the flow rate, allow the column to flow for a few minutes while you prepare the remainder of the experiment.

4. Obtain the fluorsecein dye solution and syringe from your instructor. Arrange your test tubes for collection. Weigh each test tube to obtain an initial test tube weight. Decide who will (a) inject the dye to the column, (b) collect the samples in test tubes, (c) weigh the test tubes containing water samples to measure the volume of water collected, (d) analyze the samples on the Spectronic-20 unit for absorbance, and (5) record all data from the experiment.

5. Before you start your experiment, recheck the column flow and adjust it as necessary.

6. Start your experiment by injecting 0.50 mL of dye solution into soft rubber tubing. Inject in a manner so that the dye solution enters in the narrow glass tube. This will minimize the artificial dispersion of the dye pulse caused by your injection. When the dye solution is injected, immediately start sampling. Collect approximately 5 mL in each test tube.

7. As you finish collecting each sample, weight the test tube to determine the mass of water in each sample.

8. Measure the absorbance of the dye as illustrated by your instructor.

9. Collect samples until you measure the dye pulse move through the column and the dye absorbance returns to zero. Check with your instructor to make sure you can stop the experiment.

10. To stop the experiment, close the two stopcocks simultaneously.

11. Measure the absorbance of dye in your original solution (stock solution) and in your samples.

12. Convert all of your data to a data set consisting of two data columns: (1) cumulative number of pore volumes in each sample volume and (2) dye absorbance.

Assignment. Construct a spreadsheet and plot of your data. The *x*-axis should be in pore volumes of water passing through the column and the *y*-axis should be in reduced concentrations (concentration of dye or cadmium in each pore volume divided by the original dye or cadmium concentration). Plot both data sets on the same graph.

- Determine the retardation factor for cadmium.
- Determine the K_d for cadmium based on your column experiment.
- Why is the dye plot narrower than the cadmium plot?
- Why is the height of the cadmium plot lower than the dye plot?

13.10 A FIELD STUDY OF A STREAM

Your laboratory instructor will give you plans and procedures for this exercise, since these procedures will vary greatly depending on his or her particular plans. You will be visiting a local stream and conducting several physical, chemical, and biological measurements to characterize the water quality. While playing outside and in the water is fun, keep in mind the goal at hand: evaluating the quality of the stream.

GLOSSARY OF IRIS TERMS*

This glossary contains definitions of terms used frequently in IRIS. It is intended to assist users in understanding terms utilized by the U.S. EPA in hazard and dose–response assessments. These definitions are not all-encompassing, but are useful "working definitions." It is assumed that the user has some familiarity with risk assessment and health science. For terms that are not included in this glossary, the user should refer to standard health science, biostatistics and medical textbooks and dictionaries.

Acceptable Daily Intake (ADI): The amount of a chemical a person can be exposed to on a daily basis over an extended period of time (usually a lifetime) without suffering deleterious effects.

Acute exposure: Exposure by the oral, dermal, or inhalation route for 24 hours or less.

Acute toxicity: Any poisonous effect produced within a short period of time following an exposure, usually 24–96 hours.

Adverse effect: A biochemical change, functional impairment, or pathologic lesion that affects the performance of the whole organism, or reduces an organism's ability to respond to an additional environmental challenge.

Anecdotal data: Data based on the description of individual cases rather than controlled studies.

Average daily dose (ADD): Dose rate averaged over a pathway-specific period of exposure expressed as a daily dose on a per-unit-body-weight basis. The ADD is usually expressed in terms of $mg/kg \cdot day$ or other mass–time units.

Background levels: Two types of background levels may exist for chemical substances: (a) *naturally occurring levels*: Ambient concentrations of substances present in the environment, without human influence; (b) *anthropogenic levels*: Concentrations of substances present in the environment due to human-made, non-site sources (e.g., automobiles, industries).

Benchmark dose (BMD) or concentration (BMC): A dose or concentration that produces a predetermined change in response rate of an adverse effect (called the benchmark response or BMR) compared to background.

BMDL or BMCL: A statistical lower confidence limit on the dose or concentration at the BMD or BMC, respectively.

Benchmark response (BMR): An adverse effect, used to define a benchmark dose from which an RfD (or RfC) can be developed. The change in response rate over background of the BMR is usually in the range of 5–10%, which is the limit of responses typically observed in well-conducted animal experiments.

Benign tumor: A tumor that does not spread to a secondary localization, but may impair normal biological function through obstruction or may progress to malignancy later.

Bioassay: An assay for determining the potency (or concentration) of a substance that causes a biological change in experimental animals.

*Quoted directly from http://www.epa.gov/.

A Basic Introduction to Pollutant Fate and Transport, By Dunnivant and Anders
Copyright © 2006 by John Wiley & Sons, Inc.

Bioavailability: The degree to which a substance becomes available to the target tissue after administration or exposure.

Biologically based dose–response (BBDR) model: A predictive model that describes biological processes at the cellular and molecular level linking the target organ dose to the adverse effect.

Cancer: A disease of heritable, somatic mutations affecting cell growth and differentiation, characterized by an abnormal, uncontrolled growth of cells.

Carcinogen: An agent capable of inducing cancer.

Carcinogenesis: The origin or production of a benign or malignant tumor. The carcinogenic event modifies the genome and/or other molecular control mechanisms of the target cells, giving rise to a population of altered cells.

Chronic effect: An effect that occurs as a result of repeated or long-term (chronic) exposures.

Chronic exposure: Repeated exposure by the oral, dermal, or inhalation route for more than approximately 10% of the life span in humans (more than approximately 90 days to 2 years in typically used laboratory animal species).

Chronic study: A toxicity study designed to measure the (toxic) effects of chronic exposure to a chemical.

Chronic toxicity: The capacity of a substance to cause adverse human health effects as a result of chronic exposure.

Co-carcinogen: An agent that, when administered with a carcinogen, enhances the activity of the carcinogen.

Critical concentration: An ambient chemical concentration expressed in units of $\mu g/m^3$ and used in the operational derivation of the inhalation RfC. This concentration will be the NOAEL Human Equivalent Concentration (HEC) adjusted from principal study data.

Developmental toxicity: Adverse effects on the developing organism that may result from exposure prior to conception (either parent), during prenatal development, or postnatally until the time of sexual maturation. The major manifestations of developmental toxicity include death of the developing organism, structural abnormality, altered growth, and functional deficiency.

Dose: The amount of a substance available for interactions with metabolic processes or biologically significant receptors after crossing the outer boundary of an organism. The *potential dose* is the amount ingested, inhaled, or applied to the skin. The *applied dose* is the amount presented to an absorption barrier and available for absorption (although not necessarily having yet crossed the outer boundary of the organism). The *absorbed dose* is the amount crossing a specific absorption barrier (e.g., the exchange boundaries of the skin, lung, and digestive tract) through uptake processes. *Internal dose* is a more general term denoting the amount absorbed without respect to specific absorption barriers or exchange boundaries. The amount of the chemical available for interaction by any particular organ or cell is termed the *delivered* or *biologically effective dose* for that organ or cell.

Dose–response assessment: A determination of the relationship between the magnitude of an administered, applied, or internal dose and a specific biological response. Response can be expressed as measured or observed incidence or change in level of response, percent response in groups of subjects (or populations), or the probability of occurrence or change in level of response within a population.

Dose–response relationship: The relationship between a quantified exposure (dose) and the proportion of subjects demonstrating specific biologically significant changes in incidence and/or in degree of change (response).

Epidemiology: The study of the distribution and determinants of health-related states or events in specified populations.

Estimated exposure dose (EED): The measured or calculated dose to which humans are likely to be exposed considering all sources and routes of exposure.

Excess lifetime risk: The additional or extra risk of developing cancer due to exposure to a toxic substance incurred over the lifetime of an individual.

Exposure: Contact made between a chemical, physical, or biological agent and the outer boundary of an organism. Exposure is quantified as the amount of an agent available at the exchange boundaries of the organism (e.g., skin, lungs, gut).

Exposure assessment: An identification and evaluation of the human population exposed to a toxic agent, describing its composition and size, as well as the type, magnitude, frequency, route and duration of exposure.

Hazard: A potential source of harm.

Hazard assessment: The process of determining whether exposure to an agent can cause an increase in the incidence of a particular adverse health effect (e.g., cancer, birth defect) and whether the adverse health effect is likely to occur in humans.

Hazard characterization: A description of the potential adverse health effects attributable to a specific environmental agent, the mechanisms by which agents exert their toxic effects, and the associated dose, route, duration, and timing of exposure.

Limited evidence: A term used in evaluating study data for the classification of a carcinogen by the 1986 U.S. EPA guidelines for carcinogen risk assessment. This classification indicates that a causal interpretation is credible but that alternative explanations such as chance, bias, and confounding variables could not be completely excluded.

Linear dose response: A pattern of frequency or severity of biological response that varies directly with the amount of dose of an agent.

Linearized multistage procedure: A modification of the multistage model, used for estimating carcinogenic risk, that incorporates a linear upper bound on extra risk for exposures below the experimental range.

Lower limit on effective dose 10 (LED_{10}): The 95% lower confidence limit of the dose of a chemical needed to produce an adverse effect in 10% of those exposed to the chemical, relative to control.

Lowest-observed-adverse-effect level (LOAEL): The lowest exposure level at which there are biologically significant increases in frequency or severity of adverse effects between the exposed population and its appropriate control group.

Lowest-observed effect level (LOEL or LEL): In a study, the lowest dose or exposure level at which a statistically or biologically significant effect is observed in the exposed population compared with an appropriate unexposed control group.

Malignant tumor: An abnormal growth of tissue which can invade adjacent or distant tissues.

Metastasis: The dissemination or secondary growth of a malignant tumor at a site distant from the primary tumor.

Model: A mathematical function with parameters that can be adjusted so the function closely describes a set of empirical data. A mechanistic model usually reflects observed or hypothesized biological or physical mechanisms and has model parameters with real-world interpretation. In contrast, statistical or empirical models selected for particular numerical properties are fitted to data; model parameters may or may not have real-world interpretation. When data quality is otherwise equivalent, extrapolation from mechanistic models (e.g., biologically based dose-response models) often carries higher confidence than extrapolation using empirical models (e.g., logistic model).

Modifying factor (MF): A factor used in the derivation of a reference dose or reference concentration. The magnitude of the MF reflects the scientific uncertainties of the study and database not explicitly treated with standard uncertainty factors (e.g., the completeness of the overall database). A MF is greater than zero and less than or equal to 10, and the default value for the MF is 1.

Mutagen: A substance that can induce an alteration in the structure of DNA.

Neoplasm: An abnormal growth of tissue which may be benign or malignant.

No-observed-adverse-effect level (NOAEL): The highest exposure level at which there are no biologically significant increases in the frequency or severity of adverse effect between the exposed population and its appropriate control; some effects may be produced at this level, but they are not considered adverse or precursors of adverse effects.

No-observed-effect level (NOEL): An exposure level at which there are no statistically or biologically significant increases in the frequency or severity of any effect between the exposed population and its appropriate control.

Nonlinear dose response: A pattern of frequency or severity of biological response that does not vary directly with the amount of dose of an agent.

ppb: A unit of measure expressed as parts per billion. Equivalent to 1×10^{-9}.

ppm: A unit of measure expressed as parts per million. Equivalent to 1×10^{-6}.

Prevalence: The proportion of disease cases that exist within a population at a specific point in time, relative to the number of individuals within that population at the same point in time.

Reference concentration (RfC): An estimate (with uncertainty spanning perhaps an order of magnitude) of a continuous inhalation exposure to the human population (including sensitive subgroups) that is likely to be without an appreciable risk of deleterious effects during a lifetime. It can be derived from a NOAEL, LOAEL, or benchmark concentration, with uncertainty factors generally applied to reflect limitations of the data used. Generally used in EPA's noncancer health assessments.

Reference dose (RfD): An estimate (with uncertainty spanning perhaps an order of magnitude) of a daily oral exposure to the human population (including sensitive subgroups) that is likely to be without an appreciable risk of deleterious effects during a lifetime. It can be derived from a NOAEL, LOAEL, or benchmark dose, with uncertainty factors generally applied to reflect limitations of the data used. Generally used in EPA's noncancer health assessments.

Reference value (RfV): An estimation of an exposure for [a given duration] to the human population (including susceptible subgroups) that is likely to be without an appreciable risk of adverse effects over a lifetime. It is derived from a BMDL, a NOAEL, a LOAEL, or another suitable point of departure, with uncertainty/variability factors applied to reflect limitations of the data used. [Durations include acute, short-term, longer-term, and chronic and are defined individually in this glossary.]

Risk (in the context of human health): The probability of adverse effects resulting from exposure to an environmental agent or mixture of agents.

Risk assessment (in the context of human health): The evaluation of scientific information on the hazardous properties of environmental agents (hazard characterization), the dose–response relationship (dose–response assessment), and the extent of human exposure to those agents (exposure assessment). The product of the risk assessment is a statement regarding the probability that populations or individuals so exposed will be harmed and to what degree (risk characterization).

Risk characterization: The integration of information on hazard, exposure, and dose–response to provide an estimate of the likelihood that any of the identified adverse effects will occur in exposed people.

Risk management (in the context of human health): A decision-making process that accounts for political, social, economic and engineering implications together with risk-related information in order to develop, analyze, and compare management options and select the appropriate managerial response to a potential chronic health hazard.

Slope factor: An upper bound, approximating a 95% confidence limit, on the increased cancer risk from a lifetime exposure to an agent. This estimate, usually expressed in units of proportion (of a population) affected per mg/kg · day, is generally reserved for use in the low-dose region of the dose–response relationship—that is, for exposures corresponding to risks less than 1 in 100.

Statistical significance: The probability that a result is not likely to be due to chance alone. By convention, a difference between two groups is usually considered statistically significant if chance could explain it only 5% of the time or less. Study design considerations may influence the a priori choice of a different level of statistical significance.

Subchronic exposure: Exposure to a substance spanning approximately 10% of the lifetime of an organism.

Subchronic study: A toxicity study designed to measure effects from subchronic exposure to a chemical.

Superfund: Federal authority, established by the Comprehensive Environmental Response, Compensation, and Liability Act (CERCLA) in 1980, to respond directly to releases or threatened releases of hazardous substances that may endanger health or welfare.

Susceptibility: Increased likelihood of an adverse effect, often discussed in terms of relationship to a factor that can be used to describe a human subpopulation (e.g., life stage, demographic feature, or genetic characteristic).

Susceptible subgroups: May refer to life stages, for example, children or the elderly, or to other segments of the population, for example, asthmatics or the immune-compromised, but are likely to be somewhat chemical-specific and may not be consistently defined in all cases.

Target organ: The biological organ(s) most adversely affected by exposure to a chemical, physical, or biological agent.

Teratogenic: Structural developmental defects due to exposure to a chemical agent during formation of individual organs.

Threshold: The dose or exposure below which no deleterious effect is expected to occur.

Toxicity: Deleterious or adverse biological effects elicited by a chemical, physical, or biological agent.

Toxicology: The study of harmful interactions between chemical, physical, or biological agents and biological systems.

Toxic substance: A chemical, physical, or biological agent that may cause an adverse effect or effects to biological systems.

Tumor: An abnormal, uncontrolled growth of cells. Synonym: neoplasm

Threshold limit value (TLV):   Recommended guidelines for occupational exposure to airborne contaminants published by the American Conference of Governmental Industrial Hygienists (ACGIH). TLVs represent the average concentration in mg/m^3 for an 8-hr workday and a 40-hr work week to which nearly all workers may be repeatedly exposed, day after day, without adverse effect.

Uncertainty: Uncertainty occurs because of a lack of knowledge. It is not the same as variability. For example, a risk assessor may be very certain that different people drink different amounts of water but may be uncertain about how much variability there is in water intakes within the population. Uncertainty can often be reduced by collecting more and better data, whereas variability is an inherent property of the population being evaluated. Variability can be better characterized with more data but it cannot be reduced or eliminated. Efforts to clearly distinguish between variability and uncertainty are important for both risk assessment and risk characterization.

Uncertainty/variability factor (UFs):   One of several, generally 10-fold, default factors used in operationally deriving the RfD and RfC from experimental data. The factors are intended to account for (1) variation in susceptibility among the members of the human population (i.e., interindividual or intraspecies variability); (2) uncertainty in extrapolating animal data to humans (i.e., interspecies uncertainty); (3) uncertainty in extrapolating from data obtained in a study with less-than-lifetime exposure (i.e., extrapolating from subchronic to chronic exposure); (4) uncertainty in extrapolating from a LOAEL rather than from a NOAEL; and (5) uncertainty associated with extrapolation when the database is incomplete.

Unit risk: The upper-bound excess lifetime cancer risk estimated to result from continuous exposure to an agent at a concentration of $1\,\mu g/L$ in water, or $1\,\mu g/m^3$ in air. The interpretation of unit risk would be as follows: If unit risk $= 1.5 \times 10^{-6}\,\mu g/L$, 1.5 excess tumors are expected to develop per 1,000,000 people if exposed daily for a lifetime to $1\,\mu g$ of the chemical in $1\,L$ of drinking water.

Upper bound: An plausible upper limit to the true value of a quantity. This is usually not a true statistical confidence limit.

Variability: Variability refers to true heterogeneity or diversity. For example, among a population that drinks water from the same source and with the same contaminant concentration, the risks from consuming the water may vary. This may be due to differences in exposure (i.e., different people drinking different amounts of water and having different body weights, different exposure frequencies, and different exposure durations) as well as differences in response (e.g., genetic differences in resistance to a chemical dose). Those inherent differences are referred to as variability. Differences among individuals in a population are referred to as inter-individual variability, and differences for one individual over time is referred to as intra-individual variability.

Weight-of-evidence (WOE) for carcinogenicity: A system used by the U.S. EPA for characterizing the extent to which the available data support the hypothesis that an agent causes cancer in humans. Under EPA's 1986 risk assessment guidelines, the WOE was described by categories "A through E," Group A for known human carcinogens through Group E for agents with evidence of noncarcinogenicity. The approach outlined in EPA's proposed guidelines for carcinogen risk assessment (1996) considers all scientific information in determining whether and under what conditions an agent may cause cancer in humans, and it provides a narrative approach to characterize carcinogenicity rather than categories.

LIST OF DRINKING WATER CONTAMINANTS AND MCLS*

NATIONAL PRIMARY DRINKING WATER REGULATIONS

National Primary Drinking Water Regulations (NPDWRs or primary standards) are legally enforceable standards that apply to public water systems. Primary standards protect public health by limiting the levels of contaminants in drinking water. Vist the list of regulated contaminants with links for more details.

- List of Contaminants and their Maximum Contaminant Levels (MCLs)
- Setting Standards for Safe Drinking Water to learn about EPA's standard-setting process
- EPA's Regulated Contaminant Timeline
- National Primary Drinking Water Regulations http://www.epa.gov/epahome/exitepa.htm. The complete regulations regarding these contaminants available from the Code of Federal Regulations Website

NATIONAL SECONDARY DRINKING WATER REGULATIONS

National Secondary Drinking Water Regulations (NSDWRs or secondary standards) are nonenforceable guidelines regulating contaminants that may cause cosmetic effects (such as skin or tooth discoloration) or aesthetic effects (such as taste, odor, or color) in drinking water. EPA recommends secondary standards to water systems but does not require systems to comply. However, states may choose to adopt them as enforceable standards.

- List of National Secondary Drinking Water Regulations
- National Secondary Drinking Water Regulations http://www.epa.gov/epahome/exitepa.htm. The complete regulations regarding these contaminants are available from the Code of Federal Regulations Website.

*Quoted from www.epa.org.

UNREGULATED CONTAMINANTS

These contaminants which, at the time of publication, are not subject to any proposed or promulgated national primary drinking water regulation (NPDWR), are known or anticipated to occur in public water systems, and they may require regulations under SDWA. For more information check out the list, or visit the Drinking Water Contaminant Candidate List (CCL) website.

- List of Unregulated Contaminants
- Drinking Water Contaminant Candidate List (CCL) Website
- Unregulated Contaminant Monitoring Rule (UCMR)

LIST OF CONTAMINANTS AND THEIR MCLS (EPA 816-F-02-013, July 2002)

List of Contaninants and Their MCLS (EPA 816-F-02-013, July 2002)

Contaminant	MCLG[1] (mg/L)[2]	MCL or TT[1] (mg/L)[2]	Potential Health Effects from Ingestion of Water	Sources of Contaminant in Drinking Water
Microorganisms				
Cryptosporidium	Zero	TT[3]	Gastrointestinal illness (e.g., diarrhea, vomiting, cramps)	Human and fecal animal waste
Giardia lamblia	Zero	TT[3]	Gastrointestinal illness (e.g., diarrhea, vomiting, cramps)	Human and fecal animal waste
Heterotrophic plate	n/a	TT[3]	HPC has no health effects; it is an analytic method used to measure the variety of bacteria that are common in water. The lower the concentration of bacteria in drinking water, the better maintained the water system is.	HPC measures a range of bacteria that are naturally present in the environment
Legionella	Zero	TT[3]	Legionnaire's disease, a type of pneumonia	Found naturally in water; multiplies in heating systems
Total Coliforms (including fecal coliform and *E. coli*)	Zero	5.0%	Not a health threat in itself; it is used to indicate whether other potentially harmful bacteria may be present	Coliforms are naturally present in the environment, as well as in feces; fecal coliforms and *E. coli* only come from human and animal fecal waste.

Contaminant	MCLG	MCL	Potential Health Effects	Sources of Contaminant
Turbidity	n/a	TT[3]	Turbidity is a measure of the cloudiness of water. It is used to indicate water quality and filtration effectiveness (e.g., whether disease-causing organisms are present). Higher turbidity levels are often associated with higher levels of disease-causing microorganisms such as viruses, parasites and some bacteria. These organisms can cause symptoms such as nausea, cramps, diarrhea, and associated headaches.	Soil runoff
Viruses (enteric)	Zero	TT[3]	Gastrointestinal illness (e.g., diarrhea, vomiting, cramps)	Human and animal fecal waste
Disinfection By-products				
Bromate	Zero	0.010	Increased risk of cancer	By-product of drinking water disinfection
Chlorite	0.8	1.0	Anemia; infants and young children: nervous system effects	By-product of drinking water disinfection
Haloacetic acids	n/a[6]	0.060	Increased risk of cancer	By-product of drinking water disinfection
Total Trihalomethanes (THMs) 0.10 / 0.080	None[7]	n/a[6]	Liver, kidney, or central nervous system problems; increased risk of cancer	By-product of drinking water disinfection

(Continued)

List of Contaminants and Their MCLS (EPA 816-F-02-013, July 2002)

Contaminant	MRDLG[1] (mg/L)[2]	MRDL[1] (mg/L)[2]	Potential Health Effects from Ingestion of Water	Sources of Contaminant in Drinking Water
Disinfectants				
Chloramines (as Cl_2)	MRDLG = 41	MRDL = 4.01	Eye/nose irritation; stomach discomfort, anemia	Water additive used to control microbes
Chlorine (as Cl_2)	MRDLG = 41	MRDL = 4.01	Eye/nose irritation; stomach discomfort	Water additive used to control microbes
Chlorine dioxide (as ClO_2)	MRDLG = 0.81	MRDL = 0.81	Anemia; infants and young children: nervous system effects	Water additive used to control microbes

Contaminant	MCLG[1] (mg/L)[2]	MCL or TT[1] (mg/L)[2]	Potential Health Effects from Ingestion of Water	Sources of Contaminant in Drinking Water
Inorganic Chemicals				
Antimony	0.006	0.006	Increase in blood cholesterol; decrease in blood sugar	Discharge from petroleum refineries; fire retardants; ceramics; electronics; solder
Arsenic	0.07	0.010 as of 01/23/06	Skin damage or problems with circulatory systems; may have increased risk of getting cancer	Erosion of natural deposits; runoff from orchards, runoff from glass and electronics production wastes
Asbestos	7 million fibers per liter (MFL)	7 MFL	Increased risk of developing benign intestinal polyps	Decay of asbestos cement in water mains; erosion of natural deposits
Barium	2	2	Increase in blood pressure	Discharge of drilling wastes; discharge from metal refineries; erosion of natural deposits
Beryllium	0.004	0.004	Intestinal lesions	Discharge from metal refineries and coal-burning factories; discharge from electrical, aerospace, and defense industries
Cadmium	0.005	0.005	Kidney damage	Corrosion of galvanized pipes; erosion of natural deposits; discharge from metal refineries; runoff from waste batteries and paints
Chromium (total)	0.1	0.1	Allergic dermatitis	Discharge from steel and pulp mills; erosion of natural deposits
Copper	1.3	TT[8], action Level = 1.3	Short-term exposure: gastrointestinal distress; long-term exposure: liver or kidney damage; people with Wilson's disease should consult their personal doctor if the amount of copper in their water exceeds the action level	Corrosion of household plumbing systems; erosion of natural deposits

(*Continued*)

List of Contaninants and Their MCLS (EPA 816-F-02-013, July 2002)

Contaminant	MCLG[1] (mg/L)[2]	MCL or TT[1] (mg/L)[2]	Potential Health Effects from Ingestion of Water	Sources of Contaminant in Drinking Water
Cyanide (as free cyanide)	0.2	0.2	Nerve damage or thyroid problems	Discharge from steel/metal factories; discharge from plastic and fertilizer factories
Fluoride	4.0	4.0	Bone disease (pain and tenderness of the bones); children may get mottled teeth	Water additive which promotes strong teeth; erosion of natural deposits; discharge from fertilizer and aluminum factories
Lead	zero	TT[8], action level = 0.015	*Infants and children:* Delays in physical or mental development; children could show slight deficits in attention span and learning abilities *Adults:* Kidney problems; high blood pressure	Corrosion of household plumbing systems; erosion of natural deposits
Mercury (inorganic)	0.002	0.002	Kidney damage	Erosion of natural deposits; discharge from refineries and factories; runoff from landfills and croplands
Nitrate (measured as nitrogen)	10	10	Infants below the age of six months who drink water containing nitrate in excess of the MCL could become seriously ill and, if untreated, may die; symptoms include shortness of breath and blue-baby syndrome	Runoff from fertilizer use; leaching from septic tanks, sewage; erosion of natural deposits
Nitrite (measured as nitrogen)	1	1	Infants below the age of six months who drink water containing nitrite in excess of the MCL could become seriously ill and, if untreated, may die; symptoms include shortness of breath and blue-baby syndrome	Runoff from fertilizer use; leaching from septic tanks, sewage; erosion of natural deposits

Contaminant	MCLG	MCL	Potential Health Effects	Sources of Contaminant
Selenium	0.05	0.05	Hair or fingernail loss; numbness in fingers or toes; circulatory problems	Discharge from petroleum refineries; erosion of natural deposits; discharge from mines
Thallium	0.0005	0.002	Hair loss; changes in blood; kidney, intestine, or liver problems	Leaching from ore-processing sites; discharge from electronics, glass, and drug factories
Organic Chemicals				
Acrylamide	Zero	TT[9]	Nervous system or blood problems; increased risk of cancer	Added to water during sewage/wastewater treatment
Alachlor	Zero	0.002	Eye, liver, kidney, or spleen problems; anemia; increased risk of cancer	Runoff from herbicide used on row crops
Atrazine	0.003	0.003	Cardiovascular system or reproductive problems	Runoff from herbicide used on row crops
Benzene	Zero	0.005	Anemia; decrease in blood platelets; increased risk of cancer	Discharge from factories; leaching from gas storage tanks and landfills
Benzo(a)pyrene	Zero	0.0002	Reproductive difficulties; increased risk of cancer	Leaching from linings of water storage tanks and distribution lines
Carbofuran	0.04	0.04	Problems with blood, nervous system, or reproductive system	Leaching of soil fumigant used on rice and alfalfa
Carbon tetrachloride	Zero	0.005	Liver problems; increased risk of cancer	Discharge from chemical plants and other industrial activities
Chlordane	Zero	0.002	Liver or nervous system problems; increased risk of cancer	Residue of banned termiticide
Chlorobenzene	0.1	0.1	Liver or kidney problems	Discharge from chemical and agricultural chemical factories
2,4-D	0.07	0.07	Kidney, liver, or adrenal gland problems	Runoff from herbicide used on row crops
Dalapon	0.2	0.2	Minor kidney changes	Runoff from herbicide used on rights of way
1,2-Dibromo-3-chloropropane (DBCP)	Zero	0.0002	Reproductive difficulties; increased risk of cancer	Runoff/leaching from soil fumigant used on soybeans, cotton, pineapples, and orchards
o-Dichlorobenzene	0.6	0.6	Liver, kidney, or circulatory system problems	Discharge from industrial chemical factories

(Continued)

List of Contaminants and Their MCLS (EPA 816-F-02-013, July 2002)

Contaminant	MCLG[1] (mg/L)[2]	MCL or TT[1] (mg/L)[2]	Potential Health Effects from Ingestion of Water	Sources of Contaminant in Drinking Water
p-Dichlorobenzene	0.075	0.075	Anemia; liver, kidney or spleen damage; changes in blood	Discharge from industrial chemical factories
1,2-Dichlorobenzene	Zero	0.005	Increased risk of cancer	Discharge from industrial chemical factories
1,1-Dichloroethylene	0.007	0.007	Liver problems	Discharge from industrial chemical factories
cis-1,2-Dichloroethylene	0.07	0.07	Liver problems	Discharge from industrial chemical factories
trans-1,2-Dichloroethylene	0.1	0.1	Liver problems	Discharge from industrial chemical factories
Dichloromethane	Zero	0.005	Liver problems; increased risk of cancer	Discharge from drug and chemical factories
1,2-Dichloropropane	Zero	0.005	Increased risk of cancer	Discharge from industrial chemical factories
Di(2-ethylhexyl) adipate	0.4	0.4	Weight loss, liver problems, or possible reproductive difficulties.	Discharge from chemical factories
Di(2-ethylhexyl) phthalate	Zero	0.006	Reproductive difficulties; liver problems; increased risk of cancer	Discharge from rubber and chemical factories
Dinoseb	0.007	0.007	Reproductive difficulties	Runoff from herbicide used on soybeans and vegetables
Dioxin (2,3,7,8-TCDD)	Zero	0.00000003	Reproductive difficulties; increased risk of cancer	Emissions from waste incineration and other combustion; discharge from chemical factories
Diquat	0.02	0.02	Cataracts	Runoff from herbicide use
Endothall	0.1	0.1	Stomach and intestinal problems	Runoff from herbicide use
Endrin	0.002	0.002	Liver problems	Residue of banned insecticide
Epichlorohydrin	Zero	TT[9]	Increased cancer risk, and over a long period of time, stomach problems	Discharge from industrial chemical factories; an impurity of some water treatment chemicals
Ethylbenzene	0.7	0.7	Liver or kidneys problems	Discharge from petroleum refineries
Ethylene dibromide	Zero	0.00005	Problems with liver, stomach, reproductive system, or kidneys; increased risk of cancer	Discharge from petroleum refineries

Contaminant	MCLG	MCL	Potential Health Effects	Sources of Contaminant
Glyphosate	0.7	0.7	Kidney problems; reproductive difficulties	Runoff from herbicide use
Heptachlor	Zero	0.0004	Liver damage; increased risk of cancer	Residue of banned termiticide
Heptachlor epoxide	Zero	0.0002	Liver damage; increased risk of cancer	Breakdown of heptachlor
Hexachlorobenzene	Zero	0.001	Liver or kidney problems;	Discharge from metal refineries and agricultural chemical factories
Hexachlorocyclopentadiene	0.05	0.05	Kidney or stomach problems	Discharge from chemical factories
Lindane	0.0002	0.0002	Liver or kidney problems	Runoff/leaching from insecticide used on cattle, lumber, gardens
Methoxychlor	0.04	0.04	Reproductive difficulties	Runoff/leaching from insecticide used on fruits, vegetables, alfalfa, livestock
Oxamyl (Vydate)	0.2	0.2	Slight nervous system effects	Runoff/leaching from insecticide used on apples, potatoes, and tomatoes
Polychlorinated biphenyls (PCBs)	Zero	0.0005	Skin changes; thymus gland problems; immune deficiencies; reproductive or nervous system difficulties; increased risk of cancer	Runoff from landfills; discharge of waste chemicals
Pentachlorophenol	Zero	0.001	Liver or kidney problems; increased cancer risk	Discharge from wood preserving factories
Picloram	0.5	0.5	Liver problems	Herbicide runoff
Simazine	0.004	0.004	Problems with blood	Herbicide runoff
Styrene	0.1	0.1	Liver, kidney, or circulatory System problems	Discharge from rubber and plastic factories; leaching from landfills
Tetrachloroethylene	Zero	0.005	Liver problems; increased risk of cancer	Discharge from factories and dry cleaners
Toluene	1	1	Nervous system, kidney, or liver problems	Discharge from petroleum factories
Toxaphene	Zero	0.003	Kidney, liver, or thyroid Problems increased risk of cancer	Runoff/leaching from insecticide used on cotton and cattle
2,4,5-TP (Silvex)	0.05	0.05	Liver problems	Residue of banned herbicide
1,2,4-Trichlorobenzene	0.07	0.07	Changes in adrenal glands	Discharge from textile finishing factories
1,1,1-Trichloroethane	0.20	0.2	Liver, nervous system, or circulatory problems	Discharge from metal degreasing sites and other factories
1,1,2-Trichloroethane	0.003	0.005	Liver, kidney, or immune system problems	Discharge from industrial chemical factories

(*Continued*)

List of Contaminants and Their MCLS (EPA 816-F-02-013, July 2002)

Contaminant	MCLG[1] (mg/L)[2]	MCL or TT[1] (mg/L)[2]	Potential Health Effects from Ingestion of Water	Sources of Contaminant in Drinking Water
Trichloroethylene	Zero	0.005	Liver problems; increased risk of cancer	Discharge from metal degreasing sites and other factories
Vinyl chloride	Zero	0.002	Increased risk of cancer	Leaching from PVC pipes; discharge from plastic factories
Xylenes (total)	10	10	Nervous system damage	Discharge from petroleum factories; discharge from chemical factories
Radionuclides				
Alpha particles	None[7]	Zero	15 picocuries per liter (pCi/L); increased risk of cancer	Erosion of natural deposits of certain minerals that are radioactive and may emit a form of radiation known as alpha radiation
Beta particles and photon emitters	None[7]	Zero	4 millirems per year; increased risk of cancer	Decay of natural and man-made deposits of certain minerals that are radioactive and may emit forms of radiation known as photons and beta radiation
Radium 226 and radium 228 (combined) none[7]	Zero	5 pCi/L	Increased risk of cancer	Erosion of natural deposits
Uranium	Zero	30µg/L as of 12/08/03	Increased risk of cancer, kidney toxicity	Erosion of natural deposits

[1] **Definitions:** *Maximum contaminant level (MCL)*—the highest level of a contaminant that is allowed in drinking water. MCLs are set as close to MCLGs as feasible using the best available treatment technology and taking cost into consideration. MCLs are enforceable standards. *Maximum contaminant level goal (MCLG)*—the level of a contaminant in drinking water below which there is no known or expected risk to health. MCLGs allow for a margin of safety and are nonenforceable public health goals. *Maximum residual disinfectant level (MRDL)*—the highest level of a disinfectant allowed in drinking water. There is convincing evidence that addition of a disinfectant is necessary for control of microbial contaminants. *Maximum residual disinfectant level goal (MRDLG)*—the level of a drinking water disinfectant below which there is no known or expected risk to health. MRDLGs do not reflect the benefits of the use of disinfectants to control microbial contaminants. *Treatment technique*—a required process intended to reduce the level of a contaminant in drinking water.

[2] Units are in milligrams per liter (mg/L) unless otherwise noted. Milligrams per liter are equivalent to parts per million.

[3] EPA's surface water treatment rules require systems using surface water or groundwater under the direct influence of surface water to (1) disinfect their water and (2) filter their water or meet criteria for avoiding filtration so that the following contaminants are controlled at the following levels:

472

- Cryptosporidium (as of 1/1/02 for systems serving >10,000 and 1/14/05 for systems serving <10,000) 99% removal.
- *Giardia lamblia*: 99.9% removal/inactivation.
- Viruses: 99.99% removal/inactivation.
- *Legionella*: No limit, but EPA believes that if *Giardia* and viruses are removed/inactivated, *Legionella* will also be controlled.
- Turbidity: At no time can turbidity (cloudiness of water) go above 5 nephelolometric turbidity units (NTU); systems that filter must ensure that the turbidity go no higher than 1 NTU (0.5 NTU for conventional or direct filtration) in at least 95% of the daily samples in any month. As of January 1, 2002, turbidity may never exceed 1 NTU, and must not exceed 0.3 NTU in 95% of daily samples in any month.
- HPC: No more than 500 bacterial colonies per milliliter.
- Long-Term 1 Enhanced Surface Water Treatment (Effective Date: January 14, 2005): Surface water systems or (GWUDI) systems serving fewer than 10,000 people must comply with the applicable Long-Term 1 Enhanced Surface Water Treatment Rule provisions (e.g., turbidity standards, individual filter monitoring, Cryptosporidium removal requirements, updated watershed control requirements for unfiltered systems).
- Filter Backwash Recycling; the Filter Backwash Recycling Rule requires systems that recycle to return specific recycle flows through all processes of the system's existing conventional or direct filtration system or at an alternate location approved by the state.

[4] More than 5.0% samples total coliform-positive in a month. (For water systems that collect fewer than 40 routine samples per month, no more than one sample can be total coliform-positive per month.) Every sample that has total coliform must be analyzed for either fecal coliforms or *E. coli*; if there are two consecutive TC-positive samples and if one is also positive for *E. coli* fecal coliforms, the system has an acute MCL violation.

[5] Fecal coliform and *E. coli* are bacteria whose presence indicates that the water may be contaminated with human or animal wastes. Disease-causing microbes (pathogens) in these wastes can cause diarrhea, cramps, nausea, headaches, or other symptoms. These pathogens may pose a special health risk for infants, young children, and people with severely compromised immune systems.

[6] Although there is no collective MCLG for this contaminant group, there are individual MCLGs for some of the individual contaminants:

- Trihalomethanes: bromodichloromethane (zero); bromoform (zero); dibromochloromethane (0.06 mg/L). Chloroform is regulated with this group but has no MCLG.
- Haloacetic acids: dichloroacetic acid (zero); trichloroacetic acid (0.3 mg/L). Monochloroacetic acid, bromoacetic acid, and dibromoacetic acid are regulated with this group but have no MCLGs.

[7] MCLGs were not established before the 1986 Amendments to the Safe Drinking Water Act. Therefore, there is no MCLG for this contaminant.

[8] Lead and copper are regulated by a Treatment Technique that requires systems to control the corrosiveness of their water. If more than 10% of tap water samples exceed the action level, water systems must take additional steps. For copper, the action level is 1.3 mg/L, and for lead it is 0.015 mg/L.

[9] Each water system must certify, in writing, to the state (using third-party or manufacturer's certification) that when acrylamide and epichlorohydrin are used in drinking water systems, the combination (or product) of dose and monomer level does not exceed the levels specified, as follows:

- Acrylamide = 0.05% dosed at 1 mg/L (or equivalent).
- Epichlorohydrin = 0.01% dosed at 20 mg/L (or equivalent).

National Secondary Drinking Water Regulations

National Secondary Drinking Water Regulations (NSDWRs or secondary standards) are nonenforceable guidelines regulating contaminants that may cause cosmetic effects (such as skin or tooth discoloration) or aesthetic effects (such as taste, odor, or color) in drinking water. EPA recommends secondary standards to water systems but does not require systems to comply. However, states may choose to adopt them as enforceable standards.

- For more information, read Secondary Drinking Water Regulations: Guidance for Nuisance Chemicals.

Contaminant	Secondary Standard
Aluminum	0.05–0.2 mg/L
Chloride	250 mg/L
Color	15 (color units)
Copper	1.0 mg/L
Corrosivity	Noncorrosive
Fluoride	2.0 mg/L
Foaming agents	0.5 mg/L
Iron	0.3 mg/L
Manganese	0.05 mg/L
Odor	3 threshold odor number
PH	6.5–8.5
Silver	0.10 mg/L
Sulfate	250 mg/L
Total Dissolved Solids	500 mg/L
Zinc	5 mg/L

Last updated on Thursday, March 18th, 2004.
URL: http://www.epa.gov/safewater/mcl.html#mcls

PERIODIC TABLE OF THE ELEMENTS

Periodic Table of the Elements

Key:
- Element name
- Element #
- Symbol
- Atomic Mass

Example: Carbon, 6, C, 12.01

IA	IIA	IIIB	IVB	VB	VIB	VIIB	VIII	VIII	VIII	IB	IIB	IIIA	IVA	VA	VIA	VIIA	VIIIA
H 1 — 1.01																	He 2 — 4.00
Li 3 — 6.94	Be 4 — 9.01											B 5 — 10.81	C 6 — 12.01	N 7 — 14.01	O 8 — 16.00	F 9 — 19.00	Ne 10 — 20.18
Na 11 — 22.99	Mg 12 — 24.31											Al 13 — 26.98	Si 14 — 28.09	P 15 — 30.97	S 16 — 32.07	Cl 17 — 35.45	Ar 18 — 39.95
K 19 — 39.10	Ca 20 — 40.08	Sc 21 — 44.96	Ti 22 — 47.88	V 23 — 50.94	Cr 24 — 52.00	Mn 25 — 54.94	Fe 26 — 55.85	Co 27 — 58.93	Ni 28 — 58.69	Cu 29 — 63.55	Zn 30 — 65.39	Ga 31 — 69.72	Ge 32 — 72.61	As 33 — 74.92	Se 34 — 78.96	Br 35 — 79.90	Kr 36 — 83.80
Rb 37 — 85.47	Sr 38 — 87.62	Y 39 — 88.91	Zr 40 — 91.22	Nb 41 — 92.91	Mo 42 — 95.94	Tc 43 — 99	Ru 44 — 101.07	Rh 45 — 102.91	Pd 46 — 106.42	Ag 47 — 107.87	Cd 48 — 112.41	In 49 — 114.82	Sn 50 — 118.71	Sb 51 — 121.75	Te 52 — 127.60	I 53 — 126.90	Xe 54 — 131.29
Cs 55 — 132.91	Ba 56 — 137.33	La 57 — 138.91	Hf 72 — 178.49	Ta 73 — 180.95	W 74 — 183.85	Re 75 — 186.21	Os 76 — 190.2	Ir 77 — 192.22	Pt 78 — 195.08	Au 79 — 196.97	Hg 80 — 200.59	Tl 81 — 204.38	Pb 82 — 207.2	Bi 83 — 208.98	Po 84 — 209	At 85 — 210	Rn 86 — 222
Fr 87 — 223	Ra 88 — 226	Ac 89 — 227	Rf 104 — 261	Db 105 — 262	Sg 106 — 263	Bh 107 — 262	Hs 108 — 265	Mt 109 — 266	Uun 110 — 269	Uuu 111 — 272	Uub 112 — 277						

Lanthanide Series

Ce 58 — 140.12	Pr 59 — 140.91	Nd 60 — 144.24	Pm 61 — 147	Sm 62 — 150.36	Eu 63 — 151.97	Gd 64 — 157.25	Tb 65 — 158.93	Dy 66 — 162.50	Ho 67 — 164.93	Er 68 — 167.26	Tm 69 — 168.93	Yb 70 — 173.04	Lu 71 — 174.97

Actinide Series

Th 90 — 232.04	Pa 91 — 231	U 92 — 238.03	Np 93 — 237	Pu 94 — 244	Am 95 — 243	Cm 96 — 247	Bk 97 — 247	Cf 98 — 251	Es 99 — 252	Fm 100 — 257	Md 101 — 258	No 102 — 259	Lr 103 — 260

INDEX

A Basic Introduction to Pollutant Fate and Transport, By Dunnivant and Anders
Copyright © 2006 by John Wiley & Sons, Inc.